Methods for Constructing Exact Solutions of Partial Differential Equations

Mathematical and Analytical Techniques with Applications to Engineering

Mathematical and Analytical Techniques with Applications to Engineering

Alan Jeffrey, *Consulting Editor*

Published:

Inverse Problems
A. G. Ramm

Singular Perturbation Theory
R. Johnson

Methods for Constructing Exact Solutions of Partial Differential Equations
S. V. Meleshko

Theory of Stochastic Differential Equations with Jumps and Applications
S. Rong

Forthcoming:

The Fast Solution of Boundary Integral Equations
S. Rjasanow and O. Steinbach

METHODS FOR CONSTRUCTING EXACT SOLUTIONS OF PARTIAL DIFFERENTIAL EQUATIONS

Mathematical and Analytical Techniques with Applications to Engineering

S. V. MELESHKO

Springer

Library of Congress Cataloging-in-Publication Data

Meleshko, S.V.
Methods for constructing exact solutions of partial differential equations : mathematical and analytical techniques with applications to engineering / S.V. Meleshko.
p. cm. – (Mathematical and analytical techniques with applications to engineering)
Includes index.
ISBN 0-387-25060-3 (acid-free paper) ISBN 0-387-25265-7 (e-ISBN)
1. Differential equations, Partial. 2. Differential equations, Nonlinear. I. Title. II. Series.

QA377.M445 2005
515′.353—dc22

2005042620

Printed in the United States of America.

9 8 7 6 5 4 3 2 1 SPIN 11399285

springeronline.com

Contents

Preface

Differential equations, especially nonlinear, present the most effective way for describing complex physical processes. Each solution of a system of differential equations corresponds to a particular process. Therefore, methods for constructing exact solutions of differential equations play an important role in applied mathematics and mechanics. This book aims to provide scientists, engineers, and students with an easy to follow, but comprehensive, description of the methods for constructing exact solutions of differential equations. The emphasis is on the methods of differential constraints, degenerate hodograph, and group analysis. These methods have become a necessary part of applied mathematics and mathematical physics. The book is primarily designed to present both fundamental theoretical and algorithmic aspects of these methods. The description of algorithms contains illustrative examples which are typically taken from continuum mechanics. Some sections of the book introduce new applications and extensions of these methods. For example, the sixth chapter presents integro-differential and functional differential equations, a new area of group analysis.

Nonlinear partial differential equations is a vast area. There is a great number of classical and recent results on obtaining exact solutions for this type of equations. Being both selective and comprehensive is a challenge. While I drew upon multitude of sources for this book, still many results are omitted due to space constraints. It should also be noted that the method of differential constraints is not well-known outside Russia; there are only a few books in English where the idea behind this method (without analysis) is briefly mentioned. This book is a result of an effort to introduce, at a fairly elementary level, many methods for constructing exact solutions, collected in one book. It is based on my research and various courses and lectures given to undergraduate and graduate students as well as professional audiences over the past twenty five years. The book is assembled, in a coherent and comprehensive way, from results that are scattered across many different articles and books published over the last thirty years.

The approach is analytical. The material is presented in a way that will enable the readers to succeed in their study of the subject. Introductions to theories are followed by examples. The target reader of the book are students, engineers, and scientists with diverse backgrounds and interests. For a deeper coverage of a particular method or an application the readers are referred to special-purpose books and/or scientific articles referenced in the book. The prerequisites for the study are standard courses in calculus, linear algebra, and ordinary and partial differential equations.

1. Notes to the reader

1. Analytical studies of properties of partial differential equations play an important role in applied mathematics and mathematical physics. Among them, analytical study based on the knowledge of particular classes of solutions has received a widespread attention. Each exact solution has several meanings, including an exact description of a real process in the framework of a given model, a model to compare various numerical methods, and a theory to improve the models used. This book focuses on the methods for constructing an exact solution of differential equations provided that the solution satisfies additional differential or finite constraints.

2. Most manifolds, differential equations, and other objects in the book are considered locally. All functions are assumed to be continuously differentiable a sufficient number of times. The requirement of a local study is mainly related to the inverse function theorem and the existence theorem of a local solution of an initial value problem. The local approach makes the apparatus of the study both flexible and generalizable.

3. The notion of an exact solution is not strictly defined. The concept of an exact solution is changing along with the development of mathematics. Different authors include different meaning in this notion. The exact solutions can be:

a) explicit formulae in terms of elementary functions, their quadratures, or special functions;

b) convergent series with effectively computed coefficients;

c) solutions for which the process of their finding is reduced to integration of ordinary differential equations.

The author assumes that an exact solution is a solution which has a reduced number of dependent or independent variables.

4. Particular solutions are being sought with the greatest possible functional or constant arbitrariness. Notice that any particular solution is defined by the initial differential equations and some additional analytical, geometrical, kinematic, or physical properties that lead to either the reduction of the dimension

of a problem, or the simplification of the initial equations. After finding the representation of a solution one can try to satisfy specific initial and boundary conditions by a special selection of arbitrary elements of the solution. Sometimes these methods are called half-inverse methods.

5. Compatibility analysis is one of the main techniques for constructing exact solutions. The general theory of compatibility is a special subject of algebraic analysis. Only an introduction into this theory is given in this book.

6. One of the features of a compatibility analysis is a large volume of analytical calculations. The analytical calculations include sequential executions of several algebraic operations. Since these operations are very labor intensive one has to use a computer for symbolic manipulations. Using a computer allows a considerable reduction of expense in an analytical study of systems of partial differential equations. Nowadays, obtaining new results is impossible without using a computer for analytical calculations.

2. Organization of the book

The book is divided into several chapters covering the main topics of the methods for constructing exact solutions of partial differential equations. These are united by the idea that a solution satisfies additional differential or finite constraints. For various methods the constraints are built in different ways.

The first chapter introduces methods for constructing exact solutions of partial differential equations with a single dependent function and applies these methods to studying systems of partial differential equations. For example, the Cauchy method (method of characteristics) is the main tool for finding solutions of nonlinear partial differential equations. For finding an invariant solution one has to be able to solve an overdetermined system of linear partial differential equations. Such systems can be solved by using Poison brackets. Many methods for solving differential equations along with point transformations use tangent transformations. The classical tangent transformations are the Legendre, the Hopf-Cole and the Laplace transformations. The second part of the chapter presents methods for constructing particular solutions. These methods are based on some assumptions about solutions. The assumptions are related to different representations of a solution (e.g., separation of variables, self-similar solutions, travelling waves, or partial representation) or to different requirements for a solution to satisfy such as additional functional or differential properties. The second chapter is devoted to systems of partial differential equations. If a system is written in Riemann invariants, then for homogeneous systems one obtains Riemann waves. The well-known problem of a decay of arbitrary discontinuity of a gas is solved in terms of Riemann waves. Another method that plays a very important role in gas dynamics is the hodograph method, when the hodograph is not degenerate. Presentation of self-similar solutions is given from a group analysis point of view. This way of studying self-similar solutions can also be considered as an introduction to the group analysis method. Travelling waves and solutions with linear dependence of velocity with respect to independent variables are solutions with a simple representation of the dependent variables through the independent variables. The third chapter considers the method of degenerate hodograph. This method deals with solutions that are distinguished by finite relations between the dependent variables. They form a class of solutions called multiple waves. The Riemann waves and the Prandtl–Meyer flows belong to this class of solutions. The first application of simple waves for multi-dimensional flows was made for isentropic flows of an ideal gas: simple and double waves. For double waves the Ovsiannikov theorem plays a very important role. The practical meaning of this theorem is demonstrated in the chapter by several examples. Applications of double waves in gas dynamics are followed by applications of double waves in a rigid plastic body. The chapter is completed by the study of triple waves of isentropic potential gas flows.

The fourth chapter is devoted to the method of differential constraints. Since the theory of involutive systems is the basis of the method, the first section introduces this theory. The theory of compatibility is followed by the basic definitions of the method of differential constraints. The first problem to arise in applications of the method of differential constraints is the involutiveness problem of an original system of partial differential equations with differential constraints. Since the Cartan-Khäler theorem only provides the existence of a solution for analytic systems, the existence problem of a solution for nonanalytic involutive systems appears. This problem is solved by using the notion of characteristics for an overdetermined system of partial differential equations. Characteristic curves also play the main role in defining a class of solutions that generalizes simple waves. The generalized simple waves have properties similar to simple waves. For example, the solution of the Goursat problem can be given in terms of generalized simple waves. The general study of generalized simple waves is followed by a section devoted to deriving this class of solutions for gas dynamic equations. The second part of the chapter considers applications of the method of differential constraints to systems of quasilinear equations with more than two independent variables. After the general study one finds examples of differential constraints for the system of multi-dimensional gas dynamic equations. As mentioned above, invariant solutions also can be described by differential constraints. Relations between the method of differential constraints and Lie-Bäcklund groups of transformations are studied in this chapter.

The fifth chapter presents a concise form of the basic algorithms that form the core of group analysis. The problem of finding an admitted Lie group is the first step in applications of group analysis for constructing exact solutions. The algebraic structure of the admitted Lie group introduces an algebraic structure into the set of all solutions. This algebraic structure is used to find invariant and partially invariant solutions. The main feature of these classes of solutions is that they reduce the number of independent and dependent variables. In this sense the problem of finding these solutions is simpler than the ones for the general solution. A new way of using partially invariant solutions as a means of finding exact solutions is also discussed. Finally, involving derivatives in the transformation generalizes the notion of a Lie group of point transformations and leads to the notions of Bäcklund and a group of Lie-Bäcklund transformations.

The algorithmic approach of group analysis was developed specifically for differential equations. The sixth chapter discusses an extension of group analysis for equations having nonlocal terms. As for partial differential equations, the first step involves constructing an admitted Lie group. The first section of the chapter discusses different approaches to the definition of an admitted Lie group. This discussion assists in establishing a definition of an admitted Lie group for integro-differential and functional differential equations. As for partial differential equations the main difficulty in finding an admitted Lie group consists of solving the determining equations. In contrast to partial differential equations a method for solving the determining equations depends on the nonlocal equations under study. Three different examples of solving determining equations are considered in the chapter. The last part of the chapter focuses on functional differential equations and, particularly, on delay differential equations. By example, it is shown that the method for solving determining equations for delay differential equations is similar to the one for partial differential equations.

One of the features of compatibility analysis of differential equations is the extensive analytical manipulations involved in the calculations. Computer algebra systems have become an important computational tool in analytical calculations. The goal of the seventh chapter is to demonstrate computer symbolic calculations in the study of compatibility analysis. This is demonstrated by solving the problem of linearization of a third order ordinary differential equation.

3. Acknowledgments

I am indebted to many people for inspiring my interest in this subject. During my career I have had the opportunity to work with great scientists from the mathematical schools of N.N.Yanenko and L.V. Ovsiannikov, and I am honored to consider myself to be associated with these schools. My opinions have been influenced by numerous discussions with my friends and colleagues. I would like to thank V.G. Ganzha, Yu.N. Grigoriev, N.H. Ibragimov, F.A. Murzin, V.V. Pukhnachov and V.P. Shapeev with whom I had the opportunity to carry out joint scientific projects. Discussions with L.V. Ovsiannikov, S.V. Khabirov, A.A. Talyshev, A.P. Chupakhin, E.V. Mamontov,

A.A. Cherevko and S.V. Golovin during our work on the research project "Submodels" were very stimulating.

I would like to express my appreciation to A. Jeffrey for his suggestions and comments that served to improve the quality of this book. I am indebted to him for his guidance in the preparation of this edition.

I would like to thank N. Manganaro and D. Fusco for inviting me to the University of Messina to giving lectures related to the topics of this book. These lectures became the first step in preparation of this book.

My special thanks to my friends C.P. Clements, K.J. Haller, E.R. Schultz, J.W. Ward and N.F.Samatova for their help with English corrections at different stages.

I am deeply grateful to my brother A.V. Melechko for his remarks, continuous encouragement, and support.

Nakhon Ratchasima
December 2004

Sergey V. Meleshko

Chapter 1

EQUATIONS WITH ONE DEPENDENT FUNCTION

This chapter introduces methods for constructing exact solutions of partial differential equations with one dependent function. Application of these methods is one of the steps for studying systems of partial differential equations. The methods are introduced by considering simple examples. The theory of the methods is discussed in the following chapters.

Linearity, quasilinearity, order of equations and other preliminary notions are considered in the first section. Such properties of solutions as replacement of variables and functional dependence, often used for obtaining exact solutions, are also introduced here. The next section is devoted to the Cauchy method (method of characteristics). This method is one of the main methods applied for constructing exact solutions of first order partial differential equations. The Cauchy method reduces a Cauchy problem for a partial differential equation to the Cauchy problem for a system of ordinary differential equations. This method is illustrated by the Hopf equation. The Cauchy method allows finding exact solutions with arbitrary functions. However, even knowledge of solutions with arbitrary constants can assist in constructing the general solution. This leads the reader to the solutions called complete and singular integrals. The section devoted to these solutions also contains the Lagrange-Charpit method for obtaining the complete integral. Practically, for finding any invariant solution, one has to be able to solve an overdetermined system of linear partial differential equations. For a system of quasilinear equations with a single dependent variable the problem of compatibility is solved through the concepts of Poisson brackets and complete systems.

Many methods of solving differential equations use a change of the dependent and independent variables that transforms a given differential equation into another equation with known properties. The change of variables, which also involves derivatives in the transformation, is called a tangent transformation. The classical tangent transformations such as the Legendre transformation, the Hopf-Cole transformation, and the Laplace transformations are studied in the first part of chapter 1.

The second part of the chapter is devoted to methods for constructing particular solutions. These methods are based on certain assumptions about solutions. The assumptions can be about the representation of a solution (separation of variables, self-similar solutions, travelling waves or partial representation) or they can be based on the requirements for a solution to satisfy additional functional or differential properties. The first chapter discusses functionally invariant solutions or solutions having intermediate integral.

1. Basic definitions and examples

The purpose of the section is to give introductory remarks on exact solutions of partial differential equations

$$F_k(x, u, u_{x_1}, \dots, u_{x_n}, u_{x_1x_1}, u_{x_1x_2}, \dots) = 0, \quad (k = 1, 2, \dots, m). \tag{1.1}$$

with n independent variables $x = (x_1, x_2, \dots, x_n)$ and one dependent function $u(x)$.

Definition 1.1. *A solution of equations (1.1) is a function $u(x_1, x_2, \dots, x_n)$, which being substituted into (1.1) reduces them to identities with respect to the independent variables $x_1, x_2, \dots, x_n$.*

There is also a geometrical definition of a solution, considered as a manifold.

A function $u(x_1, x_2, \dots, x_n)$ satisfying Definition 1.1 that is assumed to be sufficiently many times continuously differentiable in some domain D in R^n is called a classical solution or a genuine solution. Graphically, any solution $u = u(x_1, x_2, \dots, x_n)$ of (1.1) can be represented as a smooth surface in $R^{(n+1)}$ lying over the domain D in the $(x_1, x_2, \dots, x_n)$–hyperplane.

The maximal order of the derivatives, included in the differential equation, is called the order of this equation. If the function F_k is linear with respect to the unknown function u and its derivatives, then this equation is called a linear equation, otherwise it is called nonlinear. A nonlinear equation F_k, which is only linear with respect to the maximal order derivatives, is called a quasilinear equation.

Among the methods for constructing exact solutions of nonlinear partial differential equations that should be noted are the classical methods of finding the general solution of first order equations: the Cauchy method, complete and singular integrals, the Lagrange–Charpit method and Poisson brackets. Before giving a short introduction to these methods[1] let us consider some examples.

[1]The detail theory of these methods one can find, for example, in [32] and [163].

1.1 Replacement of the independent variables

Assume that one needs to solve the partial differential equation

$$\alpha u_x + \beta u_y = 0,$$

where α and β are constant, and $\alpha^2 + \beta^2 \neq 0$. Using the change of the independent variables

$$\xi = \beta x - \alpha y, \quad \eta = \alpha x + \beta y$$

one obtains the equation

$$(\alpha^2 + \beta^2)\omega_\eta = 0,$$

with $\omega(\xi, \eta) = u(x(\xi, \eta), y(\xi, \eta))$. The general solution of the last equation is $\omega = \omega(\xi)$. The function $\omega = \omega(\xi)$ is arbitrary. Hence, the general solution of the original equation is $u = \omega(\beta x - \alpha y)$.

Remark 1.1. *Formulae for the transformed derivatives are easily obtained by using the invariance of the differential. In fact, let us consider an arbitrary function $f(x_1, x_2, \ldots, x_n)$ and the new independent variables $\xi_i = \xi_i(x_1, x_2, \ldots, x_n)$, $(i = 1, 2, \ldots, n)$. The invariance of the differential with respect to the replacement of the independent variables means*

$$df = \sum_{i=1}^{n} f_{x_i}\, dx_i = \sum_{j=1}^{n} f_{\xi_j}\, d\xi_j. \tag{1.2}$$

Substituting the differentials

$$d\xi_j = \sum_{i=1}^{n} \frac{\partial \xi_j}{\partial x_i}\, dx_i$$

into (1.2), one obtains

$$df = \sum_{i=1}^{n} \left(\sum_{j=1}^{n} f_{\xi_j} \frac{\partial \xi_j}{\partial x_i}\right) dx_i = \sum_{i=1}^{n} f_{x_i}\, dx_i.$$

By virtue of the independence of the differentials dx_i, one finds

$$f_{x_i} = \sum_{j=1}^{n} \frac{\partial \xi_j}{\partial x_i} f_{\xi_j}, \quad (i = 1, 2, ..., n).$$

Another very well-known example where the equation is transformed to a simple form is the wave equation

$$u_{tt} - c^2 u_{xx} = 0,$$

where c is constant. Replacing the independent variables (x, t) with (ξ, η), where

$$\xi = x + ct, \ \ \eta = x - ct,$$

one obtains the general solution of the wave equation (the d'Alembert formula)

$$u = f_1(x + ct) + f_2(x - ct).$$

Here the functions f_1 and f_2 are arbitrary functions, which are defined by auxiliary initial or boundary conditions. Additional conditions (initial and boundary data) are usually related with the underlying physical problem.

The integration of some differential equations can be also simplified by including in the transformation not only the independent variables, and also some unknown functions. For example, applying the Kirchhoff transformation

$$\psi = \int_{u_0}^{u} k(\rho)d\rho,$$

to the nonlinear equation

$$div\ (k(u)\nabla u) = 0, \tag{1.3}$$

the function ψ satisfies the linear Laplace equation $\Delta\psi = 0$, which is well studied. Thus, all properties of solutions of equation (1.3) can be discussed on the basis of the solutions of the Laplace equation.

1.2 Functional dependence.

Functional dependence is often used for constructing the general solutions. For example, the partial differential equation with respect to the function $u(x, y)$

$$g_y u_x - g_x u_y = 0 \tag{1.4}$$

means, that the Jacobian $\partial(u, g)/\partial(x, y)$ vanishes. Here $g = g(x, y)$ is some given function of the independent variables x and y. The general solution of this equation is $u = \omega(g(x, y))$ with an arbitrary function $\omega(\xi)$. The proof is obtained by the replacement of the independent variables. Without loss of generality one can assume that $g_y \neq 0$. Taking

$$\xi = g(x, y), \ \ \eta = x,$$

equation (1.4) is reduced to the equation $\omega_\eta = 0$, where $u(x, y) = \omega(g(x, y), x)$. The representation

$$u = \omega \circ g$$

also gives the general solution of equation (1.4) in the more general case where $g = g(u, x, y)$.

2. The Cauchy method

One of the main tools of solving partial differential equations is the method of solving the first order nonlinear partial differential equation

$$F(x, u, p) = 0. \tag{1.5}$$

Let the initial data be given parametrically on some hypersurface

$$u = u(t), \quad x_i = x_i(t), \quad (i = 1, 2, \ldots, n). \tag{1.6}$$

Here $x = (x_1, x_2, \ldots, x_n)$ are the independent variables, $t = (t_1, t_2, \ldots, t_{n-1})$ are the parameters describing the initial values, $p = (p_1, p_2, \ldots, p_n)$, and $p_i = \partial u/\partial x_i$, $(i = 1, 2, \ldots, n)$ are partial derivatives. The functions $u(t)$, $x_i(t)$ and $F(x, u, p)$ are assumed to be sufficiently many times continuously differentiable.

Definition 1.2. *The problem of finding a solution of equation (1.5) satisfying the initial data (1.6) is called a Cauchy problem.*

The Cauchy method for constructing the solution of the Cauchy problem (1.5), (1.6) reduces this problem to finding a solution of the Cauchy problem of the system of ordinary differential equations, which is called a characteristic system,

$$\frac{dx_i}{ds} = F_{p_i}, \quad \frac{du}{ds} = p_\alpha F_{p_\alpha}, \quad \frac{dp_i}{ds} = -(F_u p_i + F_{x_i}), \ (i = 1, 2, \ldots, n), \tag{1.7}$$

with the initial data at the point $s = 0$:

$$x = x(t), \quad u = u(t), \quad p = p(t).$$

Here $x = x(t)$ and $u = u(t)$ are defined by (1.6), and summation with respect to a repeat index is assumed. The initial data $p(t)$ are found by solving equation (1.5) and the tangent conditions:

$$F(x(t), u(t), p(t)) = 0, \ u_{t_k}(t) = p_\alpha(t)\frac{\partial x_\alpha}{\partial t_k}(t), \ (k = 1, 2, \ldots, n-1). \tag{1.8}$$

As the result of solving the Cauchy problem for the characteristic system one obtains the functions $u(s, t_1, \ldots, t_{n-1})$ and $x_i(s, t_1, \ldots, t_{n-1})$, $(i = 1, 2, \ldots, n)$.

Definition 1.3. *The curve $x(s, t)$ in the space of the independent variables with fixed t, is called a characteristic.*

The solution $u = u(x)$ of the Cauchy problem (1.5), (1.6) is constructed by eliminating the parameters $s, t_1, \ldots, t_{n-1}$ from the equations $x = x(s, t)$ and $u = u(s, t)$. By virtue of the inverse function theorem for the elimination it is sufficient to require the inequality

$$\Delta(s, t_1, \ldots, t_{n-1}) \equiv \frac{\partial(x_1, x_2, \ldots, x_n)}{\partial(s, t_1, \ldots, t_{n-1})} = det \begin{pmatrix} F_{p_i} \\ \partial x_i/\partial t_k \end{pmatrix} \neq 0.$$

Theorem 1.1. *Let the initial data (1.6), (1.8) satisfy the condition*

$$\triangle(0, t_1^0, \dots, t_{n-1}^0) \neq 0 \tag{1.9}$$

at some point $t_0 = (t_1^0, \dots, t_{n-1}^0)$*. The solution* $x = x(s,t)$*,* $u = u(s,t)$*,* $p = p(s,t)$ *of the initial value problem (1.6), (1.8) of the characteristic system (1.7) gives the solution* $u(x)$ *of the Cauchy problem (1.5), (1.6) in some neighborhood of the point* $x(t_0)$*.*

Proof.

By virtue of system (1.7) one finds

$$\frac{dF}{ds} = F_{x_\alpha}\frac{dx_\alpha}{ds} + F_u\frac{du}{ds} + F_{p_\alpha}\frac{dp_\alpha}{ds} \equiv 0.$$

This means that the function $F(x(s,t), u(s,t), p(s,t))$ is an integral of system (1.7). By virtue of the choice of the initial data, one has $F(x(s,t), u(s,t), p(s,t)) = 0$. For the proof of the theorem it is enough to show that the functions p_i coincide with the derivatives $\partial u/\partial x_i$, $(i = 1, 2, \dots, n)$ of the function $u = u(x_1, x_2, \dots, x_n)$, which is recovered from the solution of the Cauchy problem (1.6)–(1.8).

Notice that the determinant of the linear system of the algebraic equations with respect to $y_1, y_2, \dots, y_n$:

$$y_\alpha x_{\alpha s} = u_s, \ \ y_\alpha x_{\alpha t_k} = u_{t_k}, \ \ (k = 1, \dots, n-1) \tag{1.10}$$

is equal to $\triangle(s, t_1, \dots, t_{n-1})$. Since $\triangle(0, t_1^0, \dots, t_{n-1}^0) \neq 0$, the determinant of system (1.10) is not equal to zero in some neighborhood of the point $(0, t_1^0, \dots, t_{n-1}^0)$. Hence, the linear system (1.10) has an unique solution. Because of the chain rule, the change of the variables $(s, t_1, \dots, t_{n-1})$ with $(x_1, \dots, x_n)$ in the function $u(s,t)$ leads to

$$\frac{\partial u}{\partial x_\alpha} x_{\alpha s} = u_s, \ \ \frac{\partial u}{\partial x_\alpha} x_{\alpha t_k} = u_{t_k}, (k = 1, \dots, n-1).$$

Hence, the solution of (1.10) is $y_i = \partial u/\partial x_i$, $(i = 1, 2, \dots, n)$.

To complete the proof of the theorem one needs to prove that the expressions $U_0 \equiv u_s - p_\alpha x_{\alpha s}$, $U_k \equiv u_{t_k} - p_\alpha x_{\alpha t_k}$, $(k = 1, \dots, n-1)$ also vanish.

In fact, by virtue of (1.7) one has $U_0 \equiv 0$ and

$$\frac{\partial U_k}{\partial s} - \frac{\partial U_0}{\partial t_k} = (F_u p_\alpha + F_{x_\alpha})\frac{\partial x_\alpha}{\partial t_k} + F_{p_\alpha}\frac{\partial p_\alpha}{\partial t_k}, \ (k = 1, \dots, n-1). \tag{1.11}$$

Since $F(x(s,t), u(s,t), p(s,t)) = 0$, the differentiation it with respect to t_k gives

$$\frac{\partial F}{\partial t_k} = F_{x_\alpha}\frac{\partial x_\alpha}{\partial t_k} + F_u\frac{\partial u}{\partial t_k} + F_{p_\alpha}\frac{\partial p_\alpha}{\partial t_k} = 0, \ \ (k = 1, 2, ..., n-1).$$

Substituting $F_{x_\alpha}\frac{\partial x_\alpha}{\partial t_k}+F_{p_\alpha}\frac{\partial p_\alpha}{\partial t_k}$ found from these equations into (1.11), they can be rewritten as follows

$$\frac{\partial U_k}{\partial s}=-F_u\left(\frac{\partial u}{\partial t_k}-p_\alpha\frac{\partial x_\alpha}{\partial t_k}\right)=-F_uU_k,\ (k=1,2,\ldots,n-1). \qquad (1.12)$$

Because of the choice of the initial data, $U_k(0,t)=0$. Because of the uniqueness of the solution of the Cauchy problem, the last equations (1.12) have the unique solution $U_k(s,t)=0$.

Comparing the expressions $U_0=0,\ U_k=0$ and system (1.10), one obtains $p_i=\partial u/\partial x_i,\ (i=1,2,\ldots,n)$. •

Remark 1.2. *Another representation of the characteristic system (1.7) is*

$$\frac{du}{p_\alpha F_{p_\alpha}}=\frac{dx_i}{F_{p_i}}=\frac{dp_i}{-(F_up_i+F_{x_i})}=ds,\ (i=1,2,\ldots,n).$$

Remark 1.3. *Let the function $F(x,u,p)$ be linear with respect to the partial derivatives (equation (1.5) is a quasilinear partial differential equation)*

$$F=a_\alpha(x,u)u_{x_\alpha}-a(x,u).$$

Since $F(x,u,p)=0$ is an integral of the characteristic system (1.7), the equation $\frac{du}{ds}=a_\alpha(x,u)p_\alpha$ in the characteristic system can be exchanged with the equation $\frac{du}{ds}=a(x,u)$. Hence, the part of the equations for the functions $x=x(s,t)$, $u=u(s,t)$ in system (1.7) forms a closed system. For these equations there is no necessity to set initial values for the variables $p_i,\ (i=1,2,\ldots,n)$. An application of the Cauchy method to such a type of equations becomes simpler.

Remark 1.4. *If the equation $F(x,u,p)=0$ is linear and homogeneous*[2]*, i.e.,*

$$F=a_\alpha(x)u_{x_\alpha},$$

the general solution of this equation has the form

$$u=\Phi(\varphi_1(x),\varphi_2(x),\ldots,\varphi_{n-1}(x)).$$

Here Φ is an arbitrary function with $n-1$ arguments, the functions $\varphi_i(x),\ (i=1,2,\ldots,n-1)$ are functionally independent solutions of this equation, and

[2] Linear homogeneous equations play a special role in solving a complete system and in using group analysis method.

they are called integrals of equation (1.5). In fact, for a linear homogeneous equation the characteristic system (1.7) is reduced to the system

$$\frac{du}{ds} = 0, \quad \frac{dx_i}{ds} = a_i(x), \ \ (i = 1, 2, \dots, n).$$

The system of ordinary differential equations

$$\frac{dx_i}{ds} = a_i(x), \quad (i = 1, 2, \dots, n).$$

only has $n - 1$ independent integrals $\varphi_i(x) = c_i$, $(i = 1, 2, \dots, n-1)$. Since $\frac{du}{ds} = 0$, the function $u(x)$ is also an integral. Hence, $u(x)$ depends on $\varphi_i(x)$, $(i = 1, 2, \dots, n-1)$:

$$u = \Phi(\varphi_1(x), \varphi_2(x), \dots, \varphi_{n-1}(x)).$$

Let us apply the Cauchy method to the equation[3]

$$\rho_t + c(\rho)\rho_x = 0, \tag{1.13}$$

where $c(\rho)$ is some function of the argument ρ. Analysis of this equation gives the majority of the basic ideas arising in studies of nonlinear hyperbolic equations: numerous physical problems are modelled by this equation. In numerical methods this equation often serves as a model equation on which those or other numerical methods are tested.

The initial data for equation (1.13) are taken on the line $t = 0$. Continuously differentiable solutions of the Cauchy problem are considered. According to the method, one needs to construct the system of characteristics, issuing from the points of the line $t = 0$. These characteristics correspond to the integrals of the characteristic system. Choosing the variable t, instead of s, as the parameter along the characteristic curves, the characteristic system takes the form

$$\frac{dx}{dt} = c(\rho), \quad \frac{d\rho}{dt} = 0. \tag{1.14}$$

Let the initial values at $t = 0$ be

$$x = \xi, \quad \rho = f(\xi), \tag{1.15}$$

where ξ is a parameter. Since the function $\rho(x,t)$ is constant on any characteristic curve, the function $c(\rho)$ is also constant on the characteristic. Thus any characteristic curve of equation (1.13) is a straight line in the (x,t)-plane

[3]If $c'(\rho) \neq 0$, the change $u = c(\rho)$ reduces equation (1.13) to the Hopf equation

$$u_t + uu_x = 0.$$

with the slope $c(\rho)$. The general solution of equation (1.13) is reduced to the construction of the family of the straight lines in the (x, t)-plane. Each of the straight lines has the slope $g(\xi) \equiv c(f(\xi))$, which is defined by the value $\rho = f(\xi)$ at the point $t = 0, \ x = \xi$. The solution of the Cauchy problem (1.14), (1.15) is

$$x = \xi + tg(\xi), \quad \rho = f(\xi). \tag{1.16}$$

This is a parametric representation of the general solution of equation (1.13). Let the function $\xi(x, t)$ be implicitly defined by the first equation (1.16). Differentiating equations (1.16) with respect to x and t , one obtains

$$1 = (1 + tg'(\xi))\xi_x, \qquad \rho_x = f'(\xi)\xi_x,$$
$$0 = (1 + tg'(\xi))\xi_t + g(\xi), \quad \rho_t = f'(\xi)\xi_t.$$

Hence,

$$\rho_t = -\frac{g(\xi)f'(\xi)}{1 + tg'(\xi)}, \quad \rho_x = \frac{f'(\xi)}{1 + tg'(\xi)},$$

and equation (1.13) is satisfied.

From the analytical representation of the solution one notes that for $F'(\xi) < 0$ the derivatives ρ_t, ρ_x can become infinite at the time $t = t_k \equiv -1/F'(\xi)$. This means that the characteristic lines cross and, since ρ has different constants $\rho = f(\xi)$ on each characteristic line, a contradictory result is obtained. Hence, a smooth solution cannot exist for all $t > 0$. The points, where the characteristics cross, is called a gradient catastrophe. The minimum value $\min(t_k(\xi))$ is called the breaking time. A smooth solution of the Cauchy problem of equation (1.13) does not exist from the moment the breaking time occurs, and thereafter the concept of a "solution" requires generalization.

3. Complete and singular integrals

Let us consider a differential equation of first order with two independent variables

$$F(x, y, u, p, q) = 0. \tag{1.17}$$

Here $F_p^2 + F_q^2 \neq 0$, and the usual notations $p = u_x, q = u_y$ are used.

Lemma 1.1. *If a family of solutions $u = f(x, y, a)$ of differential equation (1.17), depending on a parameter a, has an envelope, then this envelope is also a solution.*

Proof.

The envelope of the family $u = f(x, y, a)$ is defined by the formula

$$\psi(x, y) = f(x, y, a(x, y)),$$

where the function $a(x, y)$ is found from the equation $f_a(x, y, a) = 0$. Taking into account

$$f_a(x, y, a(x, y)) \equiv 0,$$

the derivatives of the function $\psi(x, y)$ are

$$\psi_x = f_x + f_a a_x = f_x, \ \psi_y = f_y + f_a a_y = f_y.$$

Since for any a

$$F\left(x, y, f(x, y, a), f_x(x, y, a), f_y(x, y, a)\right) = 0,$$

the function $u = \psi(x, y)$ is also a solution of equation (1.17). •

Let a family of solutions of equation (1.17) have two parameters.

Definition 1.4. *A two–parameter family $u = \phi(x, y, a, b)$ of solutions is called a complete integral of equation (1.17), if in the considered domain the rank r of the matrix*[4]

$$\begin{pmatrix} \phi_a & \phi_{ax} & \phi_{ay} \\ \phi_b & \phi_{bx} & \phi_{by} \end{pmatrix}$$

is equal to two.

Having a complete integral $u = \phi(x, y, a, b)$, one can obtain a set of solutions of equation (1.17) with one arbitrary function. In fact, assume that $a = a(x, y)$ and $b = b(x, y)$. For the function $\psi(x, y) = f(x, y, a(x, y), b(x, y))$ one finds

$$\psi_x = \phi_x + \phi_a a_x + \phi_b b_x, \quad \psi_y = \phi_y + \phi_a a_y + \phi_b b_y.$$

To use the property that the function $u = \phi(x, y, a, b)$ is a solution of equation (1.17), it is natural to require

$$\phi_a a_x + \phi_b b_x = 0, \quad \phi_a a_y + \phi_b b_y = 0. \tag{1.18}$$

If $\phi_a^2 + \phi_b^2 \neq 0$, then the determinant of the homogenous linear system (1.18) with respect to ϕ_a, ϕ_b has to be equal to zero. Hence, the Jacobian $\partial(a, b)/\partial(x, y) = 0$. Thus, for example, $b = \omega(a)$, and equations (1.18) are reduced to the equation

$$\phi_a + \phi_b \omega'(a) = 0.$$

This equation defines the envelope of the family $u = \phi(x, y, a, \omega(a))$. Finding the function $a(x, y)$ from this equation, one obtains the solution $u = \phi(x, y, a(x, y), \omega(a(x, y)))$ with one arbitrary function $\omega(a)$.

The equations

$$\phi_a = 0, \quad \phi_b = 0$$

lead to the concept of a singular integral.

---

[4]The condition $r = 2$ guarantees, that the function ϕ essentially depends on the two independent parameters.

Definition 1.5. *The envelope of a two–parameter family of solutions* $u = \phi(x, y, a, b)$, *obtained by eliminating the parameters* a *and* b *from the equations*

$$\phi_a(x, y, a, b) = 0, \quad \phi_b(x, y, a, b) = 0, \tag{1.19}$$

is called a singular integral of equation (1.17).

For some equations a singular integral can be found without knowing a two–parameter family of solutions. In fact, since a two–parameter family $u = \phi(x, y, a, b)$ of solutions satisfies equation (1.17):

$$F(x, y, \phi(x, y, a, b), \phi_x(x, y, a, b), \phi_y(x, y, a, b)) \equiv 0,$$

differentiating it with respect to the parameters a and b, and using the properties (1.19), one obtains

$$F_p\phi_{xa} + F_q\phi_{ya} = 0, \quad F_p\phi_{xb} + F_q\phi_{yb} = 0.$$

If the determinant of this linear system of algebraic equations with respect to F_p, F_q

$$det \begin{pmatrix} \phi_{ax} & \phi_{ay} \\ \phi_{bx} & \phi_{by} \end{pmatrix} \neq 0,$$

then $F_p = 0$ and $F_q = 0$. Therefore, in this case the singular integral $u = \phi(x, y, a, b)$ satisfies the equations

$$F(x, y, u, p, q) = 0, \quad F_p(x, y, u, p, q) = 0, \quad F_q(x, y, u, p, q) = 0,$$

where $p = \phi_x, \; q = \phi_y$.

To find a two–parameter family of solutions one can apply a method which uses the notion of a completely integrable system[5]. Let us consider the overdetermined system of first order equations:

$$p = f(x, y, u), \; q = g(x, y, u). \tag{1.20}$$

Definition 1.6. *An overdetermined system (1.20) is called completely integrable if the equation*

$$f_y + gf_u = g_x + fg_u \tag{1.21}$$

is identically satisfied with respect to x, y, u.

Theorem 1.2. *If system (1.20) is completely integrable, then its solution is defined and contains one arbitrary constant.*

[5]The more general case of completely integrable systems is studied in the next chapter.

Proof.

Integrating the first equation of system (1.20) with respect to x, one finds a solution $u = \phi(x, y, C(y))$. Here the function $C(y)$ is a constant of integration with respect to the independent variable x. Since $\phi_c \neq 0$, substituting it in the second equation, one obtains

$$\frac{dC}{dy} = \frac{g - \phi_y}{\phi_c}. \tag{1.22}$$

Let us show that the right hand-side in (1.22) does not depend on x. Since

$$\phi_{xc} = f_u\phi_c, \quad \phi_{xy} = f_y + f_u\phi_y,$$

one has

$$\begin{aligned}\frac{\partial}{\partial x}\left(\phi_c^{-1}(g - \phi_y)\right) &= \phi_c^{-2}\left((g_x + g_u\phi_x - \phi_{xy})\phi_c - (g - \phi_y)\phi_{xc}\right)\\ &= -\phi_c^{-1}\left((f_y + gf_u) - (g_x + qg_u)\right) = 0.\end{aligned}$$

Therefore equation (1.22) is an ordinary differential equation with respect to y, and its solution depends on one arbitrary constant, for example, b.•

Assume there is the equation

$$\Phi(x, y, u, p, q) = a \tag{1.23}$$

with a parameter a such that from this equation, and the given equation

$$F(x, y, u, p, q) = 0, \tag{1.24}$$

one can find the derivatives p and q:

$$p = f(x, y, u, a), \quad q = g(x, y, u, a)$$

such that the last overdetermined system is completely integrable. Solving this completely integrable system, one obtains a two-parameter family of solutions of the original equation (1.24): one is the parameter a and another is the constant arising when solving the totally integrable system (the parameter b).

The function $\Phi(x, y, u, p, q)$ in (1.23) can be found by the Lagrange–Charpit method. This method is based on the following idea. Since the functions (1.20) satisfy the equations

$$F(x, y, u, f(x, y, u), g(x, y, u)) \equiv 0, \quad \Phi(x, y, u, f(x, y, u), g(x, y, u)) \equiv a,$$

one can find the derivatives f_y, g_x, f_u, g_u by differentiating these equations with respect to x, y and u:

$$\begin{aligned}f_y &= \Delta^{-1}\left(-F_y\Phi_q + F_q\Phi_y\right), \quad g_x = \Delta^{-1}\left(-F_p\Phi_x + F_x\Phi_p\right),\\ f_u &= \Delta^{-1}\left(-F_u\Phi_q + F_q\Phi_u\right), \quad g_u = \Delta^{-1}\left(-F_p\Phi_u + F_u\Phi_p\right),\end{aligned}$$

where $\Delta = F_p\Phi_q - F_q\Phi_p$. Substituting these derivatives into (1.21), one has

$$\Phi_x F_p + \Phi_y F_q + \Phi_u(qF_q + pF_p) - \Phi_p(F_x + pF_u) - \Phi_q(F_y + qF_u) = 0.$$

This equation has to be satisfied when the expressions (1.20) are substituted into it. Requiring the satisfaction of this equation identically with respect to the variables x, y, u, p, q, one obtains a homogeneous quasilinear differential equation of first order for the function $\Phi(x, y, u, p, q)$. Notice that one of the integrals of this equation is $F(x, y, u, p, q)$.

The concept of complete integral can be generalized for first order equations with many independent variables $x = (x_1, x_2, \dots, x_n)$:

$$F(x, u, p) = 0. \tag{1.25}$$

Here $p = (p_1, p_2, \dots, p_n)$, $p_i = \partial u/\partial x_i$, $(i = 1, 2, \dots, n)$. Let the function $u = \phi(x_1, \dots, x_n, a_1, \dots, a_n)$ be a solution of equation (1.25), where the parameters $a = (a_1, \dots, a_n)$ are fixed.

Definition 1.7. *An n-parameter family of solutions $u = \phi(x_1, \dots, x_n, a_1, \dots, a_n)$ of equation (1.25) is called a complete integral, if the equations*

$$u - \phi(x, a) = 0, \quad p_i - \phi_{x_i}(x, a) = 0, \quad (i = 1, 2, \dots, n) \tag{1.26}$$

can be solved with respect to the parameters $a_1, \dots, a_n$. These representations of the parameters $a_1, \dots, a_n$ have to be such that substituting them into (1.25), one obtains an identity with respect to $2n$ variables, where first of these n variables are $x_1, \dots, x_n$.

Assume that $a = a(x)$, then[6] $p_i = \partial\phi/\partial x_i + (\partial\phi/\partial a_\alpha)(\partial a_\alpha/\partial x_i)$, $(i = 1, 2, \dots, n)$. In order to use the property of the n-parameter family $u = \phi(x_1, \dots, x_n, a_1, \dots, a_n)$ to be a solution, similar to equations (1.19), one can require that the functions $a_i = a_i(x)$, $(i = 1, 2, \dots, n)$ satisfy the equations

$$\frac{\partial\phi}{\partial a_\alpha}\frac{\partial a_\alpha}{\partial x_i} = 0, \quad (i = 1, 2, \dots, n). \tag{1.27}$$

Let us use a complete integral for constructing the solution of the Cauchy problem[7]

$$u = u(t_1, t_2, \dots, t_{n-1}), \quad x_i = x_i(t_1, t_2, \dots, t_{n-1}), \ (i = 1, 2, \dots, n) \tag{1.28}$$

Assume that for some

$$a_i = a_i(t_1, t_2, \dots, t_{n-1}), \quad (i = 1, 2, \dots, n) \tag{1.29}$$

[6]It is assumed summation with respect to a repeated index. Here and further, if it is not specified, the summation with respect to all values of the repeated index, which it can accept, is applied. For example, $a_\alpha x_\alpha = \sum_{\alpha=1}^n a_\alpha x_\alpha$.

[7]This method is different from the Cauchy method for solving the Cauchy problem.

one has

$$\frac{\partial \phi}{\partial a_\alpha}(x(t), a(t))\frac{\partial a_\alpha}{\partial t_j}(t) = 0, \quad (j = 1, 2, \ldots, n-1), \tag{1.30}$$

where $t = (t_1, \ldots, t_{n-1})$. Hence, for any assignments $t = t(x)$ the derivatives of the function $u(x) = \phi(x, a(t(x)))$ are

$$u_{x_j} = \frac{\partial \phi}{\partial x_j} + \frac{\partial \phi}{\partial a_\alpha}\frac{\partial a_\alpha}{\partial t_\beta}\frac{\partial t_\beta}{\partial x_j} = \frac{\partial \phi}{\partial x_j}, \quad (j = 1, 2, ..., n).$$

Thus, the solution of the Cauchy problem can be constructed in the following way. First, one finds the functions $a_i(t_1, \ldots, t_{n-1})$ from the system of equations, which consists of the $(n-1)$ equations (1.30), and the representation of the complete integral with the initial value data (1.28) substituted into it:

$$u(t) = \phi(x(t), a(t)), \quad \frac{\partial \phi}{\partial a_\alpha}(x(t), a(t))\frac{\partial a_\alpha}{\partial t_i}(t) = 0, \quad (i = 1, 2, \ldots, n-1),$$

Then from the second part of the Cauchy data (1.28):

$$x = x(t)$$

one finds $t = t(x)$. The function $u = \phi(x, a(t(x)))$ is the required solution of the Cauchy problem.

4. Systems of linear equations

This section is devoted to solving a linear system of homogeneous first order differential equations with one unknown function $u(x)$, $(x \in R^n)$:

$$X_i(u) \equiv a_{i\alpha} p_\alpha = 0, \quad (i = 1, 2, \ldots, m). \tag{1.31}$$

Here $x = (x_1, x_2, ..., x_n)$, $p_j = \partial u/\partial x_j$, the function $u(x)$ and the coefficients $a_{ij} = a_{ij}(x)$, $(i = 1, 2, \ldots, m;\ j = 1, 2, \ldots, n)$ are assumed to be sufficiently many times continuously differentiable. For the sake of simplicity it is assumed that the rank of the matrix $A = (a_{ij})$ composed of the coefficients $a_{ij} = a_{ij}(x)$, $(i = 1, 2, \ldots, m;\ j = 1, 2, \ldots, n)$ is equal to the number of the equations. This, in particular, means that $m \leq n$. Notice that with this agreement for $m = n$ there is only the trivial solution $u = const$.

Remark 1.5. *Any system of quasilinear equations with one unknown function*

$$b_{i\alpha}(x, u) p_\alpha = b_i(x, u), \quad (i = 1, 2, \ldots, m) \tag{1.32}$$

can be reduced to a linear homogeneous system of the form (1.31). For this purpose one can use an implicit representation of a solution of system (1.31), i.e., the function $Q(x, u)$ with $Q_u \neq 0$, such that $Q(x, u(x))$ is constant for *any solution $u = u(x)$ of system (1.32). The derivatives of the function $u(x)$ are defined through the derivatives of the function $Q(x,u)$:*

$$p_i = -\frac{Q_{x_i}}{Q_u}, \ (i = 1, 2, \dots, n).$$

Substituting p_i, $(i = 1, 2, ..., n)$ into system (1.32), it is reduced to the linear homogeneous system with respect to the $n+1$ independent variables $x_1, x_2, \dots, x_n, u$:

$$b_{i\alpha}(x,u)Q_{x_\alpha} + b_i(x,u)Q_u = 0, \ (i = 1, 2, \dots, m).$$

Consider the linear operators X_i with the properties

$$X_i(u_1 + u_2) = X_i(u_1) + X_i(u_2), \ X_i(u_1u_2) = u_2X_i(u_1) + u_1X_i(u_2),$$

$$X_i(X_j(u)) - X_j(X_i(u)) = [X_i(a_{j\alpha}) - X_j(a_{i\alpha})]\, p_\alpha.$$

Definition 1.8. *The operator $(X_i, X_j)(u) \equiv [X_i(a_{j\alpha}) - X_j(a_{i\alpha})]\, p_\alpha$ is called a Poisson bracket with X_i and X_j.*

It is obvious that if $u(x)$ is a solution of the equations $X_i(u) = 0$ and $X_j(u) = 0$, then it is also a solution of the equation $(X_i, X_j)(u) = 0$. Hence, new linear homogeneous equations can be produced by means of Poisson brackets. If the new equations $(X_i, X_j)(u) = 0$ are linearly independent of the equations of system (1.31), then it becomes necessary to append them to the initial system. Let m' be the number of the equations of the system composed of the equations of the initial system and the new independent equations obtained by taking all possible Poisson brackets. Because n is a finite number, there are only two possibilities: either $m' = n$ or $m' < n$. In the first case any solution is trivial. The second case leads to the following definition.

Definition 1.9. *System (1.31) is called a complete system if any Poisson bracket is linearly dependent on the equations of the initial system (1.31).*

Let $B(x) = (b_{ij}(x))$ be a nonsingular square $m \times m$ matrix. The system $Z_i(u) = 0$, $(i = 1, 2, \dots, m)$ with $Z_i = b_{i\alpha}X_\alpha$ is equivalent to system (1.31).

Lemma 1.2. *A system of linear homogeneous equations, which is equivalent to a complete system, is complete.*

Proof.

Let the system $Z_i(u) = 0$, $(i = 1, 2, \dots, m)$ be equivalent to system (1.31). Hence, there exists a nonsingular square matrix $B = (b_{ij})$ such that $Z_i(u) = b_{i\alpha}X_\alpha(u)$. Taking the Poisson bracket (Z_i, Z_j), one obtains

$$\begin{aligned}(Z_i, Z_j)(u) &= Z_i(Z_j(u)) - Z_j(Z_i(u)) \\ &= b_{i\alpha}X_\alpha(b_{j\beta}X_\beta(u)) - b_{j\alpha}X_\alpha(b_{i\beta}X_\beta(u) \\ &= \left(b_{i\alpha}X_\alpha(b_{j\beta}) - b_{j\alpha}X_\alpha(b_{i\beta})\right) X_\beta(u) + b_{i\alpha}b_{j\beta}(X_\alpha, X_\beta)(u) \\ &= h^\alpha_{ij}X_\alpha(u)\end{aligned}$$

with some functions $h_{ij}^{\alpha} = h_{ij}^{\alpha}(x)$. Since the matrix B is invertible, there exist functions $\phi_i^{\alpha} = \phi_i^{\alpha}(x)$ such that

$$X_i(u) = \phi_i^{\alpha} Z_{\alpha}(u), \ (i = 1, 2, ..., n).$$

Hence, $(Z_i, Z_j)(u)$ is linearly dependent of $Z_{\alpha}(u)$, $(\alpha = 1, 2, ..., n)$.•

By virtue of the lemma's statement, any complete system can be reduced to a complete system of the form

$$X_i(u) = p_i + \sum_{\alpha=m+1}^{n} c_{i\alpha} p_{\alpha} = 0, \ (i = 1, 2, \ldots, m), \tag{1.33}$$

where $c_{i\alpha} = c_{i\alpha}(x)$. Any Poisson bracket of the complete system (1.33) is equal to zero. In fact, the Poisson bracket of the operators (1.33) is

$$(X_i, X_j)(u) = \left(X_i(a_{j\alpha}) - X_j(a_{i\alpha})\right) p_{\alpha} = \sum_{\beta=m+1}^{n} c_{ij}^{\beta} p_{\beta}, \ (i, j = 1, 2, \ldots, m)$$

with some functions $c_{ij}^{\beta} = c_{ij}^{\beta}(x)$. Since system (1.33) is complete, there are functions $h_{ij}^{\beta}(x)$ such that

$$\sum_{\beta=m+1}^{n} c_{ij}^{\beta} p_{\beta} = h_{ij}^{\beta} X_{\beta}.$$

Since in the left-hand side there is no p_{β}, $(\beta \leq m)$, one finds that $h_{ij}^{\beta} = 0$, $(i, j = 1, 2, \ldots, m;\ \beta = 1, 2, \ldots, m)$. These equalities imply $c_{ij}^{\beta} = 0$, $(i, j = 1, 2, \ldots, m;\ \beta = m + 1, \ldots, n)$. Thus, any Poisson bracket of the complete system (1.33) is $(X_i, X_j)(u) = 0$.

Lemma 1.3. *Completeness of system (1.31) is an invariant property with respect to any invertible change of the independent variables.*

Proof.

Let $y_i = \phi_i(x)$, $(i = 1, 2, \ldots, n)$ be the new independent variables. According to the chain rule

$$\frac{\partial}{\partial x_i} = \frac{\partial \phi_{\beta}}{\partial x_i} \frac{\partial}{\partial y_{\beta}}, \ (i = 1, 2, \ldots, n). \tag{1.34}$$

Hence, to prove the lemma, one needs to show that if system (1.31) is complete, then the system

$$X_i'(u) = b_{i\beta}(y) \frac{\partial u}{\partial y_{\beta}} = 0,$$

with $b_{i\beta}(y) = a_{i\alpha}(\phi^{-1}(y))\dfrac{\partial \phi_\beta}{\partial x_\alpha}(\phi^{-1}(y))$ is also complete. Here $\phi^{-1}(y)$ is the inverse function of the function $y = \phi(x)$

$$\phi(\phi^{-1}(y)) \equiv y,\ \phi^{-1}(\phi(x)) \equiv x.$$

Let us consider the Poisson bracket

$$(X_i', X_j')(u) = \left(a_{i\alpha}\frac{\partial \phi_\gamma}{\partial x_\alpha}\frac{\partial}{\partial y_\gamma}(a_{j\delta}\frac{\partial \phi_\beta}{\partial x_\delta}) - a_{j\alpha}\frac{\partial \phi_\gamma}{\partial x_\alpha}\frac{\partial}{\partial y_\gamma}(a_{i\delta}\frac{\partial \phi_\beta}{\partial x_\delta})\right)\frac{\partial u}{\partial y_\beta}$$

$$= \left((a_{i\alpha}\frac{\partial \phi_\gamma}{\partial x_\alpha}\frac{\partial \phi_\beta}{\partial x_\delta}\frac{\partial a_{j\delta}}{\partial y_\gamma} - a_{j\alpha}\frac{\partial \phi_\gamma}{\partial x_\alpha}\frac{\partial \phi_\beta}{\partial x_\delta}\frac{\partial a_{i\delta}}{\partial y_\gamma}) + \right.$$

$$\left. (a_{i\alpha}\frac{\partial \phi_\gamma}{\partial x_\alpha}a_{j\delta}\frac{\partial}{\partial y_\gamma}(\frac{\partial \phi_\beta}{\partial x_\delta}) - a_{j\delta}\frac{\partial \phi_\gamma}{\partial x_\delta}a_{i\alpha}\frac{\partial}{\partial y_\gamma}(\frac{\partial \phi_\beta}{\partial x_\alpha})\right)\frac{\partial u}{\partial y_\beta} = 0.$$

Using (1.34), and collecting terms with the product $a_{i\alpha}a_{j\delta}$, one obtains

$$(X_i', X_j')(u) = (X_i, X_j)(u) + a_{i\alpha}a_{j\delta}\left(\frac{\partial}{\partial x_\alpha}(\frac{\partial \phi_\beta}{\partial x_\delta}) - \frac{\partial}{\partial x_\delta}(\frac{\partial \phi_\beta}{\partial x_\alpha})\right)\frac{\partial u}{\partial y_\beta}$$
$$= (X_i, X_j)(u).$$

Because system (1.31) is complete, there exists a set of functions $c_{ij}^\alpha = c_{ij}^\alpha(x)$ such that

$$(X_i, X_j)(u) = c_{ij}^\alpha X_\alpha(u).$$

This implies

$$(X_i', X_j')(u) = c_{ij}^\alpha X_\alpha'(u).$$

Thus, completeness of system (1.31) is an invariant property with respect to the change of the independent variables.•

Remark 1.6. *During the proof of the lemma, one obtained the representation of the equation $X_i(u) = 0$ after changing the independent variables $y = \phi(x)$:*

$$X_i'(u) = X_i(\phi_\alpha)\frac{\partial u}{\partial y_\alpha} = 0. \qquad (1.35)$$

Here, in the expressions of $X_i(\phi_\alpha)$, one has to substitute $x = \phi^{-1}(y)$.

The next part of the section is devoted to constructing a solution of a complete system (1.31). Without loss of generality the complete system will be considered to be in the form (1.33). The solution is obtained by sequential integration of the equations of the complete system.

Let $\phi_i(x_1, x_2, \ldots, x_n)$, $(i = 2, \ldots, n)$ be the integrals of the first equation of a complete system (1.33)

$$\frac{dx_1}{1} = \frac{dx_2}{0} = \ldots = \frac{dx_{m+1}}{c_{1m+1}} = \ldots = \frac{dx_1}{1} = \frac{dx_n}{c_{1n}}.$$

In particular, one can choose the first $m-1$ functions as $\phi_i \equiv x_i$, $(i = 2, \ldots, m)$. Assume that the function $\phi_1(x_1, x_2, \ldots, x_n)$ is such that the Jacobian $\frac{\partial(\phi_1, \ldots, \phi_n)}{\partial(x_1, \ldots, x_n)} \neq 0$. By virtue of the functional independence of the functions $\phi_i(x_1, x_2, \ldots, x_n)$, $(i = 1, 2, \ldots, n)$ one finds $X_1(\phi_1) \neq 0$. Changing the independent variables x with y:

$$y_i = \phi_i(x_1, x_2, \ldots, x_n), \ (i = 1, 2, \ldots, n),$$

one obtains

$$X_1'(u) = X_1(\phi_1)\frac{\partial u}{\partial y_1} = 0.$$

Since $X_1(\phi_1) \neq 0$, the first equation is reduced to the equation $\partial u/\partial y_1 = 0$. The remaining equations $X_i'(u)$, $(i = 2, 3, \ldots m)$ are now reduced to the form

$$\frac{\partial u}{\partial y_1} = 0, \ Y_i(u) = \frac{\partial u}{\partial y_i} + \sum_{\alpha=m+1}^{n} h_{i\alpha}\frac{\partial u}{\partial y_\alpha} = 0, \ (i = 2, \ldots, m). \tag{1.36}$$

Because the property for system to be complete is invariant with respect to changing the independent variables and reducing it to its equivalent form, system (1.36) is also complete. Since any Poisson bracket of a complete system of the form (1.33) is equal to zero, one obtains

$$(Y_1, Y_i)(u) = (\partial h_{i\alpha}/\partial y_1)\, \partial u/\partial y_\alpha = 0.$$

Hence, the factors h_{ij}, $(i = 2, \ldots, m;\ j = m+1, \ldots, n)$ in system (1.36) are independent of y_1 and, therefore, the system of equations obtained from (1.36) without the first equation, is also complete. In this system the number of the independent variables and the number of the equations is less than in (1.36). Repeating this process again m times, the general solution of system (1.33) is obtained.

5. Tangent transformations

As it has been presented, integration of some classes of differential equations can be essentially simplified by transforming them to a simpler type or to equations, solutions of which are rather well-known. These transformations can include not only independent and dependent variables, but also their derivatives

$$x' = f(x, u, p), \ u' = \phi(x, u, p), \ p' = \psi(x, u, p). \tag{1.37}$$

Here $\alpha = (\alpha_1, \alpha_2, \ldots, \alpha_n)$ is a multiindex[8], p is the vector of the partial derivatives $p_\alpha = \frac{\partial^{|\alpha|} u}{\partial x_1^{\alpha_1} \partial x_2^{\alpha_2} \ldots \partial x_n^{\alpha_n}}$. For the multiindex α the following notations are used $|\alpha| = \alpha_1 + \alpha_2 + \ldots + \alpha_n$ and $\alpha, j = (\alpha_1, \ldots, \alpha_{j-1}, \alpha_j + 1, \alpha_{j+1}, \ldots, \alpha_n)$.

[8] For $\alpha_j = 1$ and $\alpha_i = 0$, $(i \neq j)$ it is assumed that $\alpha = j$.

The transformations (1.37) are prolonged for the differentials dx, du, dp:

$$\begin{aligned} dx_i' &= \frac{\partial f_i}{\partial x_l} dx_l + \frac{\partial f_i}{\partial u} du + \frac{\partial f_i}{\partial p_\alpha} dp_\alpha, \\ du' &= \frac{\partial \phi}{\partial x_l} dx_l + \frac{\partial \phi}{\partial u} du + \frac{\partial \phi}{\partial p_\alpha} dp_\alpha, \\ dp_\gamma' &= \frac{\partial \psi_\gamma}{\partial x_l} dx_l + \frac{\partial \psi_\gamma}{\partial u} du + \frac{\partial \psi_\gamma}{\partial p_\alpha} dp_\alpha, \end{aligned}$$

where $i = 1, 2, ..., n$, the index γ is a multiindex.

Definition 1.10. *Transformation (1.37) is called a tangent transformation if it keeps the tangent conditions*

$$du - p_i dx_i = 0, \ dp_\gamma - p_{\gamma,i} dx_i = 0.$$

If the functions $\phi(x, u, p)$ and $f_i(x, u, p)$, $(i = 1, 2, ..., n)$ do not depend on derivatives, then such a transformation is called a point transformation. The tangent transformation[9], that is defined by the transformation[10] of the independent, dependent variables and the first order partial derivatives, is called a contact transformation[11]. Point and contact transformations play a special role among all tangent transformations. Their role is explained by the Bäcklund theorem, which states that if in a tangent transformation one can find a closed system[12], then such transformation is a prolongation of a point or contact transformation.

Let us consider some examples of the most familiar tangent transformations: Legendre, Hopf–Cole, Bäcklund and Laplace.

5.1 The Legendre transformation

The Legendre transformation is a classical example of a contact transformation. This transformation is defined as follows:

$$\xi_i = u_{x_i}, \ (i = 1, 2, \ldots, n), \ \omega = x_\alpha p_\alpha - u. \tag{1.38}$$

Here $\xi = (\xi_1, \xi_2, \ldots, \xi_n)$ are the new independent variables and $\omega = \omega(\xi_1, \xi_2, \ldots, \xi_n)$ is the new dependent function. For nonsingularity of the Legendre transformation one has to require

$$\det U \neq 0,$$

where U is the matrix, composed of the second order derivatives $U = (u_{x_i x_j})$, similar, $\Omega = (\omega_{\xi_i \xi_j})$.

---

[9] Which is not a point transformation.

[10] Transformations of higher order derivatives are defined through the transformations of the independent, dependent variables and first order partial derivatives by the prolongation formulae and tangent conditions.

[11] It will come up again in Chapter 5, where Lie contact transformations are discussed.

[12] Transformations of the independent, dependent variables and derivatives up to some finite order, for example, N, depend on the independent, dependent variables and derivatives up to the order N.

Taking total derivatives of (1.38) with respect to x_i, and then x_j, one finds

$$\xi_{ix_j} = u_{x_i x_j},\ u_{x_i x_\alpha}\omega_{\xi_\alpha} = x_\alpha u_{x_i x_\alpha},$$
$$u_{x_i x_j} + x_\alpha u_{x_\alpha x_i x_j} = \omega_{\xi_\alpha} u_{x_\alpha x_i x_j} + u_{x_i x_\alpha}\omega_{\xi_\alpha \xi_\beta} u_{x_\beta x_j}.$$

By virtue of nonsingularity of the matrix U, these equations give

$$x_\alpha = \omega_{\xi_\alpha},\ u_{x_i x_j} = u_{x_i x_\alpha} u_{x_j x_\beta}\omega_{\xi_\alpha \xi_\beta}$$

or the last equations can be rewritten in the matrix form $U = U\Omega U$. Hence, the transformations of the first and second order derivatives are

$$\omega_{\xi_i} = x_i,\ (i = 1, 2, ..., n),\ \Omega = U^{-1}.$$

Let us apply the Legendre transformation to the nonlinear differential equation describing a two–dimensional steady flow of a gas:

$$(c^2 - \varphi_x^2)\varphi_{xx} - 2\varphi_x\varphi_y\varphi_{xy} + (c^2 - \varphi_y^2)\varphi_{yy} = 0.$$

Here $\varphi(x, y)$ is the velocity potential, c is the sound speed which is a function of $\varphi_x^2 + \varphi_y^2$. Using the Legendre transformation

$$u = \varphi_x,\quad v = \varphi_y,\quad x = \phi_u,\quad y = \phi_v,\ \phi + \varphi = xu + yv,$$

this equation is reduced to the linear equation:

$$(c^2 - u^2)\phi_{vv} + 2uv\phi_{uv} + (c^2 - v^2)\phi_{uu} = 0. \tag{1.39}$$

The advantage of the equation (1.39) is that this equation is linear. The second advantage of (1.39) is that one can apply[13] the method of separation of variables.

5.2 The Darboux equation

Let us consider the special case of a linear hyperbolic equation of second order[14]

$$u_{xy} + \frac{m}{(x+y)}(u_x + u_y) = 0. \tag{1.40}$$

Applying the tangent transformation

$$u' = \frac{(u_x + u_y)}{(x+y)},$$

[13]For special cases of the state equation.

[14]The gas dynamics equations describing one dimensional isentropic flows of a gas can be reduced to the Darboux equation (see, for example, [147]), which in the general case has the following form

$$u_{xy} + f(x+y)\left(u_x + u_y\right) = 0.$$

For a polytropic gas with the exponent $\gamma = (2m+3)/(2m+1)$ the function $f(x+y) = m(x+y)^{-1}$.

it maps a solution $u(x, y)$ of equation (1.40) into the solution $u'(x, y)$ of the equation

$$u'_{xy} + \frac{m'}{(x+y)}(u'_x + u'_y) = 0, \quad m' = m + 1.$$

This transformation was established by Darboux and it allows changing the value of m into another m'. For example, for $m = 0$ the Darboux equation has the general solution (the d'Alambert integral):

$$u(x, y) = F(x) + G(y)$$

with arbitrary functions $F(x)$, $G(y)$. Hence, if m is a positive integer, then one obtains the general solution of the Darboux equation (1.40):

$$u(x, y) = L^m[F(x) + G(y)], \tag{1.41}$$

where L is the differential operator $L = (x + y)^{-1}(\partial/\partial x + \partial/\partial y)$. The representation (1.41) can be rewritten in the form:

$$u(x, y) = \frac{\partial^{m-1}}{\partial x^{m-1}}(\frac{\Phi(x)}{(x+y)^m}) + \frac{\partial^{m-1}}{\partial y^{m-1}}(\frac{\Psi(y)}{(x+y)^m}) \tag{1.42}$$

with some arbitrary functions $\Phi(x)$ and $\Psi(y)$. In fact, assume that $m = 1$ and $LF(x) = \Phi(x)(x + y)^{-1}$. One can show that for any integer m we have $L^m F(x) = \partial^{m-1}/\partial x^{m-1}(\Phi(x)/(x + y)^m)$. When proving this by induction one can check that

$$L^{m+1}F(x) = LL^m F(x) = L\left(\frac{\partial^{m-1}}{\partial x^{m-1}}(\frac{\Phi(x)}{(x+y)^m})\right) =$$

$$(x+y)^{-1}\left(\frac{\partial^m}{\partial x^m}(\frac{\Phi(x)}{(x+y)^m}) - m\frac{\partial^{m-1}}{\partial x^{m-1}}(\frac{\Phi(x)}{(x+y)^{m+1}})\right) = \frac{\partial^m}{\partial x^m}(\frac{\Phi(x)}{(x+y)^{m+1}}).$$

The representation (1.42) can be generalized for any real m [31].

5.3 The Hopf–Cole transformation

Another very well-known example of tangent transformation, which linearizes a differential equation, is the Hopf–Cole transformation of the Burgers equation:

$$\rho_t + \rho\rho_x = \nu\rho_{xx},$$

where ν is constant. For the sake of convenience the Hopf-Cole transformation is derived in two steps. Let $\rho = \psi_x$, then after integrating the Burgers equation with respect x, one finds

$$\psi_t + \frac{\psi_x^2}{2} = \nu\psi_{xx}.$$

Substituting $\psi = -2\nu \ln(\varphi)$ into the last equation, one obtains the heat equation

$$\varphi_t = \nu\varphi_{xx}.$$

5.4 The Bäcklund transformation

The Hopf–Cole transformation is a particular case of the Bäcklund transformation. The idea of the Bäcklund transformation[15] consists of the following. Let us consider four differential equations

$$F_j(x, t, u, p, \xi, \tau, v, q) = 0, \ \ (j = 1, 2, 3, 4). \qquad (1.43)$$

Here $u = u(x, t)$ is the function of the independent variables (x, t), p is the set of partial derivatives of the function $u(x, t)$, $v = v(\xi, \tau)$ is the function of the independent variables (ξ, τ), and q is the set of partial derivatives of the function $v(\xi, \tau)$ with respect to the independent variables (ξ, τ).

Assume that the function $u = u(x, t)$ is given and two equations of system (1.43) can be solved with respect to the variables (x, t). Substituting the variables x and t into the remaining equations of (1.43), one obtains the overdetermined system of equations

$$\widetilde{F}_j(\xi, \tau, v, q) = 0, \ \ (j = 1, 2).$$

Compatibility conditions for this overdetermined system have the form of differential equations for the function $u = u(x, t)$. If the compatibility conditions can only be expressed through the independent variables (x, t), the function $u = u(x, t)$ and its derivatives, then equations (1.43) are called the Bäcklund transformation of the functions $u(x, t)$ and $v(\xi, \tau)$.

Let us return to the Burgers equation and consider the relations

$$\xi = x, \ \ \tau = t, \ \ \varphi_x + \frac{1}{2\nu}\rho\varphi = 0, \ \ \varphi_t + \frac{1}{2}(\rho\varphi)_x = 0. \qquad (1.44)$$

Finding the derivatives φ_x and φ_t, the compatibility condition in this case consists of the equation

$$(\varphi_x)_t - (\varphi_t)_x = -\frac{\varphi}{2}(\rho_t + \rho\rho_x - \nu\rho_{xx}) = 0.$$

From another point of view, excluding the function ρ from equations (1.44), one obtains

$$\varphi_t + \frac{1}{2}(\rho\varphi)_x = \varphi_t - \nu\varphi_{xx} = 0.$$

Thus, solutions of the heat equation and the Burgers equation are related by the Bäcklund transformation.

[15] Historical review and applications of this method can be found in [72, 71].

6. A linear hyperbolic equation

The Darboux equation is a special case of a second order linear hyperbolic equation with two independent variables

$$u_{xy} + A(x, y)u_x + B(x, y)u_y + C(x, y)u = 0. \tag{1.45}$$

With equation (1.45) one can relate a pair of the functions (h, k).

Definition 1.11. *The functions*

$$h = A_x + AB - C, \quad k = B_y + AB - C$$

are called the Laplace invariants[16] *of equation (1.45).*

If one of the Laplace invariants is equal to zero, then equation (1.45) has a solution, which can be represented as a quadrature. For example, if $h = 0$, then equation (1.45) is rewritten:

$$\left(u_y + Au\right)_x + B\left(u_y + Au\right) = \left(\frac{\partial}{\partial x} + B\right)\left(u_y + Au\right) = 0.$$

Successively solving linear ordinary differential equations with respect to x and, then with respect to y, gives the general solution of equation (1.45) in the case $h = 0$. If the Laplace invariants are not equal to zero, then there is series of equations which can be transformed to equation with zero Laplace invariant. These transformations are due to Laplace. Before presenting these transformations let us justify the name "invariant" for the functions h and k.

Definition 1.12. *Two equations of the type (1.45) are equivalent with respect to the function, if they can be transformed one to another by the transformation*

$$x' = x, \quad y' = y, \quad u = \omega(x, y)u'.$$

Lemma 1.4. *Equations with the Laplace invariants (h', k') and (h, k) are equivalent with respect to the function if and only if*

$$h' = h, \quad k' = k.$$

Proof.

Substituting $u(x, y) = \omega(x, y)u'(x', y')$ into equation (1.45) one obtains for the function $u'(x, y)$ an equation of the type (1.45) with the coefficients

$$A' = A + (\ln\omega)_y, \quad B' = B + (\ln\omega)_x, \quad C' = C + \omega^{-1}\left(A\omega_x + B\omega_y + \omega_{xy}\right). \tag{1.46}$$

[16] The Laplace invariants of linear hyperbolic equation (1.45) can be found as the differential invariants of the equivalence group. This method was applied for other types of equations [75].

Calculating the Laplace invariants for the new equation, one finds

$$h' = A'_x + A'B' - C' = A_x + AB - C \ = h,$$

$$k' = B'_y + A'B' - C' = B_y + AB - C \ = k.$$

Conversely, if $h' = h, \ \ k' = k$, then

$$A'_x + A'B' - C' = A_x + AB - C, \ \ B'_y + A'B' - C' = B_y + AB - C.$$

Hence, $(A' - A)_x = (B' - B)_y$, which means the existence of a function $\omega(x, y)$ such that

$$A' - A = (\ln \omega)_y \,, \ \ B' - B = (\ln \omega)_x \,. \tag{1.47}$$

Substituting A' and B' found from these equalities into the equation $h' = h$, one finds

$$C' = C + \omega^{-1}(A\omega_x + B\omega_y + \omega_{xy}).$$

The last equation and equations (1.47) mean that the substitution $u = \omega u'$ maps equation (1.45) with the coefficients A, B and C into the equation with the coefficients A', B' and C'.•

Corollary. *Equation (1.45) is equivalent to the equation* $u_{xy} = 0$ *if and only if* $h = 0, \ k = 0$.

Apart from transformations of the type

$$x' = \alpha(x), \ \ y' = \beta(y), \ \ u = \omega(x, y)u'$$

there exist other transformations having a similar property to retain the differential structure of the equation of the type (1.45). To define these transformations let us consider the function

$$z_1 = u_y + Au.$$

Since the function u is a solution of equation (1.45), the function z_1 satisfies the equation

$$z_{1x} + Bz_1 = hu. \tag{1.48}$$

Conversely, let z_1 be a solution of equation (1.48), where $h \neq 0$ and let the function u satisfy equation (1.45). Substituting u found from (1.48) into (1.45), one finds that the function z_1 has to satisfy the equation

$$hz_{1xy} + z_{1x}(Ah - h_y) + z_{1y}Bh + z_1\left(h(AB + B_y) - Bh_y - h^2\right) = 0. \tag{1.49}$$

Hence, z_1 also satisfies an equation of the type (1.45) with the coefficients

$$A_1 = A - (\ln\ h)_y, \ \ B_1 = B, \ \ C_1 = C - A_x + B_y - B(\ln\ h)_y$$

and the Laplace invariants

$$h_1 = 2h - k - (\ln h)_{xy}, \quad k_1 = h.$$

Definition 1.13. *Equation (1.49) is called the x-Laplace transformation of equation (1.45).*

Similar the y-Laplace transformation is defined as follows: the equation

$$kz_{2xy} + z_{2x}Bk + z_{2y}(Bk - k_x) + z_2\left(k(AB + A_x) - Ak_x - k^2\right) = 0 \quad (1.50)$$

is the y-Laplace transformation

$$z_2 = u_x + Bu, \quad z_{2y} + Az_2 = ku$$

of equation (1.45). The Laplace invariants of (1.50) are

$$h_2 = k, \quad k_2 = 2k - h - (\ln k)_{xy}.$$

Direct calculations give

$$(k_1)_2 = k, \quad (h_1)_2 = h,$$
$$(k_2)_1 = k, \quad (h_2)_1 = h,$$

where the indexes show the sequence in which the Laplace transformations are taken

$$(k, h) \xrightarrow{x} (k_1, h_1) \xrightarrow{y} (k_2, h_2),$$
$$(k, h) \xrightarrow{y} (k_1, h_1) \xrightarrow{x} (k_2, h_2).$$

For example, the y-Laplace transformation of the equation with the invariants (h_1, k_1) is equivalent to the equation with (h, k).

Let us denote

$$h_0 = h, \quad h_{-1} = k.$$

The initial equation (1.45) is given by the Laplace invariants (h_0, h_{-1}). Defining the recurrence relation

$$h_{n+1} + h_{n-1} = 2h_n - \partial_x\partial_y(\ln h_n),$$

one can directly check that the x-Laplace transformation maps (h_n, h_{n-1}) into (h_{n+1}, h_n), and the y-Laplace transformation removes (h_{n+1}, h_n) into (h_n, h_{n-1}) for any integer n. Thus, one obtains the series

$$\ldots; \ (h_{-2}, h_{-3}); \ (h_{-1}, h_{-2}); \ (h_0, h_{-1}); \ (h_1, h_0); \ (h_2, h_1); \ \ldots,$$

which is called the Laplace series. The shift to the right hand side is made with the help of the x-Laplace transformation, and to the one to the left hand side is made with the help of the y-Laplace transformation. The remarkable property of this series is the fact that, if for some n the invariant $h_n = 0$, then the general solution of the initial equation is given in terms of a quadrature with two arbitrary functions, each of one argument. In fact, let $h_n = 0$ for some n. The equation with the invariants $(0, h_{n-1})$ is factorized and the general solution $u_n = u_n(x, y)$ of this equation is found by quadrature. Applying the y-Laplace transformations

$$u_{i-1} = u_{ix} + B_i u_i, \quad (i = n, n-1, \ldots, 2, 1),$$

one obtains $u_0(x, y)$, which is equivalent to the solution of the initial equation.

7. Construction of particular solutions

An uniform analytical representation of all solutions of a system of partial differential equations is ideal. This is only possible for special classes of equations. Therefore a search for particular classes of exact solutions, containing as many arbitrary functions or constants as possible presents a special interest. Notice also that by "sewing" particular solutions to each other and "duplicating" them, one can construct more general solutions.

The process of finding a particular exact solutions of a system of partial differential equations is carried out assuming additional requirements that must be satisfied by the solution. Usually a representation of the solution is assumed. The form of the representation is defined from a preliminary analysis[17] of the system of equations. It should also be noted that the vast majority of solutions were obtained by "ad hoc" methods[18]. An introduction to some of the "ad hoc" methods is given in this section.

7.1 Separation of variables

This method can be demonstrated with the heat equation

$$u_t = u_{xx}.$$

Assuming that $u(t, x) = \phi(t)\psi(x)$, and substituting it into the heat equation one obtains

$$\frac{\phi'(t)}{\phi(t)} = \frac{\psi''(x)}{\psi(x)}.$$

Since the right hand side is independent of t, and the left is independent of x, they must both be equal to the same constant $a = \pm k^2$. Depending on the sign of the constant a there are two solutions. If $a = k^2$, then $u = e^{at}(c_1 e^{-kx} + c_2 e^{kx})$. If $a = -k^2$, then $u = e^{at}\,(c_1 \sin(kx) + c_2 \cos(kx))$.

---

[17] Each method for constructing exact solutions is a special subject of studying. Some of these methods can be found, for example, in [149, 147, 130, 71, 160, 32, 163, 2] and references therein.

[18] Review of these methods can be found in [3]

For some equations to separate the independent variables one needs to perform transformations. For example, the nonlinear Klein–Gordon equation[19]

$$u_{tt} - u_{xx} = H(u). \tag{1.51}$$

is reduced to the equation

$$4u_{\xi\eta} = H(u), \tag{1.52}$$

after changing the independent variables to $\xi = t+x$, $\eta = t-x$. Assuming that a solution can be presented in the form $F(u) = f(\xi)g(\eta)$ or in the equivalent form $u = G(\zeta)$, where $\zeta = f(\xi)g(\eta)$, equation (1.52) becomes [81]:

$$4f'(\xi)g'(\eta) = H(G)/(G' + \zeta G'').$$

If the right hand side of this equation assumes the value $k(\zeta)$ such that $k(\zeta) = k_1(f)k_2(g)$, then the solution is easily separated. For example, let[20] $k(\zeta) = c_1\zeta^p$. In this case one obtains the equations

$$f' = c_2 f^p, \quad c_2 g' = c_1 g^p, \quad c_1\zeta^p = \frac{H(G)}{(\zeta G')'}.$$

Integrating these equations one finds the solution of equation (1.51) in terms of the separated variables.

It is difficult to answer a priori the question about possibility of separating the independent variables. There is an approach developed in [119] where the separation of variables is related to the infinitesimal properties of the initial equation.

7.2 Self–similar solutions

Closely related with solutions which can be obtained by the method of separation of variables are self–similar solutions[21]. Let us explain the method of obtaining self-similar solutions by using the nonlinear diffusion equation with q-dimensional spherical symmetry

$$r^{1-q}\frac{\partial(r^{q-1}c^n c_r)}{\partial r} = \frac{\partial c}{\partial t}, \quad (n > 0), \tag{1.53}$$

where (t, r) are the independent variables. Assuming that the self-similar solution $c(t, r)$ has the representation with the separated independent variables

$$c = u_0(\xi)V(\eta), \quad \xi = t, \quad \eta = r/R(t),$$

[19]This equation is applied in many physical problems. For example, if $H(u) = \sin(u)$, this equation is called the Sine–Gordon equation.

[20]It is possible to prove that this is the general case of separating $k(\zeta)$.

[21]Study of new models is usually started by considering the self-similar solutions. The theory of self–similar solutions from a group point of view is given in more detail in the next chapter.

and substituting the representation of the solution into (1.53), one has

$$\frac{R^2u_0'}{u_0^{n+1}} - (\frac{R'R}{u_0^n}) \cdot (\frac{\eta V'}{V}) = \frac{(q-1)V^nV' + \eta(V^nV')'}{\eta V}. \tag{1.54}$$

Differentiating it with respect to η, and then ξ, one obtains the equation

$$(\frac{R'R}{u_0^n})'(\frac{\eta V'}{V})' = 0. \tag{1.55}$$

If $(\frac{R'R}{u_0^n})' = 0$, the functions $u_0(t)$, $V(\eta)$ and $R(t)$ satisfy the system of ordinary differential equations

$$\begin{gathered} u_0 = kR^m,\ R' = aR^{mn-1}, \\ a(m - \frac{\eta V'}{V}) = k^n \frac{(q-1)V^nV' + \eta(V^nV')')}{\eta V}, \end{gathered} \tag{1.56}$$

where k, m, a are constants. If $a \neq 0$ and $mn \neq 2$, the solution of the first two equations (1.56) is[22]

$$R = bt^{1/(2-mn)}, \quad u_0 = kb^m t^{m/(2-mn)}. \tag{1.57}$$

In the general case the choice of the self–similar variables can be made on the basis of a dimensional analysis [149], or with the help of the scale group admitted by the initial equations [130].

7.3 Travelling waves

Travelling waves are applied in many areas of science and engineering: mostly in the problems of wave propagation in different media. A solution $u(t, x)$ of a differential equation of the form

$$u = v(x - Dt),$$

is called a travelling wave or wavefront type–solution, propagating along the axis x with a constant phase velocity D. The argument $\xi = x - Dt$ is called a phase of the wave. Usually the problem is to find a solution of an equation with the boundary conditions

$$u(-\infty, t) = u_{-\infty}, \quad u(+\infty, t) = u_{+\infty}.$$

Two such problems are considered here.

The first is a problem of a shock wave structure. The shock wave is regarded as a narrow domain containing sharp changes of the wave parameters. With

[22] Up to the shift with respect to t.

the help of travelling wave type solutions, correct relations for the strong break and estimates of the thickness of a shock layer can be obtained.

Let a smooth solution of the differential equation

$$\rho_t + c(\rho)\rho_x = \nu\rho_{xx}, \quad c(\rho) = Q'(\rho). \tag{1.58}$$

be considered. We seek a solution of (1.58) for which there exist limits: $\rho \to \rho_1$ if $x \to +\infty$ and $\rho \to \rho_2$ if $x \to -\infty$. Assume that the solution has a steady structure propagating along the x-axis with the velocity D:

$$\rho = R(\xi), \quad \xi = x - Dt.$$

The function $R(\xi)$ and the constant D have to be defined. This problem is called a problem of shock wave structure.

Substituting the representation of the solution into (1.58) and integrating, one obtains

$$Q(\rho) - D\rho + A = \nu\rho',$$

where A is a constant of the integration. If there exists a solution with $\rho_x \to 0$ as $x \to \pm\infty$, then

$$Q(\rho_1) - D\rho_1 + A = Q(\rho_2) - D\rho_2 + A = 0.$$

Defining the constant of the integration A from one of these equations and substituting it into another, one obtains

$$Q(\rho_2) - Q(\rho_1) = D(\rho_2 - \rho_1). \tag{1.59}$$

The constructed solution relates two asymptotic values ρ_1 and ρ_2 with the help of $\rho = R(x - Dt)$. The details of the wave structure between the limit values ρ_1 and ρ_2 are defined by the solution of an ordinary differential equation. In hydrodynamics this solution describes the structure of a shock wave.

Another very well-known example of the travelling waves is a solitary wave satisfied by the Korteweg–de Vries equation (KdV)

$$u_t + uu_x + Ku_{xxx} = 0, \quad K > 0. \tag{1.60}$$

Substituting the representation of the wavefront type–solution into (1.60), and integrating twice, one has

$$3Ku_\xi^2 = f(u), \tag{1.61}$$

where $f(u) \equiv -u^3 + 3Du^2 + 6Au + 6B$, A and B are constants of the integration. Assume that the roots of the polynomial $f(u)$ are α, β, γ. There are different cases according to the nature of the roots. Here only one of them is considered, where all three roots are real and satisfy the conditions $\beta = \gamma < \alpha$, i.e., $f(u) = (u - \beta)^2(\alpha - u)$, $D = (\alpha + 2\beta)/3$. Since a real and bounded solution of KdV is studied, the value of u has to belong to the interval (β, α). From (1.61) one finds

$$d\xi = \sqrt{3K}\frac{du}{(u-\beta)\sqrt{\alpha-u}}.$$

Integrating this, one obtains the solution of the KdV equation

$$u(\xi) = \beta + (\alpha-\beta)\text{sech}(\xi\sqrt{\frac{\alpha-\beta}{12K}}).$$

This solution is known as a soliton or solitary wave.

The concept of travelling waves can be generalized for the case of many independent variables $x \in R^n(x)$.

7.4 Partial representation

One of the analytical methods for studying solutions of nonlinear partial differential equations is the method of special series[1]. The special series method is applied for studying singularities of the generalized solutions of nonlinear equations. The representation of a solution in a special series requires a proof of its convergence. Solutions, for which the series is truncated, have a special interest: these solutions consist of finite sums.

Let us consider the equation of a minimum surface

$$(1+u_x^2)u_{yy} - 2u_xu_yu_{xy} + (1+u_y^2)u_{xx} = 0. \tag{1.62}$$

This equation describes behavior of a free liquid surface. A solution of this equation one can seek in the form [167]

$$u(x, y) = g(y) + xh(y).$$

Substituting the representation of the solution into equation (1.62), one obtains

$$(1+h^2)(g'' + xh'') - 2hh'(g' + xh') = 0.$$

Comparing the factors with the same degree of x, one finds

$$(1+h^2)g'' = 2hh'g', \quad g'h'' - h'g'' = 0. \tag{1.63}$$

The general solution of this system is

$$h = \tan(Ay+B), \quad g = C\,\tan(Ay+B) + D$$

with arbitrary constants A, B, C, D.

[1]Review of results obtained by this method can be found in [158].

In the general case of nonlinear differential equations with two independent variables x, t where the function $u(x, t)$ satisfies the nonlinear equation

$$(L_1 + L_2)u = f(x, t) \tag{1.64}$$

there are sufficient conditions [167], for the solution to be represented by a finite exponent series in x

$$u = \sum_{n=0}^{N} g_n(t)x^n.$$

They are as follows. Let $f(x, t)$ be a polynomial in x of a degree not higher than N. Assume that the coefficients $b_{ij}(t, x)$ of the differential operator (the linear part of the operator $L_1 + L_2$)

$$L_1 u = \sum_{i+j \leq M} b_{ij}(t, x) \frac{\partial^{i+j} u}{\partial t^i \partial x^j}$$

are polynomials in x of a degree not higher than j. The nonlinear part L_2 of the differential operator is assumed to consist of sums of the monomials

$$a_k(t) \prod_{i+j \leq M} (\frac{\partial^{i+j} u}{\partial t^i \partial x^j})^{k_{ij}},$$

in which maximum order of derivatives in x, (i.e. $\max_{k_{ij} \neq 0}(j)$) is not more than N. For each monomial one defines the value

$$\sigma_k = (\sum_{i+j \leq M} j k_{ij})/((\sum_{i+j \leq M} k_{ij}) - 1).$$

If in the operator L_2 the values $\sigma_k \geq N$ for any k, one can try to find a solution of the nonlinear differential equation (1.64) in the form

$$u = \sum_{n=0}^{N} g_n(t)x^n.$$

In [51] another approach is being developed. The representation of a solution in this approach is defined on the basis of invariant subspaces. For example, let us consider the equation

$$v_{tt} = Lv, \tag{1.65}$$

where the operator L is

$$Lv = v_{xx} + v_x^2 + v^2.$$

The space of functions $Span\{1, \cos x, \sin x\}$ is invariant with respect to the operator L:

$$L(C_1 + C_2 \cos x + C_3 \sin x) \in Span\{1, \cos x, \sin x\}.$$

Thus one can study the representation of a solution in the form:

$$v = C_1(t) + C_2(t) \cos x + C_3(t) \sin x.$$

Substituting this representation into (1.65), one has

$$\cos(x)\left(C_2'' + C_2(1 - 2C_1)\right) + \sin(x)\left(C_3'' + C_3(1 - 2C_1)\right) + C_1'' - (C_1^2 + C_2^2 + C_3^2) = 0.$$

Splitting this equation, one obtains the system of ordinary differential equations for the coefficients:

$$C_2'' = C_2(2C_1 - 1), \; C_3'' = C_3(2C_1 - 1), \; C_1'' = (C_1^2 + C_2^2 + C_3^2).$$

8. Functionally invariant solutions

One of particular classes of solutions of partial differential equations is a class of functionally invariant solutions[23].

Definition 1.14. *A solution $u(x)$, $x \in R^n$ of partial differential equation is called a functionally invariant solution, if for any function $F : R \to R$ the composition $F(u(x))$ is also a solution of the same equation.*

In this section functionally invariant solutions of linear partial differential equations of second order

$$L(u) \equiv a_{\alpha\beta}(x) u_{x_\alpha x_\beta} + b_\alpha(x) u_{x_\alpha} = 0 \tag{1.66}$$

are considered.

Let $u(x)$ be a functionally invariant solution of equation (1.66). Since $F(u)$ is also a solution of equation (1.66), one obtains

$$\begin{aligned} L(F(u)) &= F''\left(a_{\alpha\beta} u_{x_\alpha} u_{x_\beta}\right) + F'\left(a_{\alpha\beta} u_{x_\alpha x_\beta} + b_\alpha(x) u_{x_\alpha}\right) \\ &= F''\left(a_{\alpha\beta} u_{x_\alpha} u_{x_\beta}\right) = 0. \end{aligned}$$

By virtue of the arbitrariness of the function F, the functionally invariant solution $u(x)$ has to satisfy the equation

$$a_{\alpha\beta} u_{x_\alpha} u_{x_\beta} = 0. \tag{1.67}$$

[23]n [32] this class of solutions is called undisturbed waves. Review of functionally invariant solutions of partial differential equations can be found in [40]

Conversely, if $u(x)$ satisfies (1.66) and (1.67), the function $u(x)$ is a functionally invariant solution of (1.66). Thus, the problem of finding functionally invariant solutions of equation (1.66) is reduced to the problem of solving the overdetermined system of two equations (1.66) and (1.67).

Let us study functionally invariant solutions of the wave equation[24]

$$u_{tt} = u_{xx} + u_{yy}. \tag{1.68}$$

The following statement was proved in [164].

Theorem 1.3. *Any functionally invariant solution $u = \phi(x, y, t)$ of the wave equation (1.68) can be obtained as the result of solving the functional equation*

$$t\, l(\phi) + x\, m(\phi) + y\, n(\phi) = k(\phi), \tag{1.69}$$

where the functions $l(\phi), m(\phi), n(\phi), k(\phi)$ only satisfy the condition

$$l^2 = m^2 + n^2.$$

Proof.

Let $\phi(x, y, t)$ be a functionally invariant solution of (1.68). The function $\phi(x, y, t)$ has to satisfy the equation of the type (1.67), which is

$$(\phi_t)^2 = (\phi_x)^2 + (\phi_y)^2.$$

Differentiating this equation with respect to the independent variables, and making some linear transformations, one finds

$$2\phi_x\phi_y\phi_{xy} = (\phi_y)^2\phi_{xx} + (\phi_x)^2\phi_{yy},$$

$$2\phi_t\phi_x\phi_{tx} = (\phi_x)^2\phi_{tt} + (\phi_t)^2\phi_{xx},$$

$$2\phi_t\phi_y\phi_{ty} = (\phi_y)^2\phi_{tt} + (\phi_t)^2\phi_{yy}.$$

Let $\phi_t\phi_x\phi_y \neq 0$. Calculating the Jacobians

$$\frac{\partial(a_i, \phi)}{\partial(x, y)}, \quad \frac{\partial(a_i, \phi)}{\partial(x, t)}, \quad \frac{\partial(a_i, \phi)}{\partial(y, t)}, \quad (i = 1, 2),$$

where $a_1 = \phi_y/\phi_x$, $a_2 = \phi_t/\phi_x$, and substituting into them the expressions of the mixed derivatives, one obtains

$$\frac{\partial(a_i, \phi)}{\partial(x, y)} = \frac{\partial a_i}{\partial x}\phi_y - \frac{\partial a_i}{\partial y}\phi_x = 0, \quad \frac{\partial(a_i, \phi)}{\partial(x, t)} = \frac{\partial a_i}{\partial x}\phi_t - \frac{\partial a_i}{\partial t}\phi_x = 0,$$

[24] Using functionally invariant solutions of the wave equation V.I.Smirnov and S.L.Sobolev [162] explicitly solved the famous Lamb problem of finding the displacement of an elastic half-plane under the action of a concentrated impulse.

$$\frac{\partial(a_i,\phi)}{\partial(y,t)} = \frac{\partial a_i}{\partial t}\phi_y - \frac{\partial a_i}{\partial y}\phi_t = 0, \quad (i = 1, 2).$$

Hence, $a_i = a_i(\phi), \quad (i = 1, 2)$. Notice that $a_1^2 + 1 = a_2^2$, and the function $g = x + ya_1 + ta_2$ also only depends on ϕ:

$$g(\phi) = x + ya_1(\phi) + ta_2(\phi). \tag{1.70}$$

Differentiating (1.70) with respect to x, one finds $(g' - ya_1' - ta_2')\phi_x = 1$, which means that

$$g' - ya_1' - ta_2' \neq 0.$$

By virtue of the implicit function theorem, for the given functions $g(\phi), a_1(\phi)$ and $a_2(\phi)$ one can define the function $\phi(t, x, y)$ from (1.70). This proves the necessary statement of the theorem.

Introducing the notations

$$h = tl(\phi) + xm(\phi) + yn(\phi) - k(\phi),$$
$$h' = tl'(\phi) + xm'(\phi) + yn'(\phi) - k'(\phi),$$
$$h'' = tl''(\phi) + xm''(\phi) + yn''(\phi) - k''(\phi),$$

the derivatives of the function $\phi(t, x, y)$ become

$$\phi_x = -m/h', \quad \phi_y = -n/h', \quad \phi_t = -l/h',$$

$$\phi_{xx} = [m^2/h']'/h', \quad \phi_{yy} = [n^2/h']'/h', \quad \phi_{tt} = [l^2/h']'/h'.$$

Substituting them into the wave equation (1.68) one obtains the sufficient statement of the theorem.•

Remark 1.7. *The necessary statement of the theorem was obtained assuming that $\phi_t\phi_x\phi_y \neq 0$. Notice that changing the independent variables $x,\ y$ by rotating them through an angle α*

$$x' = x\cos\alpha - y\sin\alpha, \quad y' = x\sin\alpha + y\cos\alpha,$$

the product $\phi_x\phi_y$ is transformed into

$$\phi_{x'}\phi_{y'} = (\phi_x^2 - \phi_y^2)\sin\alpha\cos\alpha + \phi_x\phi_y(\cos^2\alpha - \sin^2\alpha).$$

If for all α one has $\phi_{x'}\phi_{y'} = 0$, then

$$\phi_x = 0, \quad \phi_y = 0.$$

Hence, in this case $\phi_t = 0$ and this solution corresponds to the trivial solution $\phi = const$. Also notice that if $\phi_t = 0$, then $\phi_x = \phi_y = 0$. Therefore, without loss of generality one can assume $\phi_t\phi_x\phi_y \neq 0$.

Remark 1.8. *For $m \neq 0$ the Smirnov–Sobolev formula (1.69) for functionally invariant solution can be rewritten in the form*

$$x + yf(u) + t\sqrt{1 + f^2(u)} + \psi(u) = 0.$$

8.1 Erugin's method

For finding functionally invariant solutions it is necessary to solve the overdetermined system (1.66), (1.67). The method of solving equations (1.66), (1.67) applied in [39] consists of finding a complete integral of equation (1.67), and then using this solution for integrating equation (1.66).

Let us apply this method to the wave equation. The complete integral of the equation

$$u_t^2 = u_x^2 + u_y^2, \tag{1.71}$$

has the form

$$\phi = x\alpha + y\beta + t\sqrt{\alpha^2 + \beta^2} + \gamma, \tag{1.72}$$

where α, β, γ are arbitrary constants. From the complete integral one can obtain all solutions of equation (1.71), by requiring that α, β and γ depend on one or two parameters, and by taking an envelope. Notice also that to use the advantages of the complete integral, the derivatives of the solution $u(x, y, t)$ have to be

$$u_x = \alpha, \quad u_y = \beta, \quad u_t = \sqrt{\alpha^2 + \beta^2}. \tag{1.73}$$

Differentiating these relations with respect to x, y and t, respectively, and substituting them into the wave equation (1.68), one obtains

$$\alpha_x + \beta_y = \frac{\alpha\alpha_t + \beta\beta_t}{\sqrt{\alpha^2 + \beta^2}}, \tag{1.74}$$

where the functions α, β and γ are considered as the functions of the independent variables x, y and t.

Without loss of generality, for studying the dependence of α, β and γ of one or two parameters, it is enough to consider the following two cases: either $\beta = \beta(\alpha),\ \gamma = \gamma(\alpha)$ or $\gamma = \gamma(\alpha, \beta)$.

In the first case ($\beta = \beta(\alpha),\ \gamma = \gamma(\alpha, \beta$) the function $\alpha(x, y, t)$ is found from the equation defining an envelope

$$\frac{\partial u}{\partial \alpha} = x + y\beta' + t\frac{\alpha + \beta\beta'}{\sqrt{\alpha^2 + \beta^2}} + \gamma' = 0. \tag{1.75}$$

This function has to satisfy equation (1.74), which becomes

$$\alpha_x + \beta'\alpha_y = \frac{\alpha\alpha_t + \beta\beta'\alpha_t}{\sqrt{\alpha^2 + \beta^2}}. \tag{1.76}$$

Differentiating (1.75) with respect to x, y and t, one finds

$$1 + u_{\alpha\alpha}\alpha_x = 0, \quad \beta' + u_{\alpha\alpha}\alpha_y = 0, \quad \frac{\alpha + \beta\beta'}{\sqrt{\alpha^2 + \beta^2}} + u_{\alpha\alpha}\alpha_t = 0.$$

Substituting into (1.76) the derivatives α_x, α_y and α_t found from the last equations, one gets the equation

$$1 + (\beta')^2 = (\frac{\alpha + \beta\beta'}{\sqrt{\alpha^2 + \beta^2}})^2,$$

which is equivalent to

$$\alpha\beta' - \beta = 0.$$

The general solution of the last equation is $\beta = c\alpha$, where c is an arbitrary constant. Excluding α from (1.75), and substituting it into the representation of the complete integral, one obtains

$$x + yc + t\sqrt{1 + c^2} + \phi(u) = 0$$

with the arbitrary function $\phi(u)$.

A similar study for the second case $\gamma = \gamma(\alpha, \beta)$ leads to

$$\alpha_x = -\Delta^{-1} u_{\beta\beta}, \quad \beta_x = \Delta^{-1} u_{\alpha\beta}, \quad \alpha_y = \Delta^{-1} u_{\alpha\beta}, \quad \beta_y = -\Delta^{-1} u_{\alpha\alpha},$$

$$\alpha_t = (\Delta\sqrt{\alpha^2 + \beta^2})^{-1}(\beta u_{\alpha\beta} - \alpha u_{\beta\beta}), \quad \beta_t = (\Delta\sqrt{\alpha^2 + \beta^2})^{-1}(-\beta u_{\alpha\alpha} + \alpha u_{\alpha\beta}),$$

where $\Delta = u_{\alpha\alpha} u_{\beta\beta} - u_{\alpha\beta}^2$. Hence, equation (1.74) becomes

$$\alpha^2 u_{\alpha\alpha} + 2\alpha\beta u_{\alpha\beta} + \beta^2 u_{\beta\beta} = 0$$

or after substituting into it the derivatives $u_{\alpha\alpha}, u_{\alpha\beta}, u_{\beta\beta}$

$$\alpha^2 \gamma_{\alpha\alpha} + 2\alpha\beta\gamma_{\alpha\beta} + \beta^2 \gamma_{\beta\beta} = 0.$$

The general solution of this equation is

$$\gamma = (\beta - \alpha)h(\beta/\alpha) + \psi(\beta/\alpha), \tag{1.77}$$

where $h = h(\beta/\alpha)$ and $\psi = \psi(\beta/\alpha)$ are arbitrary functions. The envelope is defined by the equations

$$x + \frac{t\alpha}{\sqrt{\alpha^2 + \beta^2}} + \gamma_\alpha = 0, \quad y + \frac{t\beta}{\sqrt{\alpha^2 + \beta^2}} + \gamma_\beta = 0. \tag{1.78}$$

Taking the linear combination of the first equation multiplied on α and second multiplied on β, and comparing this with (1.72), one finds

$$u = \gamma - \alpha\gamma_\alpha - \beta\gamma_\beta.$$

Because of the representation of γ in (1.77), one obtains $u = \psi(\beta/\alpha)$ or

$$\frac{\beta}{\alpha} = f(u).$$

Substituting the representation (1.77) into (1.78), and forming a linear combination, one finds

$$x + yf(u) + t\sqrt{1 + f^2(u)} = (1 + \beta/\alpha)h(\beta/\alpha).$$

Since the function h is arbitrary, the Smirnov–Sobolev formula is obtained

$$x + yf(u) + t\sqrt{1 + f^2(u)} + \phi(u) = 0,$$

where the functions $f(u)$, $\phi(u)$ are arbitrary.

8.2 Generalized functionally invariant solutions

Generalized functionally invariant solutions can be applied to more general linear differential equation of second order than (1.66)

$$a_{\alpha\beta}(x)u_{x_\alpha x_\beta} + b_\alpha(x)u_{x_\alpha} + c(x)u = 0. \tag{1.79}$$

Definition 1.15. *A function of the form $u(x) = g(x)F(\phi(x))$ is called a generalized functionally invariant solution if it is a solution for any function $F : R \to R$.*

Because of arbitrariness of the function F, one finds that the functions $g(x)$ and $\phi(x)$ have to satisfy the equations

$$\begin{gathered} a_{\alpha\beta}g_{x_\alpha x_\beta} + b_\alpha g_{x_\alpha} + cg = 0, \\ 2a_{\alpha\beta}g_{x_\alpha}\phi_{x_\beta} + g\left(a_{\alpha\beta}\phi_{x_\alpha x_\beta} + b_\alpha\phi_{x_\alpha}\right) = 0, \\ a_{\alpha\beta}\phi_{x_\alpha}\phi_{x_\beta} = 0. \end{gathered} \tag{1.80}$$

Notice that the case where $\phi_{x_\alpha} = 0$ for all α or $g = 0$ are trivial.

Let us study the problem of finding necessary and sufficient conditions for the existence of generalized functionally invariant solutions of the equation

$$u_{xy} + A(x, y)u_x + B(x, y)u_y + C(x, y)u = 0. \tag{1.81}$$

The problem means that the conditions for the coefficients A, B, C have to be found.

By virtue of the definition of a generalized functionally invariant solution, this solution has the form $u = g(x, y)F(\phi(x, y))$. Equations (1.80) become

$$\begin{gathered} g_{xy} + Ag_x + Bg_y + Cg = 0, \\ g_x\phi_y + g_y\phi_x + g\left(\phi_{xy} + A\phi_x + B\phi_y\right) = 0, \\ \phi_x\phi_y = 0. \end{gathered} \tag{1.82}$$

If $\phi_y = 0$, then the first two equations of (1.80) are

$$g_y + Ag = 0, \quad g_{xy} + Ag_x + Bg_y + Cg = 0.$$

Finding $g_{xy} = -(A_x g + A g_x)$ from the first equation differentiated with respect to x, and substituting it into second, one finds

$$A_x + AB - C = 0. \tag{1.83}$$

Similar for the case $\phi_x = 0$ one obtains

$$B_y + AB - C = 0. \tag{1.84}$$

Conversely, conditions (1.83) or (1.84) are sufficient for the existence of generalized functionally invariant solutions. In fact, if one looks for a solution of the form $u(x, y) = w(x, y)v(x, y)$, then

$$wv_{xy} + (w_y + Aw)v_x + (w_x + Bw)v_y + (w_{xy} + Aw_x + Bw_y + Cw)v = 0.$$

Let $A_x + AB - C = 0$. If the function $w(x, y)$ satisfies the constraint

$$w_y + Aw = 0, \tag{1.85}$$

then the function $v(x, y)$ has to satisfy the equation

$$wv_{xy} + (w_x + Bw)v_y = 0.$$

A particular class of solutions of this equation is

$$v(x, y) = F(x).$$

Hence, one finds the generalized functionally invariant solution

$$u(x, y) = w(x, y)F(x),$$

with the function $w(x, y)$, which is a solution of equation (1.85).

Similar in the case $B_y + AB - C = 0$, the function $w(x, y)$ is found from the equation

$$w_x + Bw = 0,$$

and the solution has the form $u(x, y) = w(x, y)F(y)$ with an arbitrary function $F(y)$.

Notice also that if

$$A_x + AB - C = 0, \quad B_y + AB - C = 0,$$

then the function $w(x, y)$ is defined from the compatible system of partial differential equations

$$w_y + Aw = 0, \quad w_x + Bw = 0.$$

The solution is given by the formula $u(x, y) = (\phi_1(x) + \phi_2(y))\, w(x, y)$, where the functions $\phi_1(x)$ and $\phi_2(y)$ are arbitrary.

9. Intermediate integrals

The idea of the intermediate integral method consists of reducing the order of a partial differential equation. This idea can be explained by considering a partial differential equation of second order

$$F(x, u, u_{x_1}, ..., u_{x_n}, u_{x_1x_1}, ..., u_{x_nx_n}) = 0. \tag{1.86}$$

Definition 1.16. *A first order partial differential equation*

$$v(x, u, u_{x_1}, ..., u_{x_n}, u_{x_1x_1}, ..., u_{x_nx_n}) = 0, \tag{1.87}$$

is called an intermediate integral of equation (1.86) if any solution of equation (1.87) is also a solution of equation (1.86).

Hence, with the help of intermediate integrals, solving a partial differential equation is reduced to finding solutions of an equation of smaller order.

9.1 Application to a hyperbolic second order equation

Let us study an intermediate integral of the second order quasilinear partial differential equation

$$s + pA(x, y, u) + qB(x, y, u) + C(x, y, u) = 0, \tag{1.88}$$

where the standard notations are used

$$r = u_{xx}, \quad s = u_{xy}, \quad t = u_{yy}, \quad p = u_x, \quad q = u_y.$$

An intermediate integral is sought in the form

$$v(x, y, u, p, q) = 0. \tag{1.89}$$

Let us obtain necessary conditions for existence of intermediate integrals of equation (1.88). Differentiating equation (1.89) with respect to x and y, one finds the system of linear algebraic equations for the second order derivatives

$$\begin{gathered} s + pA(x, y, u) + qB(x, y, u) + C(x, y, u) = 0, \\ v_p r + v_q s + v_u p + v_x = 0, \\ v_p s + v_q t + v_u q + v_y = 0. \end{gathered} \tag{1.90}$$

Since the general solution of equation (1.89) has a one arbitrary function, the solution of system (1.90) should also have such arbitrariness. Notice that if from a system of partial differential equations one can find all highest order derivatives, then the general solution of this system is defined up to arbitrary constants[25], i.e., there is no functional arbitrariness. Thus

$$\begin{vmatrix} 0 & 1 & 0 \\ v_p & v_q & 0 \\ 0 & v_p & v_q \end{vmatrix} = -v_p v_q = 0.$$

Let $v_p \neq 0, \quad v_q = 0$. Without loss of generality one can take $v = p+g(u,x,y)$. Using the linear dependence with respect to second order derivatives in equations (1.90), one finds

$$q(g_u - B) + g_y + Ag - C = 0.$$

Since $u(x, y)$ is an arbitrary solution of equation (1.89), from the last equation one obtains :

$$g_u - B = 0, \quad g_y + Ag - C = 0. \tag{1.91}$$

Notice that the last equation of system (1.90) is a linear combination of the first and second equations. System (1.91) is an overdetermined system for the function $g(x, y, u)$. From the compatibility condition $g_{uy} = g_{yu}$, one gets

$$AB + B_y + A_u g - C_u = 0.$$

If $A_u = 0$, then

$$B_y + AB - C_u = 0$$

and in this case, these conditions are also sufficient for the existence of an intermediate integral. In fact, these conditions guarantee that the system of equations (1.91) for the function $g(x, y, u)$ comprises a complete system. Solutions (1.91) are defined up to one arbitrary function of one argument.

If $A_u \neq 0$, then

$$g = \frac{B_y + AB - C_u}{A_u}.$$

Substituting the function $g(x, y, u)$ into (1.91), one has

$$\left(\frac{B_y + AB - C_u}{A_u}\right)_u - B = 0, \quad \left(\frac{B_y + AB - C_u}{A_u}\right)_y + A\frac{B_y + AB - C_u}{A_u} - C = 0.$$

These conditions provide the existence of the intermediate integral for equation (1.88).

The case $v_p = 0$ is studied in similar fashion.

Remark 1.9. *If equation (1.88) is a linear homogeneous equation ($A_u = 0$, $B_u = 0$, and $C = h(x, y)u$), then the conditions of the existence of an intermediate integral mean that one of the Laplace invariants is equal to zero.*

[25]This statement is the consequence of the Cartan-Käller theorem of compatibility theory.

However, the intermediate integral method can be applied to a more general class of equations.

Remark 1.10. *An intermediate integral is a differential constraint for a solution of the initial differential equation. This differential constraint is constructed with the requirement that any solution of the differential constraint must be a solution of the initial equation. The method of differential constraints can be considered as a further generalization of the intermediate integral method for constructing exact solutions. Applications of the method of differential constraints to second order partial differential equations can be found in [160], and references therein.*

9.2 Application to the gas dynamic equations

The one dimensional gas dynamic equations in Lagrange variables are

$$\tau_t - u_q = 0, \ \ u_t + p_q = 0, \ \ S_t = 0,$$

where t is the time, q is the mass Lagrange coordinate, u is the velocity, p is the pressure, S is the entropy of a gas, and $\tau = V(p, S)$ is the state equation.

The third equation gives the integral

$$S = S(q).$$

The first and second equations allow the introduction of the functions $\varphi_1(t, q)$ and $\varphi_2(t, q)$ such that

$$d\varphi_1 = \tau\, dq + u\, dt, \ \ d\varphi_2 = u\, dq - p\, dt.$$

Following [106], let us consider the function $\xi(t, q)$ instead of the function $\varphi_2(t, q)$

$$d\xi = d\varphi_2 + d\,(pt) = u\, dq + t\, dp. \tag{1.92}$$

Assuming $p_t \neq 0$, one can choose the new independent variables (p, q). Since of (1.92)

$$u = \xi_q, \ \ t = \xi_p.$$

Hence, taking the differentials

$$du = \xi_{qq} dq + \xi_{pq} dp, \ \ dt = \xi_{pq} dq + \xi_{pp} dp,$$

one obtains

$$d\varphi_1 = \varphi_{1p}\, dp + \varphi_{1q}\, dq = \tau\, dq + u\, dt = (\tau + \xi_q \xi_{pq}) dq + \xi_q \xi_{pp} dp.$$

Since $d\varphi_1$ is a differential, the mixed derivatives of the function φ_1 are equal

$$(\tau + \xi_q \xi_{pq})_p = (\xi_q \xi_{pp})_q.$$

This gives the Monge-Ampere equation

$$\xi_{qq}\xi_{pp} - \xi_{pq}^2 + F^2(p, q) = 0, \tag{1.93}$$

where $F^2(p, q) = -V_p(p, S(q)) \neq 0$.

Let us study the problem: what functions $F(p, q)$ allow equation (1.93) to have an intermediate integral.

The intermediate integral has the form

$$Q(p, q, \xi, \xi_p, \xi_q) = 0. \tag{1.94}$$

Differentiating this equation with respect to p and q, one finds

$$\begin{aligned} Q_{\xi_p}\xi_{pp} + Q_{\xi_q}\xi_{pq} + Q_\xi\xi_p + Q_p = 0, \\ Q_{\xi_p}\xi_{pq} + Q_{\xi_q}\xi_{qq} + Q_\xi\xi_q + Q_q = 0. \end{aligned} \tag{1.95}$$

Since the variables p and q are symmetrically involved in equation (1.93), assume $Q_{\xi_p} \neq 0$. Without loss of generality one can account $Q_{\xi_p} = 1$. Substituting into (1.93) the derivatives ξ_{pp} and ξ_{pq} found from (1.95)

$$\xi_{pp} = Q_{\xi_q}\xi_{qq} + Q_q - Q_p, \ \xi_{pq} = -Q_{\xi_q}\xi_{qq} - Q_q,$$

one has

$$\xi_{qq}(Q_\xi Q_{\xi_q}\xi_q + Q_\xi\xi_p + Q_{\xi_q}Q_q + Q_p) + (Q_\xi\xi_q + Q_q)^2 - F^2 = 0.$$

If the coefficient with ξ_{qq} is not equal to zero, then all second order derivatives are defined and there is no functional arbitrariness. Hence, these coefficients have to vanish

$$\begin{aligned} Q_\xi Q_{\xi_q}\xi_q + Q_\xi\xi_p + Q_{\xi_q}Q_q + Q_p = 0, \\ Q_\xi\xi_q + Q_q = \varepsilon F. \end{aligned}$$

where $\varepsilon = \pm 1$. Substituting $Q = \xi_p + \eta(p, q, \xi, \xi_p, \xi_q) = 0$, and taking some linear combinations, one gets

$$\begin{aligned} \eta_{\xi_q}\varepsilon F - \eta\eta_\xi + \eta_p = 0, \\ \eta_\xi\xi_q + \eta_q = \varepsilon F. \end{aligned}$$

Introducing the function $G = G(\eta, p, q, \xi, \xi_q)$ such that $G_\eta \neq 0$ and $G(\eta(p, q, \xi, \xi_q), p, q, \xi, \xi_q) = 0$, the last system is reduced to the system of linear homogeneous equations

$$\begin{aligned} X_1 G = G_{\xi_q}\varepsilon F - \eta G_\xi + G_p = 0, \\ X_2 G = G_\xi\xi_q + G_q + \varepsilon F G_\eta = 0. \end{aligned}$$

This system can be studied by means of Poisson brackets. Taking the Poisson bracket $[X_1, X_2]G = 0$, one obtains

$$X_3 G = 2F G_\xi - F_q G_{\xi_q} + F_p G_\eta = 0,$$

and then the Poisson brackets $[X_1, X_3]G = 0,\ [X_2, X_3]G = 0$ are

$$X_4G = G_\eta(2FF_{pp} - 3F_p^2) - G_{\xi_q}(2FF_{pq} - 3F_pF_q) = 0,$$
$$G_\eta(2FF_{pq} - 3F_pF_q) - G_{\xi_q}(2FF_{qq} - 3F_q^2) = 0.$$

Since $G_\eta \neq 0$, the determinant of this linear system with respect to the derivatives G_η and G_{ξ_q} has to be equal to zero

$$(2FF_{pp} - 3F_p^2)(2FF_{qq} - 3F_q^2) - (2FF_{pq} - 3F_pF_q)^2 = 0. \tag{1.96}$$

Assuming $2FF_{pp} - 3F_p^2 = 0$, one finds $2FF_{pq} - 3F_pF_q = 0$. The general solution of these equations is

$$F(p, q) = (c_1 p + g(q))^{-2},$$

where $g(q)$ is an arbitrary function and c_1 is constant.

Let $2FF_{pp} - 3F_p^2 \neq 0$. The Poisson bracket $[X_1, X_4]G = 0$ gives one more equation

$$F_{ppq}(2FF_{pp} - 3F_p^2) - F_{ppp}(2FF_{pq} - 3F_pF_q) + 4F_{pp}(F_pF_{pq} - F_{pp}F_q) = 0. \tag{1.97}$$

By virtue of equations (1.96) and (1.97) the other Poisson brackets are equal to zero. Hence, the system $X_iG = 0,\ (i = 1, 2, 3, 4)$ is a complete system. The function $F(p, q)$ has to satisfy the overdetermined system of equations (1.96), (1.97). This system of equations is involutive[26].

For example, for a polytropic gas

$$F^2 = H(q)p^\kappa,\ \kappa = -\frac{\gamma + 1}{\gamma}.$$

Equation (1.97) is satisfied identically, equation (1.96) becomes

$$H'' = \frac{(\kappa + 5)(H')^2}{(\kappa + 4)H}.$$

Without loss of generality one can assume that $H = q^{-(\kappa+4)}$. In this case $F = \varepsilon q^{-2}(p/q)^{\kappa/2}$, and a solution of the equations $X_iG = 0,\ (i = 1, 2, 3, 4)$ can be chosen in the form

$$G = u + p\eta - qu_q - \varphi, \tag{1.98}$$

where the function $\varphi = \varphi(p/q)$ satisfies the relation $\varphi'(\lambda) = -\varepsilon\lambda^{\kappa/2}$.

[26] From equations (1.96) and (1.97) one can define the derivatives of third order F_{ppq}, F_{pqq}, F_{qqq}. The next prolongation does not give new equations.

Remark 1.11. *In [106, 99, 178] there are given the following solutions of the involutive system (1.96), (1.97):* $F = H(c_1 p + c_2 q)$ *and* $F = (q + c_2)^{-2} H\left(\frac{p+c_1}{q+c_2}\right)$ *with an arbitrary function* H *and constant* c_1, c_2.

Remark 1.12. *Applications of the method of differential constraints to equation (1.93) can be found in [160], and the references therein. Equation (1.93) has also been studied by the group analysis method [83].*

Chapter 2

SYSTEMS OF EQUATIONS

In the previous chapter the reader was acquainted with equations having one unknown function. However, almost all partial differential equations can be reduced to a system of quasilinear equations.

The first section of this chapter is devoted to explaining classical knowledge related to systems of quasilinear partial differential equations. The main definitions of quasilinear systems, as well as the notions of characteristics and relations along characteristics are considered in this section.

Systems written in Riemann invariants play an important role in continuum mechanics. In particular, homogeneous systems have solutions, called Riemann waves, where only one Riemann invariant is changed. The well-known problem of the decay of arbitrary discontinuity is solved in terms of Riemann waves. Chapter 2 also contains an application of Riemann waves for describing one-dimensional motion of an elastic-plastic material.

Another method playing a very important role in gas dynamics is the hodograph method. For some problems this method allows linearization. If a hodograph is degenerate, such solutions form a class of solutions called solutions with degenerate hodograph. This class of solutions is considered in the next chapter. Here it is worth mentioning that solutions with degenerate hodograph have a group-invariant nature: they are partially invariant solutions. Another class of solutions which also has group-invariant nature is the class of self-similar solutions. This class of invariant solutions, very well-known in continuum mechanics, is based on the analysis of dimensions of studied quantities. The approach used in the book for introducing self-similar solutions is related to admitted scale groups. This way of studying self-similar solutions can also be considered as an introduction to group analysis method. The theory of self-similar solutions is followed by applications to one-dimensional gas dynamics, in particular to the problem of an intense gas explosion.

Among the approaches using a simple representation of the dependent variables through the independent variables are travelling waves and solutions with linear dependence of velocity with respect to spatial (all or part) independent variables.

The chapter ends with a general study of completely integrable systems.

1. Basic definitions

Most mathematical problems in science are described by systems of partial differential equations. The vast majority of these problems are reduced to problems involving the study of systems of quasilinear partial differential equations. In this section one can find sufficient notations for further reading[1] related with these types of systems.

Let $x = (x_1, x_2, \dots, x_n) \in R^n$ be the independent variables and $u = (u_1, u_2, \dots, u_n), \in R^m$ be the dependent functions.

Definition 2.1. *A system of first order partial differential equations of the form*

$$\sum_{\alpha=1}^{n}\sum_{\beta=1}^{m} a_{i\beta}^{\alpha}(x,u)\frac{\partial u_\beta}{\partial x_\alpha} = f_i(x,u), \ \ (i = 1, 2, \dots, N) \tag{2.1}$$

is called a system of quasilinear partial differential equations.

Any system of quasilinear partial differential equations can be written in the matrix form

$$\sum_{\alpha=1}^{n} A_\alpha \frac{\partial u}{\partial x_\alpha} = f, \tag{2.2}$$

where A_α are rectangle $N \times m$ matrices with the entries $a_{i\beta}^{\alpha}$ (i is the row number, β is the column number), and the vector-columns

$$f = (f_1, f_2, \dots, f_m)^t, \ \ \partial u/\partial x_\alpha = (\partial u_1/\partial x_\alpha, \partial u_2/\partial x_\alpha, \dots, \partial u_m/\partial x_\alpha)^t.$$

System (2.1) or (2.2) is called a determined system if $N = m$, and overdetermined if $N > m$. If the vector f and the matrices A_α, $(\alpha = 1, 2, ..., n)$ do not depend on the independent variables $x = (x_1, x_2, \dots, x_n)$, then the system is called an autonomous system. If the vector $f = 0$, then it is called a homogeneous system.

The matrix

$$A(\xi) = \sum_{\alpha=1}^{n} A_\alpha \xi_\alpha,$$

[1]For more detail study of properties of systems of quasilinear partial differential equations one can read, for example, [147].

associated with system (2.2) is called the characteristic matrix. Here $\xi = (\xi_1, \xi_2, ..., \xi_n) \in R^n$.

Definition 2.2. *A vector ξ is called a normal characteristic vector of system (2.2) at the point (x, u), if*

$$\det(A(\xi)) = 0. \tag{2.3}$$

Equation (2.3) is called the characteristic equation.

For any normal characteristic vector ξ there exists a left eigenvector $l(\xi) = (l_1(\xi), l_2(\xi), \ldots, l_m(\xi))$ of the matrix $A(\xi)$

$$l(\xi)A(\xi) = 0. \tag{2.4}$$

Let $\eta \in R^n$ be a unit fixed vector, then any vector ξ can be decomposed

$$\xi = z\eta + \sigma,$$

where σ is a vector orthogonal to η.

Definition 2.3. *System (2.2) is called a hyperbolic system at a point (x, u), if there exists a vector η, such that for any vector σ, which is orthogonal to η, the characteristic equation*

$$\det(A(z\eta + \sigma)) = 0$$

has m real roots with respect to z and the set of eigenvectors l^k, $(k = 1, 2, \ldots, m)$ corresponding to these roots, composes a basis in R^m. A system is called a hyperbolic system in a domain D, if it is hyperbolic at any point $(x, u) \in D$ (the vector η can depend on (x, u)). A direction defined by the vector η is called a hyperbolic direction.

Let a solution $u = \phi(x)$ of system (2.2) be given. The point $(x, u) = (x, \phi(x))$ is defined by the point x.

Definition 2.4. *A surface $\Gamma \subset R^n(x)$ such that a tangent hyperplane at any $x \in \Gamma$ has a characteristic direction is called a characteristic surface of system (2.2) on this solution, or simply a characteristic.*

The problem of finding a characteristic given by the equation $h(x) = 0$ is reduced to the problem of solving the equation

$$\det(A(\nabla h)) = 0.$$

This equation is a first order partial differential equation. The characteristics of this equation are called bicharacteristics of system (2.2).

Remark 2.1. *If one of the matrices A_i, $(i = 1, 2, ..., n)$ is nonsingular, then without loss of generality one can assume that A_1 is the unit matrix.*

Remark 2.2. *The concept of characteristic surfaces plays an important role in the process of constructing solutions. Their allocations give information about the qualitative behavior of the solution, and about the correctness of initial and boundary value problems. Moreover, they play a crucial role in defining the so called "simple" solutions. Sewing together simple solutions one can find more "complicated" solutions. The classical example of this construction is the solution of the decay of an arbitrary discontinuity problem in gas dynamics.*

2. Riemann invariants

In this section the Riemann invariants for a hyperbolic system of quasilinear partial differential equations with two independent variables[2]

$$\frac{\partial u}{\partial x_1} + A\frac{\partial u}{\partial x_2} = f. \tag{2.5}$$

are considered. The direction of hyperbolicity of system (2.5) is assumed $\eta = (1, 0)$. Hence, the matrix A has m real eigenvalues $\lambda_k(x, u)$ with the left eigenvectors $l^k(x, u)$, $(k = 1, \ldots, m)$ of the matrix A. The eigenvectors $l^k(x, u)$, $(k = 1, \ldots, m)$ form a basis for R^m. Hence, the matrix $L = (l_j^k)$ comprising the coordinates $l_j^k(x, u)$ of the vectors $l^k(x, u)$ is nonsingular. If the characteristic, corresponding to the eigenvalue $\lambda_k(x, u)$ of system (2.5) is defined by the equation $h_k = x_2 - \phi_k(x_1) = 0$, then the function $\phi_k(x_1)$ satisfies the equation $d\phi_k/dx_1 = \lambda_k$.

Multiplying system (2.5) by the left eigenvectors $l^k(x, u)$, one obtains:

$$l^k(x, u)\left(\frac{\partial u}{\partial x_1} + \lambda_k(x, u)\frac{\partial u}{\partial x_2}\right) = l^k f, \quad (k = 1, 2, \ldots, m). \tag{2.6}$$

Let each of the differential forms

$$\omega_k(x, u, du) = l_\alpha^k \, du_\alpha, \quad k = 1, 2, \ldots, m$$

with fixed x_1, x_2 have the integrating factor[3] $\mu_k(x, u)$, i. e.,

$$\mu_k(x, u)\omega_k(x, u, du) = \frac{\partial r_k(x, u)}{\partial u_\alpha} du_\alpha, \quad (k = 1, 2, \ldots, m).$$

Hence, after multiplying equations (2.6) by μ_k, they take on the form

$$D_{x_1} r_k + \lambda_k D_{x_2} r_k = g_k(x, u), \quad (k = 1, 2, \ldots, m),$$

---

[2] A notation of Riemann invariants in the case of more than two independent variables is given in the next Chapter.

[3] In the case $m = 2$ the integrating factor always exists. This is not so for $m > 2$.

where $g_k = \mu_k f_k + r'_{kx_1} + \lambda_k r'_{kx_2}$, D_{x_i} is the total derivative operator with respect to x_i, and the value r'_{kx_i} is the partial derivative of the function r_k with respect to x_i with fixed values of the dependent variables $u_1, u_2, \ldots, u_m$. Since the Jacobian

$$\frac{\partial(r_1, \ldots, r_m)}{\partial(u_1, \ldots, u_m)} = \det(L) \neq 0,$$

changing the dependent variables $(u_1, \ldots, u_m)$ into the new dependent functions $(r_1, \ldots, r_m)$, one reduces system (2.5) to the system of quasilinear partial differential equations

$$\frac{\partial r_k}{\partial x_1} + \lambda_k(x, u)\frac{\partial r_k}{\partial x_2} = g_k, \quad (k = 1, 2, \ldots, m). \tag{2.7}$$

The values r_k are called Riemann invariants and system (2.7) is a system written in invariants. If system (2.5) is homogeneous and autonomous ($A = A(u)$, $f = 0$), then the system written in invariants (2.7) is also homogeneous. If $g_k = 0$ in (2.7), then the Riemann invariant r_k is constant along the characteristic $\frac{dx}{dt} = \lambda_k$.

Let a system written in Riemann invariants (2.7) be homogeneous $g_i = 0$, $(i = 1, 2, \ldots, m)$. Assume that all Riemann invariants except one, for example r_k, are constant, i. e.,

$$r_i = c_i, \quad (1 \leq i \leq m, \ i \neq k).$$

The Riemann invariant r_k satisfies the equation

$$\frac{\partial r_k}{\partial x_1} + \lambda_k(\hat{x}, r_k)\frac{\partial r_k}{\partial x_2} = 0,$$

where $\lambda_k(\hat{x}, r_k) = \lambda_k(x, c_1, \ldots, c_{k-1}, r_k, c_{k+1}, \ldots, c_m)$. Such a solution is called a simple wave or a Riemann wave.

2.1 The problem of stretching an elastic–plastic bar

The equations, governing a one–dimensional time dependent motion of an elastic–plastic material with the state equation $\sigma = f(\varepsilon)$, $f' = \phi^2(\varepsilon) > 0$, are

$$v_t = \sigma_x, \quad \varepsilon_t = v_x. \tag{2.8}$$

Taking linear combinations of the first equation and second multiplied by $\phi(\varepsilon)$, one obtains

$$(v_t + \phi(\varepsilon)\varepsilon_t) - \phi(\varepsilon)(v_x + \phi(\varepsilon)\varepsilon_x) = 0,$$
$$(v_t - \phi(\varepsilon)\varepsilon_t) + \phi(\varepsilon)(v_x - \phi(\varepsilon)\varepsilon_x) = 0.$$

Hence, the Riemann invariants and characteristic eigenvalues are

$$r_1 = v + F(\varepsilon), \quad \lambda_1 = -\phi(\varepsilon),$$
$$r_2 = v - F(\varepsilon), \quad \lambda_2 = \phi(\varepsilon),$$

where $F(\varepsilon) = \int_{\varepsilon_0}^{\varepsilon} \phi(\eta)d\eta$. The original system of partial differential equations (2.8) becomes

$$\frac{\partial r_1}{\partial t} + \lambda_1(r_1, r_2)\frac{\partial r_1}{\partial x} = 0,$$
$$\frac{\partial r_2}{\partial t} + \lambda_2(r_1, r_2)\frac{\partial r_2}{\partial x} = 0,$$

The dependent variables are recovered through the Riemann invariants

$$v = \frac{r_1 + r_2}{2}, \quad F(\varepsilon) = \frac{r_1 - r_2}{2}.$$

Let us pose the problem of stretching a semi-infinite bar. Assume that at the initial time $t = 0$ an elastic–plastic half–infinitely long bar is in the unperturbed state:

$$v(x, 0) = 0, \varepsilon(x, 0) = \varepsilon_0, \quad x \geq 0.$$

The end of the bar $x = 0$ at the time $t = 0$ starts stretching with the velocity

$$v(0, t) = -u(t). \tag{2.9}$$

It is assumed that $u(0) = 0$ and $u'(t) \geq 0$. The problem is to find the loading wave propagating in the bar.

The solution of this problem can be constructed with the help of Riemann waves.

Since the Riemann invariants are constant along their own characteristics, the solution in the domain, joining to the initial data $V_1 = \{(x, t) \mid 0 \leq t \leq \infty, \ 0 \leq x \leq t\phi(\varepsilon_0)\}$, is constant: $v = 0, \varepsilon = \varepsilon_0$. Because the Riemann invariant r_1 is constant along the characteristics $dx/dt = -\phi$ (which cross the characteristic curve $x = t\phi(\varepsilon_0)$) and have a constant value on the characteristic $x = t\phi(\varepsilon_0)$, this invariant is also constant in the domain V_2, joining to V_1. This means, that in this domain one obtains the Riemann wave: $r_1 = v + F(\varepsilon) = F(\varepsilon_0)$. In this Riemann wave the other Riemann invariant r_2 constant along the characteristics $dx/dt = \phi$, which are straight lines. Hence, the solution in the domain V_2 is defined by the values $v = -u(t)$ and $F(\varepsilon) = F(\varepsilon_0) + u(t)$ at the point $(x_0(t), t)$, where $x_0(t) = -\int_0^t u(\tau)\,d\tau$. The relation $u'(t) \geq 0$ provides the condition that characteristics $\frac{dx}{dt} = \phi$ intersect in the domain V_2. If the condition $u'(t) \geq 0$ is broken, this leads to the formation of a gradient catastrophe. The relation $u(0) = 0$ gives a nonsingular Riemann wave. If $u(0) > 0$, then the part of the domain V_2 is occupied by the rarefaction Riemann wave. This part is bounded by the characteristic $x = t\phi(\varepsilon_1)$, where $F(\varepsilon_1) = F(\varepsilon_0) + u(0)$. The deformation ε in this domain is defined by the equation $\phi(\varepsilon) = \frac{x}{t}$. The characteristics $\frac{dx}{dt} = \phi$ issue from the origin $(x, t) = (0, 0)$.

3. Hodograph method

The basic idea of the hodograph method consists of interchanging the role of the dependent and independent variables. For some classes of equations this method reduces them to a system of linear partial differential equations. The essence of the hodograph method is described by the system of the equations, governing two–dimensional irrotational isentropic flows of a gas ($\nu = 0, 1$):

$$u_y - v_x = 0, \ (u^2 - c^2)u_x + 2uvu_y + (v^2 - c^2)v_y = \frac{\nu}{y}c^2 v, \tag{2.10}$$

where (u, v) is the velocity, c is the sound speed, which is expressed through the value $q^2 = u^2 + v^2$. The plane $R^2(u, v)$ is called a hodograph plane, and (u, v) are the hodograph variables.

Assume that the Jacobian[4] $\triangle = \frac{\partial(u,v)}{\partial(x,y)} \neq 0$. Choosing (u, v) as the new independent variables, one can find

$$x = x(u, v), \quad y = y(u, v).$$

Differentiating these relations with respect to x and y, one obtains

$$1 = x_u u_x + x_v v_x, \quad 0 = y_u u_x + y_v v_x,$$
$$0 = x_u u_y + x_v v_y, \quad 1 = y_u u_y + y_v v_y.$$

Since $\triangle \neq 0$, one finds

$$u_x = \triangle y_v, \quad v_x = -\triangle y_u, \quad u_y = -\triangle v_v, \quad v_y = \triangle x_u. \tag{2.11}$$

The two-dimensional gas dynamics equations (2.10) may then be written

$$x_v - y_u = 0, \ (u^2 - c^2)y_v + 2uvx_v + (v^2 - c^2)x_u = \frac{\nu}{y}c^2 v(x_u y_v - x_v y_u). \tag{2.12}$$

The first equation of (2.12) leads to the existence of a potential: a function $\phi = \phi(u, v)$ such that

$$x = \phi_u, \quad y = \phi_v.$$

The second equation of (2.12) becomes the equation for the function ϕ

$$(u^2 - c^2)\phi_{vv} + 2uv\phi_{uv} + (v^2 - c^2)\phi_{uu} = \frac{\nu c^2 v}{\varphi_v}(\phi_{uu}\phi_{vv} - \phi_{uv}^2). \tag{2.13}$$

Equation (2.13) assumes a specially simple form in the case $\nu = 0$ in the polar coordinates ($u = q\cos\theta, v = q\sin\theta$):

$$(M^2 - 1)\phi_{\theta\theta} - q^2\phi_{qq} + (M^2 - 1)q\phi_q = 0, \tag{2.14}$$

where $M = q/c$ is the Mach number. The most important property of equation (2.14) consists of its linearity. Thus the hodograph transformation can simplify

[4]Vanishing of the Jacobian $\triangle$ defines a class of solutions which are called solutions with degenerate hodograph. A detailed study of these solutions is given in the next Chapter.

the original system of equations. It should be also noted that equation (2.14) allows finding solutions with separated variables q and θ. In fact, substituting the representation of the solution

$$\phi = h(\theta)g(q)$$

into (2.14), one obtains

$$\frac{h''}{h} - \frac{q}{g(M^2-1)}\left(g'' - g'(M^2-1)\right) = 0.$$

Alongside the complete interchange of the dependent and independent variables, sometimes it is convenient to make only a partial change of them. For example, for the equations of a stationary boundary layer

$$\frac{\partial\psi}{\partial y}\frac{\partial^2\psi}{\partial x\partial y} - \frac{\partial\psi}{\partial x}\frac{\partial^2\psi}{\partial y^2} = U\frac{dU}{dx} + \nu\frac{\partial^3\psi}{\partial y^3}, \quad U = U(x),$$

applying the transformation from the independent variables (x, y) to the Prandtl–Mises variables: $\xi = x, \eta = \psi(x, y)$, one obtains the equations

$$\frac{\partial u^2}{\partial\xi} - \frac{dU^2}{d\xi} = 2\nu u\frac{\partial}{\partial\eta}(u\frac{\partial u}{\partial\eta}),$$

where $u = 1/(\partial y/\partial\eta)$. Introducing the variable $2z = U^2 - u^2$, one finds the equation of a boundary layer in the Mises form

$$\frac{\partial z}{\partial\xi} = \nu\sqrt{U^2 - 2z}\frac{\partial^2 z}{\partial\eta^2}.$$

4. Self–similar solutions

4.1 Definitions and basic properties

One of the modelling stages of a problem in continuum mechanics is the dimensional analysis of the quantities of the variables involved. This analysis also allows forming representations of solutions, which are called self-similar solutions. One example of a self-similar solution was presented in the first chapter. Here the main definitions and properties of self–similar solutions are given. Since the basis for dimensional analysis is a scale group, the given approach is based on the concept of an admitted scale group.

Let $(x_1, \ldots, x_n)$ and $(u_1, \ldots, u_m)$ be the independent and dependent variables.

Definition 2.5. *A transformation $h_a : R^{n+m} \to R^{n+m}$ of the form*

$$x_i' = x_i \prod_{\alpha=1}^{r}(a_\alpha)^{\lambda_\alpha^i}, \quad u_k' = u_k \prod_{\alpha=1}^{r}(a_\alpha)^{\mu_\alpha^k}, \qquad (i = 1, \ldots, n; \;\; k = 1, \ldots, m) \tag{2.15}$$

is called a scale group H^r of transformations of the space $R^{n+m}(x, u)$. The variables a_α $(\alpha = 1, \ldots, r)$ are called its parameters.

It is natural to require

$$\equiv rank \begin{pmatrix} \lambda_1^1 & \ldots & \lambda_1^n & \mu_1^1 & \ldots & \mu_1^m \\ \ldots & \ldots & \ldots & \ldots & \ldots & \ldots \\ \lambda_r^1 & \ldots & \lambda_r^n & \mu_r^1 & \ldots & \mu_r^m \end{pmatrix} = r. \tag{2.16}$$

Otherwise, introducing new scale parameters, it is possible to reduce the number r.

Under action of the transformation (2.15) the first order derivatives are transformed according to the formulae

$$\frac{\partial u_k'}{\partial x_i'} = \frac{\partial u_k}{\partial x_i} \prod_{\alpha=1}^{r} (a_\alpha)^{\mu_\alpha^k - \lambda_\alpha^i}. \tag{2.17}$$

Similar formulae are valid for higher order derivatives.

Definition 2.6. *The group of transformations, consisting of the transformations of the independent, dependent variables (2.15) and the derivatives (2.17), is called a prolonged group of H^r.*

For any function $F : R^{n+m} \to R$ the total derivative with respect to the parameter a_j is

$$\frac{\partial F(x', u')}{\partial a_j} = \frac{1}{a_j}\left(\sum_{\alpha=1}^{n} \lambda_j^\alpha x_\alpha' \frac{\partial F(x', u')}{\partial x_\alpha'} + \sum_{\gamma=1}^{m} \mu_j^\gamma u_\gamma' \frac{\partial F(x', u')}{\partial u_\gamma'}\right). \tag{2.18}$$

Definition 2.7. *The linear differential operator*

$$\zeta_j \partial = \sum_{\alpha=1}^{n} \lambda_j^\alpha x_\alpha \frac{\partial}{\partial x_\alpha} + \sum_{\gamma=1}^{m} \mu_j^\gamma u_\gamma \frac{\partial}{\partial u_\gamma}$$

is called an infinitesimal operator of the group H^r.

Definition 2.8. *A function $F : R^{n+m} \to R$ is called an invariant of the group H^r, if for any transformation $h_a \in H^r$:*

$$F(x', u') = F(x, u). \tag{2.19}$$

Theorem 2.1. *A function $F : R^{n+m} \to R$ is an invariant of the group H^r if and only if*

$$\zeta_j \partial F = 0, \quad (j = 1, \ldots, r). \tag{2.20}$$

Proof.
If the function $F(x,u)$ is invariant, then $F(x',u') = F(x,u)$. Differentiating it with respect to the parameter a_j, $(j = 1,\dots,r)$, and setting the parameters $a_i = 1$, $(i = 1,2,...,r)$, one obtains (2.20). Conversely, if (2.20) is valid, then by virtue of (2.18) one has $\partial F(x',u')/\partial a_j = 0$, $(j = 1,\dots,r)$. This means that $F(x',u')$ does not depend on the parameters a_j, $(j = 1,\dots,r)$. Since for $a_i = 1$, $(i = 1,2,...,r)$ its value is equal to $F(x,u)$, hence, $F(x',u') = F(x,u)$. Therefore F is an invariant of the group H^r.•

Theorem 2.2. *For a scale group H^r of the space $R^{n+m}(x,u)$ with the condition $r < n+m$ there exist $n+m-r$ independent invariants. They are the monomials*

$$J^k(x,u) = \prod_{\alpha=1}^{n}(x_\alpha)^{\theta_\alpha^k}\cdot\prod_{\beta=1}^{m}(u_\beta)^{\sigma_\beta^k},\quad (k = 1,\dots,m+n-r). \tag{2.21}$$

Proof.
By virtue of the criterion, the monomial

$$J(x,u) = \prod_{\alpha=1}^{n}(x_\alpha)^{\theta_\alpha}\cdot\prod_{\beta=1}^{m}(u_\beta)^{\sigma_\beta}$$

is an invariant of the group H^r, if and only if

$$(\zeta_j\partial)J = J(\sum_{\alpha=1}^{n}\lambda_j^\alpha\theta_\alpha + \sum_{\gamma=1}^{m}\mu_j^\gamma\sigma_\gamma) = 0,\quad (j = 1,\dots,r). \tag{2.22}$$

Because of (2.16), the system of linear algebraic equations (2.22) with $n+m$ unknown θ_α, σ_γ has $n+m-r$ linearly independent solutions. Let the solutions be

$$(\theta_1^k,\dots,\theta_n^k,\sigma_1^k,\dots,\sigma_m^k),\quad (k = 1,\dots,m+n-r). \tag{2.23}$$

Then the monomials (2.21) with exponents (2.23) are also independent. By virtue of the exponent representation of the monomials (2.21), it is enough to prove the independence of their exponent.

Assume that there exist constants c_k, $(k = 1,\dots,m+n-r)$, at least one of them is not equal to zero $(c_1^2 + c_2^2 + \dots + c_{m+n-r}^2 \neq 0)$, and for which one has the equality

$$\prod_{k=1}^{n+m-r}(J^k)^{c_k} = 1.$$

This is only possible if

$$\sum_{k=1}^{n+m-r}\theta_j^k c_k = 0,\quad \sum_{k=1}^{n+m-r}\sigma_\gamma^k c_k = 0,\quad (j = 1,\dots,n;\ \gamma = 1,\dots,m),$$

But by virtue of the linear independence of (2.23), one obtains $c_k = 0$, $(k = 1, \ldots, m+n-r)$. This contradicts to the assumption $c_1^2 + c_2^2 + \ldots + c_{m+n-r}^2 \neq 0$.•

Definition 2.9. *A manifold assigned by the equations* $\phi_k(x, u) = 0$, $(k = 1, 2, \ldots, l)$, *is called an invariant manifold of the group* H^r, *if*

$$\phi_k(x', u') = 0, (k = 1, 2, \ldots, l).$$

The concept of a scale group is closely related with the dimensional analysis theory of physical quantities [130].

Each physical quantity ϕ is characterized by a unit of its measurement E and a numerical value $|\phi|$, hence, $\phi = |\phi|E$. Let E_α, $(\alpha = 1, \ldots, r)$ be some independent units of measurements that E is expressed through them $E = \prod_{\alpha=1}^{r} E_\alpha^{\lambda_\alpha}$. The value E is called the dimension of the physical quantity ϕ in the terms of the units E_α and it is denoted $[\phi]$.

If one changes the scales of the units E_i into the new units E_i'

$$E_i = a_i E_i', \quad (i = 1, \ldots, r),$$

the numerical value of the physical quantity ϕ in the new units is

$$\phi = |\phi|[\phi] = |\phi|(\prod_{\alpha=1}^{r} (a_\alpha)^{\lambda_\alpha})(\prod_{\alpha=1}^{r} (E_\alpha')^{\lambda_\alpha}) = |\phi|'[\phi]'.$$

This means that $|\phi|' = |\phi|(\prod_{\alpha=1}^{r} (a_\alpha)^{\lambda_\alpha})$, i.e., the change of the numerical values of the physical quantity ϕ is similar with the scale group H^r. When constructing exact solutions by the dimensional analysis theory one can use the theory of invariant solutions with respect to the scale group H^r. This theory is explained next.

Definition 2.10. *A scale group* H^r *is said to be an admitted by a system of partial differential equations if the manifold assigned by this system is invariant with respect to the prolonged group* H^r.

Nonsingular invariant solutions are constructed as follows. First, one finds all of the independent invariants J^k, $(k = 1, \ldots, m+n-r)$. They should be such that it is possible to solve m of them (for example, J^k, $(k = 1, \ldots, m)$) with respect to all dependent variables. A sufficient condition for this is the inequality

$$\frac{\partial(J^1, \ldots, J^m)}{\partial(u^1, \ldots, u^m)} = (\prod_{k=1}^{m} J^k / \prod_{\alpha=1}^{m} u_\alpha) \det \begin{pmatrix} \sigma_1^1 & \ldots & \sigma_m^1 \\ \ldots & \ldots & \ldots \\ \sigma_1^m & \ldots & \sigma_m^m \end{pmatrix} \neq 0.$$

Without loss of generality one can let $\sigma^i_j = \delta_{ij}$ $(i, j = 1, \dots, m)$. The remaining invariants J^{m+k}, $(k = 1, \dots, n-r)$ can be chosen depending only of the independent variables (the case where $r = n$ is also possible). Hence,

$$rank \begin{pmatrix} \lambda^1_1 & \dots & \lambda^n_1 \\ \dots & \dots & \dots \\ \lambda^1_r & \dots & \lambda^n_r \end{pmatrix} = r \le n.$$

After obtaining the independent invariants J^k, $(k = 1, \dots, m+n-r)$, one supposes the dependence of the first invariants J^k, $(k = 1, \dots, m)$ of the remaining, i.e.,

$$J^k = \varphi_k(J^{m+1}, \dots, J^{m+n-r}),\ (k = 1, \dots, m).$$

Since the invariants J^k, $(k = 1, \dots, m)$ can be solved with respect to all dependent variables u^i, $(i = 1, ..., m)$, defining the dependent variables from the last equations, one obtains a representation of the invariant solution. Substituting the representation of the functions u^i, $(i = 1, ..., m)$ into the initial system of partial differential equations, one obtains the system of equations for the unknown functions φ_k, $(k = 1, \dots, m)$. This system involves a smaller number of independent variables.

Definition 2.11. *An invariant solution of an admitted scale group H^r is called a self–similar solution*[5].

Let us apply the theory to the nonlinear diffusion equation[6]

$$r^{1-q}\frac{\partial(r^{q-1}c^n c_r)}{\partial r} = \frac{\partial c}{\partial t}, \quad (n > 0) \tag{2.24}$$

The dependent and independent variables are scaled as follows

$$c = a^{\alpha_1}\widehat{c},\ r = a^{\alpha_2}\widehat{r},\ t = a^{\alpha_3}\widehat{t}.$$

Equation (2.24) in the new variables becomes

$$a^{n\alpha_1 - 2\alpha_2 + \alpha_3}\widehat{r}^{1-q}\frac{\partial}{\partial\widehat{r}}(\widehat{r}^{q-1}\widehat{c}^n\frac{\partial\widehat{c}}{\partial\widehat{r}}) - \frac{\partial\widehat{c}}{\partial\widehat{t}} = 0.$$

If equation (2.24) is invariant with respect to scaling, then it is necessary that

$$n\alpha_1 - 2\alpha_2 + \alpha_3 = 0.$$

For $\alpha_3 \ne 0$ the combinations $J^1 = ct^{-\alpha_1/\alpha_3}$, $J^2 = rt^{-\alpha_2/\alpha_3}$ are invariant with respect to this scaling. Assuming

$$ct^{-\alpha_1/\alpha_3} = V(rt^{-\alpha_2/\alpha_3}),$$

one obtains the representation of an invariant (self–similar) solution of (1.56).

---

[5] With a group point of view such solutions are called self–similar solutions in narrow sense [130].
[6] Self-similar solutions of this equation were considered in the previous chapter.

4.2 Self–similar solutions in an inviscid gas

The one-dimensional gas dynamics equations are considered to illustrate the method of constructing self-similar solutions. Notice that the solution of the problem of a strong explosion in a gas was found with the help of self–similar solutions.

The system of equations describing a one-dimensional motion of a gas is

$$\begin{aligned} &\frac{\partial u}{\partial t} + v\frac{\partial u}{\partial x} + \frac{1}{\rho}\frac{\partial p}{\partial x} = 0, \\ &\frac{\partial \rho}{\partial t} + \frac{\partial \rho u}{\partial x} + \nu\frac{\rho u}{x} = 0, \\ &\frac{\partial}{\partial t}\left(\frac{p}{\rho^\gamma}\right) + v\frac{\partial}{\partial x}\left(\frac{p}{\rho^\gamma}\right) = 0. \end{aligned} \tag{2.25}$$

Here γ is the exponent of the adiabatic curve, ν characterizes a geometrical structure of the problem: $\nu = 0$ for plane flows, $\nu = 1$ for cylindrical flows, and $\nu = 2$ for spherical flows. Because the number of the independent variables $n = 2$, by virtue of the condition $n \geq r$ only two cases are possible: either $r = 1$ or $r = 2$.

Let $r = 1$, and define the one-parameter scale group H^1 admitted by system (2.25) with the equations ($a \equiv a_1$):

$$u' = ua^{\mu^1}, \ \rho' = \rho a^{\mu^2}, \ p' = pa^{\mu^3}, \ t' = ta^{\lambda^1}, \ x' = xa^{\lambda^2}.$$

For the transformations that remain invariant the manifold, assigned by equations (2.25), one obtains

$$\mu^1 = \lambda^2 - \lambda^1, \ \ \mu^3 = \mu^2 + 2(\lambda^2 - \lambda^1).$$

Further it is assumed that $(\lambda^1)^2 + (\lambda^2)^2 \neq 0$. Since $m + n - r = 4$, when forming the independent invariants $J^k = t^{\theta_1^k} x^{\theta_2^k} u^{\sigma_1^k} \rho^{\sigma_2^k} p^{\sigma_3^k}$, $(k = 1, 2, 3, 4)$ it is enough to find independent solutions of the linear equation

$$\lambda^1\theta_1^k + \lambda^2\theta_2^k + (\lambda^2 - \lambda^1)\sigma_1^k + \mu^2\sigma_2^k + (\mu^2 + 2(\lambda^2 - \lambda^1))\sigma_3^k = 0.$$

If $\lambda^1 = 0$, one can choose the following independent invariants

$$J^1 = ux^{-1}, \ \ J^2 = \rho x^{-\alpha}, \ \ J^3 = px^{-(\alpha+2)}, \ J^4 = t, \ \ (\alpha = \mu^2/\lambda^2).$$

The self–similar solution has the representation[7]

$$u = x\phi_1(t), \ \ \rho = x^\alpha\phi_2(t), \ \ p = x^{\alpha+2}\phi_3(t).$$

The system of equations (2.25) becomes

$$\begin{aligned} &\phi_1' + \phi_1^2 + (\alpha + 2)\phi\phi_2^{\gamma-1} = 0, \\ &\phi_2' + (\alpha + 1 + \nu)\phi_1\phi_2 = 0, \\ &\phi' + (\alpha + 2 - \gamma\alpha)\phi_1\varphi = 0, \end{aligned}$$

[7]Such solutions are called solutions with a linear profile of velocity.

where $\phi \equiv \phi_3/\phi_2^\gamma$. The solution of this system of equations can be found in terms of quadratures.

If $\lambda^1 \neq 0$, the invariants can be chosen as follows

$$J^1 = ut^{\alpha+1}, \quad J^2 = \rho t^\beta, \quad J^3 = pt^{2(\alpha+1)+\beta}, \quad J^4 = xt^\alpha,$$

where $\alpha = -\lambda^2/\lambda^1, \ \beta = -\mu^2/\lambda^1$. The self–similar solution in this case has the representation

$$u = t^{-(1+\alpha)}\phi_1(\lambda), \quad \rho = t^{-\beta}\phi_2(\lambda), \quad p = t^{-(\beta+2(\alpha+1))}\phi_3(\lambda),$$

where $\lambda = xt^\alpha$. There is another equivalent representation

$$u = \frac{x}{t}U(\lambda), \quad \rho = x^{-(k+3)}t^{-s}R(\lambda), \quad p = x^{-(k+1)}t^{-(s+2)}P(\lambda). \qquad (2.26)$$

Substituting these expressions in equations (2.25), one obtains the system of ordinary differential equations for the functions $U, \ R, \ P$

$$\lambda[(\alpha + U)U' + \tfrac{P'}{R}] = -U^2 + U + (k+1)\tfrac{P}{R},$$
$$\lambda[U' + (\alpha + U)\tfrac{R'}{R}] = s + (k - \nu + 2)U,$$
$$\lambda(\alpha + U)[\tfrac{P'}{P} - \gamma\tfrac{R'}{R}] = s(1-\gamma) + 2 + [k(1-\gamma) + 1 - 3\gamma]U.$$

This system of equations is split into two equations of the form

$$\lambda\frac{dU}{d\lambda} = f_1(U, Z), \ \lambda\frac{dR}{d\lambda} = Rf_2(U, Z), \qquad (2.27)$$

and one differential equation

$$\frac{dZ}{dU} = \frac{Z}{U+\alpha}\frac{M(Z,U)}{N(Z,U)},$$

where $Z = P/R$. Notice that the last equation in the case where $k = -3, \ s = -(2 + \alpha(\nu + 3))$ has the integral

$$ZU + (U + \alpha)(\frac{1}{2}U^2 + \frac{1}{\gamma - 1}Z) = 0. \qquad (2.28)$$

The values of the constant α and β in each particular self–similar solution are chosen by analysis of the initial parameters of the problem.

4.3 An intense explosion in a gas

The problem of a strong explosion in a gas is formulated as follows[8].

[8]Detailed analysis of this problem can be found in [149] and references therein.

At the moment $t = 0$ in an undisturbed gas ($u_1 = 0$) with the initial density ρ_1 and zero pressure $p_1 = 0$ at the center of symmetry ($x = 0$) an explosion occurs, i.e., a final energy E_0 is instantly released. The areas of the disturbed and undisturbed parts of a gas are separated by a shock wave. The gas dynamics values in front of the wave $\rho_1,\ p_1,\ u_1$ and behind it $\rho_2,\ p_2,\ u_2$ are related by the conditions across the shock wave (the Hugoniot relations):

$$\rho_2(u_2 - D) = \rho_1(u_1 - D),$$
$$p_2 + \rho_2(u_2 - D)^2 = p_1 + \rho_1(u_1 - D)^2,$$
$$\frac{\gamma}{(\gamma-1)}\frac{p_2}{\rho_2} + \frac{(u_2-D)^2}{2} = \frac{\gamma}{(\gamma-1)}\frac{p_1}{\rho_1} - \frac{(u_1-D)^2}{2}.$$

These relations express the conservation of mass, impulse and energy laws on the shock wave $x = x_b(t)$. Here $D = dx_b/dt$ is the velocity of the shock wave propagation. From the Hugoniot relations one can obtain the values of the density, pressure and velocity behind the front of the wave

$$\rho_2 = \frac{(\gamma+1)}{(\gamma-1)}\rho_1,\quad p_2 = \frac{2}{(\gamma+1)}\rho_1 D^2,\quad u_2 = \frac{2}{(\gamma+1)}D.$$

The disturbed part of the gas is located in the interval $(0, x_b)$. Because the energy of the volume of the gas is equal to $\rho v^2/2 + p/(\gamma - 1)$, according to the conservation of energy

$$\frac{2\nu\pi}{(\nu+1)}\int_0^{x_b}[\rho u^2/2 + p/(\gamma-1)]x^\nu dx = E_0. \tag{2.29}$$

The motion of the gas after the explosion ($t > 0$) is defined by the dimension of the parameters $E_0,\ \rho_1,\ x$ and t. They have the dimensions

$$[E_0] = ML^\nu T^{-2},\quad [\rho_1] = ML^{-3},\quad [x] = L,\quad [t] = T.$$

Here M is the dimension of the mass, L is the dimension of the length and T is the dimension of the time. In this problem there is only one dimensionless variable parameter

$$\lambda = (E/\rho_1)^{-\frac{1}{(\nu+3)}} x t^{-\frac{2}{(\nu+3)}},$$

where $E = hE_0$ with some constant h. This constant is chosen to scale the variable λ. Assume that the disturbed part of the gas is governed by the self-similar solution of the type (2.26). The front of the shock wave is $x_b = \lambda_* t^{-\alpha}(E/\rho_1)^{\frac{1}{(\nu+3)}}$. Without loss of generality one can let $\lambda_* = 1$. According to the dependence of the variable λ of t and x, in (2.26) one has to choose $\alpha = -2/(\nu + 3)$. Analyzing the dimension of the density $[\rho]$ it also follows that $\beta = 0$. Therefore when seeking the solution of the problem of an intense explosion, one can try to find it in the class of self–similar solutions (2.26) with $s = 0, k = -3$:

$$\rho = \rho_1 R(\lambda),\quad u = \frac{x}{t}U(\lambda),\quad p = \rho_1\frac{x^2}{t^2}P(\lambda).$$

Notice that for these parameters there is the integral (2.28). Because the coordinate of the shock wave front is $x_b = t^{-\alpha}(E/\rho_1)^{\frac{1}{(\nu+3)}}$, then $D = -\alpha x_b t^{-1}$. This gives the initial data at $\lambda = 1$ (on the front of the shock wave)

$$U(1) = -\frac{2\alpha}{(\gamma+1)}, \; R(1) = \frac{(\gamma+1)}{(\gamma-1)}, \; P(1) = \frac{2\alpha^2}{(\gamma+1)}, \; Z(1) = \frac{2\alpha^2(\gamma-1)}{(\gamma+1)^2}.$$

Equation (2.29) in the self-similar variables is reduced to the equation

$$\frac{2\nu\pi E}{(\nu+1)} \int_0^1 \lambda^{2+\nu} \left(RU^2 + \frac{P}{(\gamma-1)} \right) d\lambda = E_0.$$

This equation serves for finding the constant

$$h = \frac{2\nu\pi}{(\nu+1)} \int_0^1 \lambda^{2+\nu} \left(RU^2 + \frac{P}{(\gamma-1)} \right) d\lambda.$$

Therefore, the solution of intense explosion can be found in the form of quadratures of (2.27).

5. Solutions with a linear profile of velocity

Among the approaches for obtaining classes of exact solutions in continuum mechanics there is the method where the velocity vector is required to be linear with respect to the spatial independent variables, with respect to all or their part of them. Such solutions were studied a long time ago by Dirichlet, Dedekind and Riemann. One example of such a solution was obtained in the previous section[9]. Assuming linearity of the velocity with respect to some independent variables, one usually obtains polynomial equations with respect to them. Splitting these equations leads to an overdetermined system of partial differential equations. The main problem in these studies is the compatibility problem for the overdetermined system of equations.

Let us consider the equations describing the isentropic motion of a polytropic gas. For the sake of simplicity[10] we consider the two-dimensional case

$$\frac{d\theta}{dt} + \kappa\theta\, div(u) = 0, \quad \frac{du_i}{dt} + \frac{\partial\theta}{\partial x_i} = 0, \quad (i = 1, 2). \tag{2.30}$$

We will assume that the velocity vector has the representation

$$u_i = m_{i\alpha}(t) x_\alpha, \quad (i = 1, 2). \tag{2.31}$$

The last two equations of (2.30) define the derivatives

$$\frac{\partial\theta}{\partial x_i} = -a_{i\alpha} x_\alpha, \quad (i = 1, 2), \tag{2.32}$$

[9]Applications of this method to the Navier-Stokes equations can be found in [159]

[10]Similar, but more cumbersome, one can obtain solutions in the three-dimensional case.

where

$$a_{ii} = m'_{ii} + m^2_{ii} + m_{12}m_{21}, \ (i = 1, 2),$$

$$a_{12} = m'_{12} + m_{12}(m_{11} + m_{22}), \ a_{21} = m'_{21} + m_{21}(m_{11} + m_{22}).$$

The condition

$$\partial^2\theta/\partial x_1 \partial x_2 = \partial^2\theta/\partial x_2 \partial x_1$$

gives $a_{12} = a_{21}$. Integrating (2.32) with respect to x_1 and x_2, one finds

$$\theta = -\frac{1}{2} a_{\alpha\beta} x_\alpha x_\beta + \phi(t), \tag{2.33}$$

where $\phi(t)$ is an arbitrary function of integration. Substituting the expression for θ and the velocity in the first equation of (2.30), one obtains the squared form with respect to x_1 and x_2

$$b_{\alpha\beta} x_\alpha x_\beta - b = 0 \tag{2.34}$$

with the coefficients

$$b_{11} = a'_{11} + 2m_{11}a_{11} + 2m_{21}a_{12} + \kappa(m_{11} + m_{22})a_{11},$$

$$b_{22} = a'_{22} + 2m_{22}a_{22} + 2m_{12}a_{12} + \kappa(m_{11} + m_{22})a_{22},$$

$$b_{12} = b_{21} = a'_{12} + m_{12}a_{11} + m_{21}a_{22} + (\kappa + 1)(m_{11} + m_{22})a_{12},$$

$$b = -2[\phi' + \kappa(m_{11} + m_{22})\phi].$$

Factors of the squared form (2.34) are independent of the independent variables x_1 and x_2. Hence, splitting equation (2.34), one has a system of five ordinary differential equations for five functions: $\phi(t)$ and $m_{ij}(t)$, $(i, j = 1, 2)$.

More general class of solutions is obtained if one assumes linearity only with respect to one independent variable.

6. Travelling waves

The idea of a travelling wave was presented in the previous chapter. The concept of travelling waves can be generalized for many independent variables $x \in R^n$ and many dependent variables $u \in R^m$. In this section the generalization for the multidimensional case is given.

Let Lx be a linear form of the independent variables

$$Lx = L_\alpha x_\alpha.$$

The representation of a solution is assumed in the form

$$u(x) = v(Lx),$$

where the function $v(\xi)$ depends on one independent variable ξ. The value ξ is called a phase of the wave. Fixing the phase $\xi = Lx$, one obtains the front of the wave, where the values of the dependent variables are constant. Hence, the front of the wave is a plane propagating in the space R^n. Usually one of the independent variables plays the role of time t (for example, $t = x_n$), the phase of the wave is represented as $\xi = \eta y - Dt$, where $\eta \in R^{n-1}$ is a unit vector, $y = (x_1, x_2, ..., x_{n-1})$, and ηy is a scalar product.

Definition 2.12. *A solution $u(x_1, x_2, \ldots, x_n)$ is called an r–multiple travelling wave, if it has the representation*

$$u(x) = v(Lx),$$

where Lx be a vector, which coordinates are linear forms of the independent variables

$$(Lx)_i = L_{i\alpha}x_\alpha, \ (i = 1, 2, ..., r, \ r < n).$$

The variables $\xi = Lx \in R^r$ are called parameters of the wave.

Here the vector function $v(\xi)$ depends on a smaller number of the independent variables ξ. Fixing all but one of the components of the vector ξ defines a wave propagating in the subspace of R^n of the dimension[11] $n - (r - 1)$. Equations for the function $v(\xi)$ are obtained by substituting the representation of the solution into the initial system of equations.

Notice that for an r-multiple travelling wave the rank of the Jacobi matrix of the dependent variables with respect to the independent variables is less or equal than r. This provides an idea for further generalization of the travelling wave concept. The generalization is achieved by rejecting the requirement that the parameters of the wave are linear forms. These solutions are called solutions with a degenerate hodograph, and they are studied in the next chapter.

Let us apply the method to two-dimensional flows of a fluid, described by the Navier-Stokes equations

$$\begin{gathered} u_t + uu_x + vu_y + p_x = u_{xx} + u_{yy}, \\ v_t + uv_x + vv_y + p_y = v_{xx} + v_{yy}, \\ u_x + u_y = 0. \end{gathered}$$

Without loss of generality one can assume that the travelling wave type solution has the representation

$$u = u(\xi), \ v = v(\xi), \ p = p(\xi), \ \xi = x + ay + Dt.$$

Substituting the representation of the solution into the Navier-Stokes equations, one obtains

$$\begin{gathered} p' + (u + av + D)u' = (a^2 + 1)u'', \\ ap' + (u + av + D)v' = (a^2 + 1)v'', \\ u' + av' = 0. \end{gathered} \tag{2.35}$$

[11] Without loss of generality it is assumed that the rank of the matrix $L = (L_{\alpha\beta})$ is equal to k.

Taking the linear combination of the first equation and the second equation multiplied by the constant a, and using the third equation, one finds

$$p' = 0.$$

Integrating the third equation of (2.35), one gets

$$u = av + c,$$

where c is constant. The second equation of (2.35) becomes

$$(c + D)v' = (a^2 + 1)v''.$$

Notice that by virtue of a Galilean transformation and a rotation one can assume that $c = 0$ and $a = 0$.

7. Completely integrable systems

One class of overdetermined systems, for which the problem of compatibility is solved, is the class of completely integrable systems. One particular case of such a system was considered in the first chapter. Here the theory of completely integrable systems is developed in the general case.

Definition 2.13. *A system*

$$\frac{\partial z^i}{\partial a^j} = f^i_j(a, z), \ (i = 1, 2, \ldots, N; \ j := 1, 2, \ldots, r) \qquad (2.36)$$

is called a completely integrable if it has a solution for any initial values a_o, z_o in some open domain D.

Lemma 2.1. *Any system of the type (2.36) is completely integrable if and only if the equalities*

$$\frac{\partial f^i_j}{\partial a^\beta} + f^\gamma_\beta \frac{\partial f^i_j}{\partial z^\gamma} = \frac{\partial f^i_\beta}{\partial a^j} + f^\gamma_j \frac{\partial f^i_\beta}{\partial z^\gamma}, \qquad (2.37)$$
$$(i = 1, 2, , \ldots, N; \ \beta, j = 1, 2, \ldots, r)$$

are identically satisfied with respect to the variables $(a, z) \in D$.

Proof.

Let $z = z(a)$ be a solution of the initial value problem

$$z(a_o) = z_o.$$

Calculating the derivatives

$$\frac{\partial}{\partial a^j}\left(\frac{\partial z^i}{\partial a^\beta}\right) = \frac{\partial}{\partial a^\beta}\left(\frac{\partial z^i}{\partial a^j}\right)$$

at the point a_o, one obtains

$$\frac{\partial f_j^i}{\partial a^\beta}(a_o,z_o)+f_\beta^\gamma(a_o,z_o)\frac{\partial f_j^i}{\partial z^\gamma}(a_o,z_o)=\frac{\partial f_\beta^i}{\partial a^j}(a_o,z_o)+f_j^\gamma(a_o,z_o)\frac{\partial f_\beta^i}{\partial z^\gamma}(a_o,z_o).$$

Since the initial values a_o, z_o are arbitrary, the first part of the lemma is proven.

For proving the second part of the lemma, the theorems of existence, uniqueness and continuity with respect to parameters of a solution of an initial value problem are applied. Let e be an arbitrary vector in the closed unit ball $\overline{B_1(0)}$, and pose the problem

$$\frac{\partial v^i}{\partial t}(t,e)=e^\alpha f_\alpha^i(a_o+te,v(t,e)),\;\; v^i(0)=z_o^i,\;\; (i=1,2,,\ldots,N). \quad (2.38)$$

This problem has a unique solution $v(t,e)$, which is defined in the maximal interval $t\in(0,\tau_e)$. Direct calculations show that the function $u(t,e)=v(\lambda t,e)$ is a solution of the problem

$$\frac{\partial u^i}{\partial t}(t,e)=\lambda e^\alpha f_\alpha^i(a_o+\lambda te,u(t,e)),\;\; u^i(0)=z_o^i,\;\; (i=1,2,,\ldots,N).$$

Because of the uniqueness of the solution for $\lambda e\in\overline{B_1(0)}$, the vector function $u(t,e)=v(t,\lambda e)$. Hence, $v(t,\lambda e)$ is defined in the interval $t\in(0,\tau_{\lambda e})$, and $\lambda\tau_{\lambda e}=\tau_e$.

Set $\tau_*=\inf_e\tau_e$, and assume that $\tau_*=0$. This means that for any $\varepsilon>0$ there exists a vector $e\in\overline{B_1(0)}$ such that $\tau_*=0\le\tau_e<\varepsilon$. Choosing $\varepsilon=1/k$ one constructs a sequence of the vectors $\{e_k\}$ such that $\tau_{e_k}\to 0$. Because the unit ball $\overline{B_1(0)}$ is a compact, there exists a convergent subsequence $\{e_{k_l}\}\to e_*\in\overline{B_1(0)}$. For the vector e_* the solution of the problem (2.38) is defined in the interval $(0,\tau_{e_*})$, where $\tau_{e_*}>0$. Because of the continuity with respect to the parameter e, there exists an interval $(0,\tau)$ and a neighborhood U_{e_*} of the vector e_* such that $0<\tau<\tau_{e_*}$ and for any $e\in U_{e_*}$ the solution $v(t,e)$ is defined in the interval $(0,\tau)$. This contradicts to the condition that $\tau_{e_{k_l}}\to 0$. Hence, $\tau_*>0$.

Set $\tau=\min(1,\tau_*)$. The functions $v^i(t,a-a_o)$ are defined for any $a\in B_\tau(a_o)$ and $t\in(0,1]$. Let us prove this statement.

If $\tau_*\ge 1$, this follows from the definition of τ_*. In fact, in this case $\tau=1$, and for any $a\in\overline{B_\tau(a_o)}$ the functions $v^i(t,e)$ with $e=a-a_o\in\overline{B_1(0)}$ are defined in the interval $(0,\tau)$, where $\tau\ge\tau_*\ge 1$. If $\tau_*<1$, then $\tau=\tau_*$. For any vector $a\in B_{\tau_*}(a_o)$ such that $a-a_o\ne 0$ the vector $e=\lambda^{-1}(a-a_o)$, with $\lambda=|a-a_o|^{-1}$, belongs to the ball $\overline{B_1(0)}$. Notice that $\lambda e\in\overline{B_\tau(0)}\subset\overline{B_1(0)}$. Since $\tau_{\lambda e}=\lambda^{-1}\tau_e\ge\lambda^{-1}\tau_*=\lambda^{-1}\tau\ge 1$, then for any $a\in B_\tau(a_o)$ the values $v^i(1,a-a_o)$ are defined. Thus, the functions $v^i(t,a-a_o)$ are defined at the point $t=1$ for any $a\in B_\tau(a_o)$.

Let us show that the functions $z^i(a) = v^i(1, a - a_o)$, $\forall a \in B_\tau(a_o)$ satisfy the equations

$$R^i_j(a) = \frac{\partial z^i}{\partial a^j}(a) - f^i_j(a, z(a)) = 0, \ \forall a \in B_\tau(a_o).$$

For this purpose the functions $S^i_j(t) = tR^i_j(a_o + te)$, where $e = a - a_o \in B_\tau(0)$, are studied.

First one can obtain some useful identities. Notice that

$$z^i(a_o + te) = v^i(1, te) = v^i(t, e).$$

Differentiating these equalities with respect to t, and using equations (2.38), one has

$$e^\beta \frac{\partial z^i}{\partial a^\beta}(a_o + te) = e^\beta f^i_\beta(a_o + te, z(a_o + te)),$$

which at the point $t = 1$ become

$$(a^\beta - a^\beta_o)\frac{\partial z^i}{\partial a^\beta}(a) = (a^\beta - a^\beta_o) f^i_\beta(a, z(a)).$$

Differentiating these relations with respect to the coordinate a^j, one obtains

$$\frac{\partial z^i}{\partial a^j}(a) + (a^\beta - a^\beta_o)\frac{\partial^2 z^i}{\partial a^\beta a^j}(a) =$$
$$f^i_j(a, z(a)) + (a^\beta - a^\beta_o)\left(\frac{\partial f^i_\beta}{\partial a^j}(a, z(a)) + \frac{\partial f^i_\beta}{\partial z^\gamma}(a, z(a))\frac{\partial z^\gamma}{\partial a^j}(a, z(a))\right).$$

Substituting the expressions $\frac{\partial z^i}{\partial a^j}(a) = f^i_j(a, z(a)) + R^i_j(a)$ into these equations, they become

$$(a^\beta - a^\beta_o)\frac{\partial^2 z^i}{\partial a^\beta a^j}(a) =$$
$$-R^i_j(a) + (a^\beta - a^\beta_o)\left(\frac{\partial f^i_\beta}{\partial a^j}(a, z(a)) + \frac{\partial f^i_\beta}{\partial z^\gamma}(a, z(a))\frac{\partial z^\gamma}{\partial a^j}(a, z(a))\right).$$

Because of the definition $S^i_j(t) = tR^i_j(a_o + te)$, at the point $a = a_o + te$ the last equations are

$$te^\beta\left(\frac{\partial^2 z^i}{\partial a^j \partial a^\beta}(a_o + te)\right) = -R^i_j(a_o + te) + te^\beta \frac{\partial f^i_\beta}{\partial a^j}(a_o + te) +$$
$$e^\beta\left(S^\gamma_j(t) + tf^\gamma_j(a_o + te)\right)\frac{\partial f^i_\beta}{\partial z^\gamma}(a_o + te). \tag{2.39}$$

From another point of view, differentiating the relations $tR^i_j(a_o + te) = S^i_j(t)$ with respect to t, one obtains

$$\frac{dS^i_j}{dt} = R^i_j + te^\beta\left(\frac{\partial^2 z^i}{\partial a^j \partial a^\beta} - \frac{\partial f^i_j}{\partial a^\beta} - (R^\gamma_\beta + f^\gamma_\beta)\frac{\partial f^i_j}{\partial z^\gamma}\right).$$

Using (2.39), these equations are reduced to the equations

$$\frac{dS^i_j(t)}{dt} = -te^\beta\left(\frac{\partial f^i_j}{\partial a^\beta} + (R^\gamma_\beta + f^\gamma_\beta)\frac{\partial f^i_j}{\partial z^\gamma}\right) + e^\beta\left(t\frac{\partial f^i_\beta}{\partial a^j} + \left(S^\gamma_j + tf^\gamma_j\right)\frac{\partial f^i_\beta}{\partial z^\gamma}\right).$$

Changing tR^i_j with $S^i_j(t)$, one finds

$$\frac{dS^i_j(t)}{dt} = -e^\beta S^\gamma_\beta\frac{\partial f^i_j}{\partial z^\gamma} + e^\beta S^\gamma_j\frac{\partial f^i_\beta}{\partial z^\gamma} + te^\beta\left(\frac{\partial f^i_\beta}{\partial a^j} + f^\gamma_j\frac{\partial f^i_\beta}{\partial z^\gamma} - \frac{\partial f^i_j}{\partial a^\beta} - f^\gamma_\beta\frac{\partial f^i_j}{\partial z^\gamma}\right)$$

Because of the relations (2.37), one finds that the functions $S^i_j(t)$ satisfy the problem

$$\frac{dS^i_j}{dt} = -e^\beta\left(S^\gamma_\beta\frac{\partial f^i_j}{\partial z^\gamma} - S^\gamma_j\frac{\partial f^i_\beta}{\partial z^\gamma}\right), \quad S^i_j(0) = 0.$$

By virtue of the uniqueness of the solution of this problem, one finds that $S^i_j(t) = 0$ or $R^i_j(a) = 0$.

Chapter 3

METHOD OF THE DEGENERATE HODOGRAPH

The vast majority of exact solutions in continuum mechanics have been obtained by the method of the degenerate hodograph. This method deals with solutions which are distinguished by finite relations between the dependent variables. Solutions with degenerate hodograph form a class of solutions called multiple waves. Riemann waves and Prandtl–Meyer flows are the simplest solutions of this class[1]. The main problem with the theory of multiple waves is obtaining a compatible system of equations in the space of the dependent and independent variables.

The chapter starts by giving the main definitions and basic facts of the theory. Simple waves of systems with two independent variables are closely related to the Riemann invariants. Attempts to generalize the notion of Riemann invariants to equations with more than two independent variables are discussed. One of these approaches deals with simple integral elements.

The simplest case of multiple waves is the case of simple waves. The first application of simple waves for multi-dimensional flows was made for isentropic flows of an ideal gas. From a group analysis point of view a multiple wave is a partially invariant solution. For example, a simple wave is a partially invariant solution with the defect one; the defect of a double wave is equal to two. In the theory of partially invariant solutions there is the problem of reducibility to a smaller defect. Solutions, irreducible to invariant, take a special place among partially invariant solutions. This is related to the fact that the problem of compatibility for an invariant multiple wave is easier than for a partially invariant multiple wave. Hence, the problem of reducibility arises. There are few theorems which state sufficient conditions of reducibility. One of them

[1]Applications of the method of degenerate hodograph to the gas dynamic equations can be found in [160] and the references therein.

is the Ovsiannikov theorem. This theorem provides restrictions on systems of partial differential equations when describing irreducible double waves. The practical meaning of this theorem is demonstrated with several examples in this chapter. The Ovsiannikov theorem is also an imprescriptible part of the classification problem of double waves.

Applications of multiple waves in multi–dimensional gas dynamics admit the possibility that the degenerate hodograph method can also be applied to the theory of plasticity, where so far only simple waves have been applied. Applications of double waves to gas dynamics are followed by applications of double waves to rigid plastic bodies.

The chapter ends with the study of triple waves of isentropic potential gas flows.

1. Basic definitions

Let us consider an autonomous system of quasilinear equations

$$\sum_{\alpha=1}^{n} G_\alpha(u)\frac{\partial u}{\partial x_\alpha} = f(u). \tag{3.1}$$

Here $x=(x_1, x_2, \ldots, x_n)$ are the independent variables, $u^i=u^i(x_1, x_2, \ldots, x_n)$, $(i = 1, 2, \ldots, m)$ are the unknown functions, G_α are matrices with the elements $a_{ij}^\alpha(u)$, $(i = 1, 2, \ldots, N;\ j = 1, 2, \ldots, m;\ \alpha = 1, 2, \ldots, n)$. Originally the method of degenerate hodograph was applied to homogeneous systems ($f = 0$). Unless otherwise stated, homogeneous systems are considered in this section.

Definition 3.1. *A solution of system (3.1) for which the rank of the Jacobi matrix in a domain $G \subset R^n(x)$ satisfies the condition*

$$rank\left(\frac{\partial(u^1, u^2, \ldots, u^m)}{\partial(x_1, x_2, \ldots, x_n)}\right) = r$$

is called a multiple wave of the rank r.

A multiple wave is called a simple wave if $r = 1$, a double wave if $r = 2$ and a triple wave if $r = 3$. The value $r = 0$ corresponds to uniform flow with constant u^i, $(i = 1, 2, \ldots, m)$. The value $r = n$ corresponds to the common case of nondegenerate solutions. Multiple waves of all ranks $r \le n - 1$ form a class of solutions with a degenerate hodograph.

The singularity of the Jacobi matrix means that the functions $u^i(x)$, $(i = 1, 2, \ldots, m)$ are functionally dependent (the hodograph is degenerate), and the number of functional constraints is equal to $m - r$, i.e.,

$$u^i = \Phi_i(\lambda^1, \lambda^2, \ldots, \lambda^r), \quad (i = 1, 2, \ldots, m), \tag{3.2}$$

with some functions $\lambda^1(u), \lambda^2(u), \ldots, \lambda^r(u)$, which are called parameters of the wave. The solutions with a degenerate hodograph generalize the travelling wave type solutions. In an r–multiple travelling wave the wave parameters are linear forms of the independent variables, contrary to an r–multiple waves where the wave parameters are some unknown functions. To find an r–multiple wave one needs to substitute the representation (3.2) into system (3.1). The obtained overdetermined system of differential equations for the wave parameters $\lambda^i(x)$, $(i = 1, 2, \ldots, r)$ formed in this way must then be studied for compatibility. These compatibility conditions are equations for the functions $\Phi_i(\lambda^1, \lambda^2, \ldots, \lambda^r)$, $(i = 1, 2, \ldots, m)$.

The main problem of the theory of solutions with degenerate hodograph involves obtaining a closed system of equations in the space of the dependent variables (hodograph), in establishing the arbitrariness of the general solution, and in defining a flow in the physical space.

A homogeneous ($f = 0$) system (3.1) is not changed by the transformations

$$x_i' = ax_i + b_i, \quad (i = 1, 2, \ldots, n), \tag{3.3}$$

which forms a Lie group G^{n+1} of transformations[2]. From a group analysis point of view any r–multiple wave is a partially invariant solution with respect to this group of transformations[3]. The solutions, irreducible to invariant, take a special place among all partially invariant solutions. This is related with the circumstance that the problem of constructing invariant multiple waves is much easier than the problem of constructing partially invariant solutions. Namely, wave parameters for an invariant r–multiple wave can be chosen only from the following two types (up to equivalence transformations). The first type of the waves has the wave parameters

$$\lambda^1 = x_{n-r+1}, \; \lambda^2 = x_{n-r+2}, \; \ldots, \; \lambda^r = x_n,$$

and for the second type the wave parameters are

$$\lambda^1 = x_{n-r}/x_n, \; \lambda^2 = x_{n-r+1}/x_n, \; \ldots, \; \lambda^r = x_{n-1}/x_n.$$

The equivalence transformations are defined by the linear mapping of the independent variables $x' = Vx + c$ with a nonsingular square $n \times n$ matrix V and a constant vector c.

Moreover, the analysis of compatibility for partially invariant solutions is more difficult than for invariant solutions. Thus, it is worthwhile to find out a priori a form of irreducible waves. In the general case this problem is difficult[4]. The practical significance of these conditions are as follows. In the process of forming compatibility conditions for the wave parameters, it is necessary to set a veto on the reduction. It should be also noted that the notation of "irreducible"

---

[2]The Lie group of transformations admitted by system (3.1) can be wider than the group G^{n+1} (3.3).

[3]A group analysis approach to solutions with a degenerate hodograph can be found in [130].

[4]There are only some sufficient conditions of reducibility [113, 129].

used in this chapter means solutions that are irreducible to solutions invariant with respect to the group G^{n+1} (3.3).

The study of a solution with a degenerate hodograph requires the investigation of an overdetermined system for the wave parameters. Usually the analysis of these overdetermined systems is difficult. Therefore, additional assumptions about solution need to be applied. Originally geometrical and kinematic conditions were required: either the rectilinearity of level lines or of the potentiality of the flows. Other restrictions were constructed on the base of the algebraic structure of system (3.1), related to the so called simple integral elements of the system.

Because, in any case, in the study of solutions with degenerate hodograph one needs to analyze overdetermined systems, the classification of such solutions with respect to functional arbitrariness of the Cauchy problem is more natural from the compatibility theory point of view. One of the classes of such solutions is a class of multiple waves where the overdetermined system for the wave parameters has solutions with functional arbitrariness. This class has the property that after reducing the overdetermined system to an involutive system[5], the rank of the Jacobi matrix composed of the equations of this system with respect to the highest order derivatives is not equal to the number of all of the highest order derivatives.

2. Remarks on multiple waves and Riemann invariants

One class of restrictions for multiple waves was suggested on the basis of algebraic structure[6] of system (3.1). These restrictions are related to the simple integral elements of system (3.1).

Definition 3.2. *If there exist nonzero vectors $p = (p_1, p_2, \ldots, p_m) \in R^m$ and $q = (q_1, q_2, \ldots, q_n) \in R^n$ such that $a^{\alpha}_{i\beta} p_\beta q_\alpha = 0$, then the matrix $P = p \otimes q$ is called a simple integral element of system (3.1), and the vectors p, q are called characteristic vectors: a vector p in the hodograph space R^m, and q in R^n.*

The name "a characteristic vector" is related with the following property of system (3.1). If $P = p \otimes q$ is a simple integral element, then

$$\text{rank } (a^{j}_{i\beta} p_\beta) < n, \ \ \text{rank } (q_\alpha G_\alpha) < m.$$

If there exists a function $u(x)$, satisfying the relations $\partial u^i / \partial x_j = p_i q_j$ with the simple integral element $P = p \otimes q$, then $u(x)$ is a simple wave of system (3.1). In this approach an r–multiple wave is generated by simple waves, appropriate to simple integral elements $P_k = p^k \otimes q^k$, $(k = 1, 2, \ldots, r)$.

[5]It can be done after finite number of prolongations [24].

[6]See, for example, [92].

Let $P_k = p^k \otimes q^k$, $(k = 1, 2, \dots, r)$ be a finite set of simple integral elements with the linearly independent vectors $p^k = (p_1^k, p_2^k, \dots, p_m^k)$ for which there exists a function $u(x)$ such that

$$\frac{\partial u^i}{\partial x_j} = \xi_\alpha p_i^\alpha q_j^\alpha,$$

where $\xi_k = \xi_k(x, u)$, $(k = 1, 2, ..., r)$. Because of linearity and homogeneity of system (3.1) with respect to the derivatives, the function $u(x)$ is a solution of system (3.1).

Definition 3.3. *If the wave parameters $R_k(x)$, $(k = 1, 2, \dots, r)$ of a solution $u = U(R_1, R_2, ..., R_r)$ of system (3.1) satisfy the conditions*

$$p_i^k = \frac{\partial u^i}{\partial R_k}, \; q_j^k = \frac{\partial R_k}{\partial x_j}, \; (k = 1, \dots, r; \; i = 1, \dots, m; \; j = 1, \dots, n)$$

with the simple integral elements $P_k = p^k \otimes q^k$, $(k = 1, 2, \dots, r)$, then such solution is called a Riemann wave and the parameters R_k of the wave are called generalized Riemann invariants.

Thus an r–multiple (Riemann) wave requires additional restrictions on a solution, leading to more restrictive conditions than requiring only the functional arbitrariness of a solution.

3. Simple waves

The simplest class of multiple waves is the class of simple waves. This class of solutions is applied to many problems in continuum mechanics.

3.1 General theory

According to the definition of a simple wave this class of solutions has the representation

$$u^i = u^i(\lambda), \; (i = 1, 2, \dots, m), \tag{3.4}$$

where $\lambda = \lambda(x_1, x_2, \dots, x_n)$ is a wave parameter.

Substituting (3.4) into the original homogeneous system (3.1), one has the overdetermined homogeneous system of quasilinear differential equations for the function $\lambda = \lambda(x_1, x_2, \dots, x_n)$:

$$c_{i\alpha}(\lambda)\frac{\partial \lambda}{\partial x_\alpha} = 0, \; (i = 1, 2, \dots, m) \tag{3.5}$$

with the coefficients $c_{ik} = a_{i\beta}^k u'_\beta$, where the prime means the derivative of the functions (3.4) with respect to the wave parameter λ.

The structure of the solution of system (3.5) depends on the matrix C, which is composed of the coefficients $c_{ik}(\lambda)$. System (3.5) has nontrivial solutions if

$$r \equiv rank\ C < min(n, m).$$

In this case, without loss of generality, system (3.5) can be written in the form

$$\frac{\partial \lambda}{\partial x_i} = \sum_{\alpha=r+1}^{n} b_{i\alpha}(\lambda)\frac{\partial \lambda}{\partial x_\alpha}, \quad (i = 1, 2, \ldots, r). \tag{3.6}$$

Equations for the mapping $u = u(\lambda)$, which guarantee the existence of a non-trivial simple wave are called equations of the simple wave.

The description of all solutions of system (3.6) is given by the following theorem.

Theorem 3.1. *The general solution of system (3.6) is defined implicitly by the equation*

$$\lambda = f(x_{r+1} + \sum_{\alpha=1}^{r} x_\alpha b_{\alpha r+1}, \ldots, x_n + \sum_{\alpha=1}^{r} x_\alpha b_{\alpha n}), \tag{3.7}$$

where $f : R^{n-r} \to R$ is an arbitrary mapping.

Proof of the theorem consists of finding consecutively the general solutions of the equations of system (3.6). For example, the integrals of the first equation $(i = 1)$ are

$$x_2, x_3, \ldots, x_r, y_1, \ldots, y_{n-r}$$

where $y_j = x_{r+j} + x_1 b_{1j}, \ (j = 1, 2, \ldots, n-r)$. Hence, the general solution of the first equation is represented by the formula

$$\lambda = f(x_2, x_3, \ldots, x_r, y_1, \ldots, y_{n-r})$$

with an arbitrary function $f : R^{n-1} \to R$. By substituting the derivatives

$$\begin{aligned}\frac{\partial \lambda}{\partial x_1} &= \frac{1}{\Delta}(\sum_{\alpha=1}^{n-r}\frac{\partial f}{\partial y_\alpha}b_{1\alpha}), \quad \frac{\partial \lambda}{\partial x_i} = \frac{1}{\Delta}\frac{\partial f}{\partial x_i}, \quad \frac{\partial \lambda}{\partial x_k} \\ &= \frac{1}{\Delta}\frac{\partial f}{\partial y_k}, \quad (i = 2, \ldots, r; \ k = r+1, \ldots, n)\end{aligned}$$

into the remaining equations of system (3.6), they are reduced to

$$\frac{\partial f}{\partial x_i} = \sum_{\alpha=1}^{n-r}\frac{\partial f}{\partial y_\alpha}b_{i\alpha}(f) \ (i = 2, \ldots, r),$$

where $\Delta = 1 - \sum_{\alpha=1}^{n-r} x_1 b'_{1\alpha}(\partial f/\partial y_\alpha)$. Thus, the function $f(x_2, \ldots, x_r, y_1, \ldots, y_{n-r})$ satisfies a similar system of equations as the function $\lambda(x)$, but with a smaller number of the independent variables. Repeating this process, one finely obtains (3.7).•

A surface in R^n, where $\lambda = const$, is called a level surface. The level surface of a simple wave is an intersection of $n - r$ planes.

The most frequently occurring case in applications is with $r = n - 1$. If $r = n - 1$, the representation of a simple wave of system (3.5) is given by the formula

$$\sum_{\alpha=1}^{n} x_\alpha \Delta_\alpha(\lambda) = F(\lambda), \tag{3.8}$$

where $F(\lambda)$ is an arbitrary function, Δ_i, $(i = 1, 2, \ldots, n)$ are known $(n-1)$–order minors of the matrix C, which are functions of λ.

3.2 Isentropic flows of a gas

The system describing three–dimensional isentropic motion of an inviscid gas is

$$\frac{du_i}{dt} + \frac{\partial\theta}{\partial x_i} = 0, \quad (i = 1, 2, 3), \tag{3.9}$$
$$\frac{d\theta}{dt} + \kappa\theta \operatorname{div} u = 0.$$

Here (u_1, u_2, u_3) is the velocity vector, $\theta = c^2/\kappa$, $\kappa = \gamma - 1$, c is the sound speed, γ is the polytropic exponent of the gas, $d/dt \equiv \partial/\partial t + u_\alpha \partial/\partial x_\alpha$.

Taking the wave parameter $\lambda = \theta$, equations (3.5) become

$$u_i' \frac{d\theta}{dt} + \frac{\partial\theta}{\partial x_i} = 0, \quad (i = 1, 2, 3), \tag{3.10}$$
$$\frac{d\theta}{dt} + \kappa\theta u_\alpha' \frac{\partial\theta}{\partial x_\alpha} = 0.$$

Here $x_4 = t$, and the matrix C is:

$$C = \begin{pmatrix} u_1u_1' + 1 & u_2u_1' & u_3u_1' & u_1' \\ u_1u_2' & u_2u_2' + 1 & u_3u_2' & u_2' \\ u_1u_3' & u_2u_3' & u_3u_3' + 1 & u_3' \\ u_1 + \kappa\theta u_1' & u_2 + \kappa\theta u_2' & u_3 + \kappa\theta u_3' & 1 \end{pmatrix}.$$

The existence of a nontrivial simple wave requires that det $(C) = 0$. This equation is reduced to

$$(u_1')^2 + (u_2')^2 + (u_3')^2 = 1/(\kappa\theta). \tag{3.11}$$

The last equation guarantees that $\det(C) = 3$. It allows system (3.10) to be rewritten in the form:

$$\frac{\partial\theta/\partial x_1}{u_1'} = \frac{\partial\theta/\partial x_2}{u_2'} = \frac{\partial\theta/\partial x_3}{u_3'} = -\frac{\partial\theta/\partial t}{(1 + u_\alpha u_\alpha')}$$

with the general integral (3.8):

$$x_\alpha u_\alpha' - t(1 + u_\alpha u_\alpha') = F(\theta).$$

Assuming the value $\tau = 2\sqrt{\theta/\kappa}$ instead of the wave parameter θ, the general solution of (3.11) can be defined by the formulae[7]

$$u_1(\tau) = \int \sin\psi(\tau)\cos\phi(\tau)\,d\tau, \quad u_2(\tau) = \int \sin\psi(\tau)\sin\phi(\tau)\,d\tau,$$

$$u_3(\tau) = \int \cos\psi(\tau)\,d\tau,$$

where $\psi(\tau),\ \phi(\tau)$ are arbitrary functions of τ.

Special cases of simple waves are obtained under additional assumptions. In particular, for a steady simple wave ($\partial\theta/\partial t = 0$) there is

$$1 + u_\alpha u'_\alpha = 0. \tag{3.12}$$

Integration of the last equation gives the Bernoulli integral

$$u_\alpha u_\alpha + 2\theta = const.$$

The level surfaces of the functions u_i and θ are the planes in the space of the independent variables x_1, x_2, x_3:

$$x_\alpha u'_\alpha = F(\theta). \tag{3.13}$$

Applications of simple waves in classical gas dynamics are very well–known. They are: the Riemann waves of one–dimensional unsteady motion of a gas and the Prandtl–Meyer waves of a steady two–dimensional flow. The more general is the solution, the more complex is the problem that can be solved. For example, a steady three–dimensional simple wave generalizes the Prandtl–Meyer wave: using this more general simple wave one can construct a flow over developable surface in the three-dimensional case [19]. In fact, a developable surface Γ in the space $R^3(x)$ is parametrically given by the equations

$$x_i = q_i(s) + \nu p_i(s), \ (i = 1, 2, 3), \tag{3.14}$$

where $q_i(s),\ p_i(s)$ are some functions satisfying the equations

$$\begin{vmatrix} p_1 & p_2 & p_3 \\ p'_1 & p'_2 & p'_3 \\ q'_1 & q'_2 & q'_3 \end{vmatrix} = 0, \ (p_1, p_2, p_3) \times (q'_1, q'_2, q'_3) \neq 0. \tag{3.15}$$

By virtue of (3.15) a normal direction to the surface Γ is

$$(n_1, n_2, n_3) = \alpha(p_2 q'_3 - p_3 q'_2,\ p_3 q'_1 - p_1 q'_3,\ p_1 q'_2 - p_2 q'_1),$$

where α is a scale constant.

---

[7] This solution was given in [176].

Theorem 3.2. *The flow over an arbitrary, sufficiently smooth developable surface* Γ, *which is not a plane, can be described by a simple wave.*

Proof.

For the proof it is sufficient to find the functions $u_i(\theta)$, $(i = 1, 2, 3)$, satisfying equations (3.11)–(3.13) and the impermeability conditions on the surface Γ:

$$u_\alpha n_\alpha = 0. \tag{3.16}$$

The last equation means that the gas is not flowing through the surface Γ. Because the surface Γ is not a plane and the normal direction to it only depends on the parameter s, this parameter can be defined from (3.16)

$$s = \phi(u_1, u_2, u_3).$$

Hence, the surface Γ is described by two parameters θ and ν. By virtue of arbitrariness of the parameter ν, equation (3.13), considered on Γ, is split into the two equations

$$u'_\alpha \hat{p}_\alpha = 0, \tag{3.17}$$

$$u'_\alpha \hat{q}_\alpha = F. \tag{3.18}$$

Here $\hat{p}_i(u) = p_i(\phi(u)), \hat{q}_i(u) = q_i(\phi(u))$. Notice that equation (3.18) serves to find the function $F(\theta)$ (if the functions $u_i(\theta)$, $(i = 1, 2, 3)$ are known).

Thus, for finding the simple wave $u_i(\theta)$, $(i = 1, 2, 3)$ describing the flow over developable surface Γ, it is enough to prove the solvability of the Cauchy problem for the system of ordinary differential equations

$$u'_\alpha \hat{p}_\alpha(u) = 0, \; u_\alpha u'_\alpha = -1, \tag{3.19}$$

$$u'_\alpha u'_\alpha = 1/(\kappa\theta). \tag{3.20}$$

Let the initial data $u_i = u_i^0$, $(i = 1, 2, 3)$ at $\theta = \theta_0$ for the Cauchy problem of the system (3.19), (3.20) satisfy the conditions

$$(\hat{p}_1(u^0), \hat{p}_2(u^0), \hat{p}_3(u^0)) \times (u_1^0, u_2^0, u_3^0) \neq 0.$$

Without loss of generality it is possible to assume that $\Delta_3(u^0) \neq 0$. Here $\Delta_1(u) = u_2\hat{p}_3(u) - u_3\hat{p}_2(u)$, $\Delta_2(u) = u_3\hat{p}_1(u) - u_1\hat{p}_3(u)$, $\Delta_3(u) = u_1\hat{p}_2(u) - u_2\hat{p}_1(u)$. From equations (3.19) one obtains

$$u'_1 = (\Delta_1 u'_3 - \hat{p}_2)/\Delta_3, \; u'_2 = (\Delta_2 u'_3 + \hat{p}_1)/\Delta_3. \tag{3.21}$$

After substituting (3.21) into (3.20), there is the squared algebraic equation with respect to u'_3:

$$(u'_3)^2(\Delta_\alpha \Delta_\alpha) + 2u'_3(\Delta_2\hat{p}_1 - \Delta_1\hat{p}_2) - \Delta_3^2/(\kappa\theta) = 0.$$

Since the discriminant of the equation is not negative, the system (3.19), (3.20) can be solved with respect to the derivatives u_1', u_2', u_3'. This means that the system (3.19), (3.20) can be written in the canonical form for which there exists an unique solution of the Cauchy problem in some neighborhood of $\theta = \theta_0$.•

4. Double waves

For a double wave a parametric representation has the form

$$u_i = u_i(\lambda, \mu), \ (i = 1, 2, \dots, m) \tag{3.22}$$

with the wave parameters $\lambda = \lambda(x)$, $\mu = \mu(x)$. The result of substituting representation (3.22) into the homogeneous system (3.1) is the overdetermined system

$$A_\alpha(u_\lambda \lambda_\alpha + u_\mu \mu_\alpha) = 0. \tag{3.23}$$

where $\lambda_i = \partial\lambda/\partial x_i$, $\mu_i = \partial\mu/\partial x_i$, $(i = 1, 2, \dots, n)$. System (3.23) needs to be studied for compatibility. In the general case, it is impossible to analyze compatibility. As already mentioned, the compatibility of invariant double waves is easier to study. Therefore it is useful to find the form of double waves which are reducible to invariant double waves. In this section some sufficient conditions for the reduction of a double wave to an invariant solution are given [113, 129].

4.1 Homogeneous $2n - 1$ equations

Assume that the number of independent equations for the wave parameters obtained in the process of establishing the consistency of a double wave type solution is equal to $2n - 1$. Here n is the number of the independent variables. Such a double wave is described by the following theorem [129].

Theorem 3.3. *If in the homogeneous system of quasilinear equations (3.1) the number of independent equations $N = 2n - 1$, then the double wave is an invariant double wave. In this wave the wave parameters can be chosen (up to equivalence transformations) from one of these two types: either $\lambda = x_1$, $\mu = x_2$ or $\lambda = x_1/x_3$, $\mu = x_2/x_3$.*

Proof.

Since the total number of the derivatives $\lambda_i \equiv \partial\lambda/\partial x_i$, $\mu_i \equiv \partial\mu/\partial x_i$, $(i = 1, 2, \dots, n)$ is equal to $2n$, then, without loss of generality, the system specified in the theorem can be written in the form

$$\lambda_i = p_i(\lambda, \mu)\,\lambda_1, \ \ \mu_i = q_i(\lambda, \mu)\,\lambda_1, \ \ (i = 1, 2, \dots, n), \tag{3.24}$$

where $p_1 = 1$ is used for convenience. The proof of the theorem involves making consecutive transformations of system (3.24) to simpler forms.

First of all note that the invertible change of the wave parameters

$$\lambda' = \phi(\lambda, \mu), \ \ \mu' = \psi(\lambda, \mu) \tag{3.25}$$

transforms the factors of system (3.24) to

$$p_i' = \frac{\phi_\lambda p_i + \phi_\mu q_i}{\phi_\lambda + \phi_\mu q_1}, \quad q_i' = \frac{\psi_\lambda p_i + \psi_\mu q_i}{\phi_\lambda + \phi_\mu q_1}, \quad (i = 1, 2, \ldots, n).$$

This means it is possible to transform system (3.24) by the change (3.25) to a system with $q_1' = 0$. For this purpose it is necessary to choose ψ such that

$$\psi_\lambda + \psi_\mu q_1 = 0.$$

The property $q_1 = 0$ is retained in any change of the type (3.25) if $\psi_\lambda = 0$. Further simplifications are carried out with the simultaneous preservation of the equalities $q_1 = 0$ and $q_2 = 1$. There only take place for transformations where $\psi_\lambda = 0, \ \ \phi_\lambda = \psi_\mu$, i.e.,

$$\lambda' = \lambda\psi'(\mu) + \omega(\mu), \quad \mu' = \psi(\mu). \tag{3.26}$$

Because $\mu_1 = 0$ and $\mu_2 = \lambda_1$, differentiation (3.24) with respect to x_1 gives $\lambda_{11} = 0$ and

$$\lambda_{i1} = p_{i\lambda}\lambda_1^2, \quad q_{i\lambda} = 0, \quad (i = 2, \ldots, n).$$

Differentiating the first part of these equations with respect to x_1, one obtains $p_{i\lambda\lambda} = 0, \ \ (i = 2, \ldots, n)$. Therefore, system (3.24), by the change (3.26), can further be simplified to a system with $q_1 = 0, \ q_2 = 1$ and $p_2 = 0$. Forming mixed derivatives by differentiating (3.24), one has

$$\lambda_{ij} = \lambda_1^2(p_{i\lambda}p_j + p_{j\lambda}p_i + p_{i\mu}q_j) = \lambda_1^2(p_{i\lambda}p_j + p_{j\lambda}p_i + p_{j\mu}q_i),$$

$$\mu_{ij} = \lambda_1^2(q_i q_{j\mu} + p_{i\lambda}q_j) = \lambda_1^2(q_j q_{i\mu} + p_{j\lambda}q_i), \quad (j, i = 2, \ldots, n).$$

Comparing the left-hand and right-hand sides, and because of $\lambda_1 \neq 0$, one gets

$$p_{i\mu}q_j = p_{j\mu}q_i, \quad (p_{i\lambda} - q_{i\mu})q_j = (p_{j\lambda} - q_{j\mu})q_i, \quad (j, i = 2, \ldots, n).$$

By virtue of the relations $p_2 = 0$ and $q_2 = 1$, and setting $j = 2$ in the last equations,

$$p_{i\mu} = 0, \quad p_{i\lambda} = q_{i\mu}, \quad (i = 2, \ldots, n).$$

The general solution of these equations and the equations $q_{i\lambda} = 0$, obtained before, is

$$p_i = \lambda A_i + B_i, \quad q_i = \mu A_i + C_i, \quad (i = 2, \ldots, n),$$

where A_i, B_i, C_i are arbitrary constants, only satisfying the relations

$$A_1 = 0, \ B_1 = 1, \ C_1 = 0, \ A_2 = 0, \ B_2 = 0, \ C_2 = 1.$$

Further simplifications can be made with the help of linear invertible transformations of the independent variables. Firstly, the transformation

$$x_1' = B_\alpha x_\alpha, \;\; x_2' = C_\alpha x_\alpha, \;\; x_i' = x_i \;\; (i = 3, \ldots, n) \tag{3.27}$$

is used. System (3.24) after this transformation becomes the form

$$\lambda_2 = 0, \;\; \lambda_i = A_i \lambda \lambda_1, \;\; \mu_1 = 0, \;\; \mu_2 = \lambda_1, \;\; \mu_i = A_i \mu \mu_2, \;\; (i = 3, \ldots, n). \tag{3.28}$$

If all $A_i = 0, \;\; (i = 1, 2, \ldots, n)$, the general solution of system (3.28) is

$$\lambda = K x_1 + L, \;\; \mu = K x_2 + M, \;\; (K \neq 0)$$

with arbitrary constants K, L, M. Shifting and scaling the independent variables, this solution is reduced to

$$\lambda = x_1, \quad \mu = x_2.$$

If at least one of the $A_i, \;\; (i = 1, 2, \ldots, n)$ is not equal to zero, for example, $A_3 \neq 0$, then the linear invertible transformation of the independent variables

$$x_1' = x_1, \;\; x_2' = x_2, \;\; x_3' = A_\alpha x_\alpha, \;\; x_i' = x_i, \;\; (i = 4, \ldots, n)$$

reduces system (3.28) to

$$\lambda_2 = 0, \;\; \lambda_3 = \lambda \lambda_1, \;\; \mu_1 = 0, \;\; \mu_2 = \lambda_1, \;\; \mu_3 = \mu \mu_2, \;\; \lambda_i = \mu_i = 0, \quad (i = 4, \ldots, n).$$

The general solution of this system is

$$\lambda = \frac{x_1 + L}{K - x_3}, \;\; \mu = \frac{x_2 + M}{K - x_3}$$

with arbitrary constants K, L, M. Changing the independent variables, the solution is reduced to: $\lambda = x_1/x_3, \;\; \mu = x_2/x_3$. •

The practical meaning of the last theorem is demonstrated by considering a double wave type solution of a plane irrotational isentropic flow of a polytropic gas

$$\frac{du_i}{dt} + \frac{\partial \theta}{\partial x_i} = 0, \;\; (i = 1, 2), \tag{3.29}$$

$$\frac{d\theta}{dt} + \kappa \theta \mathrm{div}\, u = 0, \quad \frac{\partial u_1}{\partial x_2} - \frac{\partial u_2}{\partial x_1} = 0.$$

Substituting $\theta(u_1, u_2)$ into (3.29) the following system, consisting of the four quasilinear differential equations, is obtained:

$$S_i \equiv \frac{du_i}{dt} + \theta_\alpha \frac{\partial u_\alpha}{\partial x_i} = 0, \;\; (i = 1, 2), \tag{3.30}$$

$$S_3 \equiv \psi_\alpha \frac{\partial u_\alpha}{\partial x_\alpha} + 2\theta_1 \theta_2 \frac{\partial u_2}{\partial x_1} = 0, \;\; S_4 \equiv \frac{\partial u_1}{\partial x_2} - \frac{\partial u_2}{\partial x_1} = 0,$$

where $\theta_i \equiv \partial\theta/\partial u_i$, $\psi_i = \theta_i^2 - \kappa\theta$, $(i = 1, 2)$. Taking the total derivatives D_i, $(i = 0, 1, 2)$ with respect to the independent variables x_i, $(x_0 \equiv t)$, and substituting the derivatives $\partial u_i/\partial x_j$, $(i = 1, 2, 3;\ j = 0, 1, 2)$, found from system (3.30) through the parametric derivatives $\partial u_i/\partial x_1$, $(i = 1, 2)$, into the equation

$$D_0 S_3 + (u_\alpha + \theta_\alpha) D_\alpha S_3 - \psi_\alpha D_\alpha S_\alpha - 2\theta_1\theta_2 D_1 S_2 = 0,$$

one obtains the homogeneous quadratic form with respect to the derivatives $\partial u_i/\partial x_1$, $(i = 1, 2)$:

$$M b_{\alpha\beta} \frac{\partial u_\alpha}{\partial x_1} \frac{\partial u_\beta}{\partial x_1} = 0, \tag{3.31}$$

where

$$b_{ii} = -\psi_i^2,\ b_{21} = b_{12} = (\psi_1\psi_2 - 2\theta_1^2\theta_2^2),$$
$$M = \psi_2(1 + \theta_{11}) - 2\theta_1\theta_2\theta_{12} + \psi_1(1 + \theta_{22}).$$

If at least one of the coefficients $M b_{ij}$, $(i, j = 1, 2)$ is not equal to zero, equation (3.31) gives a fifth quasilinear homogeneous equation. By virtue of the Ovsiannikov theorem, such a solution is an invariant solution. Thus, for the irreducible double wave, $M b_{ij} = 0$, $(i, j = 1, 2)$. Hence,

$$M = 0. \tag{3.32}$$

It is possible to prove that the condition $M = 0$ provides involution of system (3.29) with two arbitrary functions of a one argument. Notice that equation (3.32) was obtained in [137] from another point of view: namely, requiring by solutions of the double wave type to have functional arbitrariness.

4.2 Systems of four quasilinear homogeneous equations with 3 independent variables

The case of double waves with $n = 3$ independent variables, where the wave parameters $u = (\lambda, \mu)$ satisfy the four first order homogeneous quasilinear equations

$$H_1 u_1 + H_2 u_2 + H_3 u_3 = 0, \tag{3.33}$$

is studied in this section. Here $\lambda_i = \partial\lambda/\partial x_i$, $\mu_i = \partial\mu/\partial x_i$, $u_i = (\lambda_i, \mu_i)$, $(i = 1, 2, 3)$, $H_1(u)$, $H_2(u)$, $H_3(u)$ are 4×2 matrices satisfying the condition

$$rank(H_1, H_2, H_3) = 4.$$

The requirement for a solution to be a double wave means that

$$rank \begin{pmatrix} \lambda_1 & \lambda_2 & \lambda_3 \\ \mu_1 & \mu_2 & \mu_3 \end{pmatrix} = 2. \tag{3.34}$$

4.2.1 Equivalence transformations

The property of system (3.33) to be homogeneous and autonomous is invariant with respect to the following transformations:

a) any invertible change of the wave parameters $\lambda' = L(\lambda, \mu)$, $\mu' = M(\lambda, \mu)$;

b) any non-singular linear transformation of the independent variables.

By using these equivalence transformations, and because of the double wave condition (3.34), it can be shown that any system (3.33) of four independent equations can be reduced to one of the following forms: either

$$\lambda = x_3, \quad \mu = \mu(x_1, x_2), \quad a(\lambda, \mu)\mu_1 + b(\lambda, \mu)\mu_2 + 1 = 0 \tag{3.35}$$

or

$$u_3 = Au_1, \quad u_2 = Bu_1, \tag{3.36}$$

where $A = (a_{ij}(u))$, $B = (b_{ij}(u))$ are 2×2 square matrices.

Let us prove it. Without loss of generality one can assume that $rank(H_3) = 2$. In fact, let $rank(H_3) = 0$, then $H_3 = 0$ and $\det(H_1, H_2) = 4$. This gives $u = u(x_3)$, which contradicts to the condition for a solution to be a double wave. If $rank(H_3) = 1$, then without loss of generality one can assume that in the matrix H_3 only the first row is nonzero. Hence, the rank of the matrix $(\overline{H}_1, \overline{H}_2)$ is equal to 3, where matrices $\overline{H}_i$ are composed of the matrices H_i without the first row. This is only possible if the rank of one of the matrices $\overline{H}_1, \overline{H}_2$ is equal to 2. Exchanging the independent variables, this case is transformed to the case where $rank(H_3) = 2$. Hence, it only remains to study the case $rank(H_3) = 2$.

Let $rank(H_3) = 2$. System (3.33) can be rewritten as

$$u_3 = \overline{H}_1 u_1 + \overline{H}_2 u_2, \quad \overline{\overline{H}}_1 u_1 + \overline{\overline{H}}_2 u_2 = 0$$

with some square matrices $\overline{H}_1, \overline{H}_2, \overline{\overline{H}}_1, \overline{\overline{H}}_2$. The matrices $\overline{\overline{H}}_1, \overline{\overline{H}}_2$ satisfy the condition

$$rank(\overline{\overline{H}}_1, \overline{\overline{H}}_2) = 2.$$

If the determinant of one of the matrices $\overline{\overline{H}}_1, \overline{\overline{H}}_2$ is not equal to zero, then (3.33) can be rewritten in the form (3.36).

If both matrices are singular

$$\det(\overline{\overline{H}}_1) = 0, \ \det(\overline{\overline{H}}_2) = 0,$$

then without loss of generality one can assume that

$$\overline{\overline{H}}_2 = \begin{pmatrix} 1 & b \\ 0 & 0 \end{pmatrix}.$$

Changing the independent variables to

$$x_1' = x_1, \ x_2' = \beta x_1 + x_2,$$

the second part of system $\overline{\overline{H}}_1 u_1 + \overline{\overline{H}}_2 u_2 = 0$ is transformed to

$$\overline{\overline{H}}_1 u_1 + (\overline{\overline{H}}_2 + \beta \overline{\overline{H}}_1) u_2 = 0.$$

Here the derivatives $u_1,\ u_2$ are with respect to the new independent variables $x_1',\ x_2'$. If there exists a β such that

$$\det\left(\overline{\overline{H}}_2 + \beta \overline{\overline{H}}_1\right) \neq 0,$$

then system (3.33) can be rewritten in the form (3.36). Thus it remains to study the case

$$\det\left(\overline{\overline{H}}_2 + \beta \overline{\overline{H}}_1\right) = \beta(h_{22} - bh_{21}) = 0,$$

where $h_{ij},\ (i, j = 1, 2)$ are the entries of the matrix $\overline{\overline{H}}_1$. Since $rank(\overline{\overline{H}}_1, \overline{\overline{H}}_2) = 2$, the value $h_{21} \neq 0$. In this case the system $\overline{\overline{H}}_1 u_1 + \overline{\overline{H}}_2 u_2 = 0$ is reduced to the equations

$$\lambda_1 + b\mu_1 = 0,\ \lambda_2 + b\mu_2 = 0.$$

Changing the dependent variables to $\lambda' = L(\lambda, \mu),\ \mu' = \mu$ with

$$dL(\lambda, \mu) = a(\lambda, \mu)\,(d\lambda + b(\lambda, \mu)\,d\mu)\,,$$

the last equations are transformed to the equations

$$\lambda_1 = 0,\ \lambda_2 = 0.$$

Thus, system (3.33) is transformed to the system

$$\lambda_1 = 0,\ \lambda_2 = 0,\ \lambda_3 = a_1\mu_1 + b_1\mu_2,\ \mu_3 = a_2\mu_1 + b_2\mu_2 \tag{3.37}$$

with some functions $a_i = a_i(\lambda, \mu),\ a_1^2 + b_1^2 \neq 0$. This means that $\lambda = f(x_3)$. Note that if

$$\det\begin{pmatrix} a_1 & b_1 \\ a_2 & b_2 \end{pmatrix} \neq 0,$$

this system can be reduced to a system of the form (3.36). In the case that

$$\det\begin{pmatrix} a_1 & b_1 \\ a_2 & b_2 \end{pmatrix} = 0,$$

since of $a_1^2 + b_1^2 \neq 0$ the last two equations of system (3.37) are reduced to the equations

$$\lambda_3 = a_1\mu_1 + b_1\mu_2,\ \overline{a}\lambda_3 + \mu_3 = 0.$$

By changing the dependent variables $\lambda' = L(\lambda),\ \mu' = M(\lambda, \mu)$ with

$$dM(\lambda, \mu) = c(\lambda, \mu)\,(d\mu + \overline{a}(\lambda, \mu)\,d\lambda)$$

system (3.37) is transformed to system (3.35). Here $L(\lambda)$ is the inverse function of the function $f(x_3)$, i.e., $x_3 = L(f(x_3))$. Further analysis is related to studying solutions of systems (3.35) and (3.36).

4.2.2 Solution of system (3.35)

Differentiating the last equation of system (3.35) with respect to x_3 one obtains the equations

$$a_\lambda \mu_1 + b_\lambda \mu_2 = 0 \qquad (3.38)$$
$$a_{\lambda\lambda} \mu_1 + b_{\lambda\lambda} \mu_2 = 0$$

These equations are linear and homogeneous equations for the derivatives μ_1, μ_2. Because of the double wave condition the discriminant of this system $\Delta_1 \equiv ab_\lambda - ba_\lambda$ is equal to zero. Without loss of generality it is assumed that $a \neq 0$. Hence, $b = g(\mu)a$ with some function $g = g(\mu)$. Therefore, the function $\mu = \mu(x_1, x_2)$ satisfies the equation

$$\mu_1 + g(\mu)\mu_2 = -1.$$

If $g'(\mu) = 0$, the solution of system (3.35) is an invariant solution. If $g'(\mu) \neq 0$, then the last equation after changing the dependent variable $\mu' = g(\mu)$, is reduced to the equation

$$\mu'_1 + \mu'\mu'_2 = f(\mu'),$$

where $f(\mu') = -g'(g^{-1}(\mu'))$.

Thus the theorem is proved.

Theorem 3.4. *All systems (3.35) with solutions irreducible to invariant are reduced by equivalent transformations to the equation*

$$\mu_1 + \mu\mu_2 = f(\mu),$$

where $\lambda = x_3, \quad \mu = \mu(x_1, x_2)$.

4.2.3 Solutions of system (3.36)

Taking the total derivatives D_i with respect to x_i, $(i = 1, 2, 3)$ in the expression

$$D_2(u_3 - Au_1) - D_3(u_2 - Bu_1) = 0,$$

one obtains the equations

$$Gu_{11} - C < u_1, u_1 >= 0. \qquad (3.39)$$

Here $G = AB - BA$ is a 2×2 matrix with the entries

$$g_{11} = -g_{22} = \Delta, \quad g_{12} = a_{12}(b_{22} - b_{11}) - b_{12}(a_{22} - a_{11}),$$
$$g_{21} = -a_{21}(b_{22} - b_{11}) + b_{21}(a_{22} - a_{11}),$$

$\Delta = a_{12}b_{21} - a_{21}b_{12}$, C is a bilinear mapping, where its coordinates are determined by the entries of the matrices A, B and their derivatives with respect to λ and μ.

If $\det(G) \neq 0$, all second order derivatives λ_{ij} and μ_{ij}, $(i, j = 1, 2, 3)$ are defined. They are expressed by homogeneous quadratic forms with respect to λ_1 and μ_1. Equating the mixed derivatives

$$\lambda_{ijk} = \lambda_{\sigma(ijk)}, \;\; \mu_{ijk} = \mu_{\sigma(ijk)},$$

one obtains the cubic homogeneous forms

$$\sum_{\alpha=0}^{3} c_\alpha^k(\lambda, \mu)\lambda_1^\alpha \mu_1^{3-\alpha} = 0, \;\; (k = 1, 2, ..., 8),$$

with coefficients which depend on the entries of the matrices A, B and their derivatives. Here $\sigma(ijk)$ is an arbitrary permutation of ijk, $(i, j, k = 1, 2, 3)$. If at least one of the coefficients of these cubic forms is not equal to zero, one obtains a fifth homogeneous equation, which means according to the Ovsiannikov theorem that this solution is invariant. Thus, for solutions irreducible to invariant, it is necessary to equate all coefficients of the cubic forms to zero

$$c_\alpha^k(\lambda, \mu) = 0, \;\; (k = 1, 2, ..., 8; \;\alpha = 0, 1, 2, 3).$$

These equations give the compatibility conditions. Analysis of these conditions is cumbersome and is not presented here. Notice that in this case a solution of system (3.36) is only defined up to two constants.

Further study is devoted to systems, which have irreducible solutions with functional arbitrariness. Hence, one needs to assume $det\ G = 0$.

The equation

$$\begin{aligned} det\ G = a_{12}a_{21}(b_{22} - b_{11})^2 + b_{12}b_{21}(a_{22} - a_{11})^2 - \Delta^2 \\ -(a_{12}b_{21} + a_{21}b_{12})(b_{22} - b_{11})(a_{22} - a_{11}) = 0 \end{aligned} \qquad (3.40)$$

is a quadratic polynomial equation with respect to $(b_{22} - b_{11})$.

If $a_{12}a_{21}\Delta \neq 0$, the discriminant of equation (3.40) cannot be negative, so

$$\left((a_{22} - a_{11})^2 + 4a_{12}a_{21}\right)\Delta^2 \geq 0.$$

This implies $(a_{22}-a_{11})^2+4a_{12}a_{21} \geq 0$. Under this condition the matrix A has real eigenvalues. The case where the matrix A has real eigenvalues is studied later.

If $a_{12}a_{21}\Delta = 0$, then either $a_{12}a_{21} = 0$ (in this case the matrix A has real eigenvalues) or

$$\Delta = 0, \quad a_{12}a_{21} \neq 0 \qquad (3.41)$$

The last relations lead to

$$b_{22} = b_{11} + b_{21}\xi, \;\; \xi = a_{21}^{-1}(a_{22} - a_{11}),$$

which give $G = 0$. In the case where $G = 0$, equations (3.39) consist of two homogeneous quadratic forms with respect to $u_1 = (\lambda_1, \mu_1)$:

$$C < u_1, u_1 >= 0$$

If at least one of the coefficients of these quadratic forms is non–zero, it gives fifth independent first order homogeneous autonomous quasilinear equation. Because of the reduction theorem, such solutions are reduced to invariant. Hence, for irreducible solutions, one has $C = 0$. Since $a_{12}a_{21} \neq 0$ (by assumption), equations (3.40), (3.41) and $C = 0$ yield the relations

$$\begin{gathered} b_{21} = \frac{a_{21}b_{12}}{a_{12}}, \quad b_{22} = \xi + b_{11}, \quad \frac{\partial b_{12}}{\partial \lambda} - \frac{\partial b_{11}}{\partial \mu} + \frac{b_{12}}{a_{12}}\left(\frac{\partial a_{11}}{\partial \mu} - \frac{\partial a_{12}}{\partial \lambda}\right) = 0, \\ \frac{\partial b_{11}}{\partial \lambda} - \frac{a_{21}}{a_{12}}\frac{\partial b_{12}}{\partial \mu} + \xi\frac{\partial b_{11}}{\partial \mu} + \frac{b_{12}}{a_{12}}\left(\frac{a_{21}}{a_{12}}\frac{\partial a_{12}}{\partial \mu} - \frac{\partial a_{11}}{\partial \lambda} - \xi\frac{\partial a_{11}}{\partial \mu}\right) = 0. \end{gathered} \tag{3.42}$$

Under these conditions, system (3.42) is an involutive system for the coefficients b_{12}, b_{11}, and its solution is defined within two arbitrary functions of a one argument. Notice that the coefficients a_{11}, a_{22} are also arbitrary functions.

Theorem 3.5. *System (3.36) with det $G = 0$ can have solutions irreducible to invariant if either the matrix A has real eigenvalues, or the coefficients of the matrices A and B satisfy the conditions (3.42).*

Remark 3.1. *The double waves considered in applications known to the author are of the type (3.36) with the matrix A having real eigenvalues. This property is not explicitly pointed out there. It is a consequence of the following. The classified double waves involve the hodograph transformation $x_1 = P(\lambda, \mu, x_3)$, $x_2 = Q(\lambda, \mu, x_3)$, followed by obtaining a second order degenerate algebraic equation for $\partial P/\partial x_3$ and $\partial Q/\partial x_3$, which is split into the product of two linear forms. It can be shown[8] that this is only possible if the matrix A has real eigenvalues.*

Let the matrix A have real eigenvalues. If the Jordan matrix of the matrix A is diagonal, system (3.36) can be transformed by equivalence transformations to a system of the type (3.36) with a diagonal matrix A:

$$a_{12} = a_{21} = 0.$$

In this case the matrix

$$G = (a_{22} - a_{11})\begin{pmatrix} 0 & -b_{12} \\ b_{21} & 0 \end{pmatrix}.$$

For a matrix A with a triangular Jordan matrix, system (3.36) can be transformed by equivalence transformations to a system of the type (3.36) with a

[8] An approach with the hodograph transformation will be used to study double waves of the Prandtl-Reuss equations.

triangular matrix A:

$$a_{11} = a_{22}, \ a_{21} = 0, \ a_{12} \neq 0.$$

Hence, in this case

$$G = a_{12} \begin{pmatrix} b_{21} & b_{22} - b_{11} \\ 0 & -b_{21} \end{pmatrix}.$$

Since $\det(G) = 0$, one has to study only two cases, either $rank(G) = 0$ or $rank(G) = 1$.

Assume that $rank(G) = 0$, i.e., $G = 0$. For systems with irreducible solutions it implies $C = 0$. System (3.36) is involutive with two arbitrary functions of a one argument. The equations $C = 0$ are restrictions for the entries of the matrices A and B. If the matrix A is triangular, these restrictions are a particular case of (3.42), where $\xi = 0$. If the matrix A is diagonal, the relations $G = 0$ give $(a_{22} - a_{11})(b_{12}^2 + b_{21}^2) = 0$. In the case $b_{12}^2 + b_{21}^2 \neq 0$ the equations $C = 0$ lead to $a_{11} = a_{22} = const$. Using the change of the independent variables

$$x_1' = x_1 + a_{11}x_3, \ x_2' = x_2, \ x_3' = x_3$$

a solution of system (3.36) is reduced to the invariant solution

$$\lambda = \lambda(x_1', x_2'), \ \mu = \mu(x_1', x_2').$$

Hence, for systems with irreducible solutions $b_{12} = b_{21} = 0$. In this case[9] the equations $C = 0$ are

$$\begin{aligned} (a_{22} - a_{11})\tfrac{\partial b_{11}}{\partial \mu} - (b_{22} - b_{11})\tfrac{\partial a_{11}}{\partial \mu} &= 0, \\ (a_{22} - a_{11})\tfrac{\partial b_{22}}{\partial \lambda} - (b_{22} - b_{11})\tfrac{\partial a_{22}}{\partial \lambda} &= 0. \end{aligned} \tag{3.43}$$

Let $rank(G) = 1$. For a triangular matrix A the coefficients of the matrix B satisfy the conditions $b_{21} = 0$ and $b_{22} - b_{11} \neq 0$. Since $b_{22} - b_{11} \neq 0$, the matrix B can be reduced to a diagonal form. Exchanging the independent variables x_2 and x_3, this case is transformed to the case with a diagonal matrix A. Hence, it is enough to study the case with a diagonal matrix A. For a diagonal matrix A the condition $rank(G) = 1$ implies $a_{22} - a_{11} \neq 0$, $b_{12}^2 + b_{21}^2 \neq 0$ and $b_{12}b_{21} = 0$. Without loss of generality one can assume that $b_{12} = 0$ and $b_{21} \neq 0$. The first equation (3.39) becomes

$$\lambda_1 b_{21} \frac{\partial a_{11}}{\partial \mu} + \mu_1(a_{22} - a_{11}) \left(\frac{\partial a_{11}}{\partial \mu}(b_{22} - b_{11}) - \frac{\partial b_{11}}{\partial \mu} \right) = 0.$$

For systems with irreducible solutions the coefficients of this homogeneous linear form with respect to λ_1 and μ_1 have to be equal to zero, that gives

$$a_{11} = a_{11}(\lambda), \quad b_{11} = b_{11}(\lambda).$$

[9]This class of double waves is studied in [43].

The second equation in (3.39) becomes

$$\lambda_{11} = a\lambda_1^2 + b\lambda_1\mu_1, \tag{3.44}$$

where

$$\eta = b_{21}(a_{22} - a_{11}), \quad a = \eta^{-1}\left(b_{21}\frac{\partial a_{11}}{\partial \lambda} - (a_{22} - a_{11})\frac{\partial b_{21}}{\partial \lambda}\right),$$

$$b = \eta^{-1}\left((b_{22} - b_{11})\frac{\partial a_{22}}{\partial \lambda} - b_{21}\frac{\partial a_{22}}{\partial \mu} - (a_{22} - a_{11})\frac{\partial b_{22}}{\partial \lambda}\right).$$

Without loss of generality one can set $a''_{11} = 0$, since for $a'_{11} \neq 0$ by the equivalence transformation $\lambda' = a_{11}(\lambda),\ \mu' = \mu$ the system can be transformed to the case with $a_{11} = \lambda$.

For $b \neq 0$, the derivative μ_{11} can be found from the equations

$$\begin{aligned} D_3(\lambda_{11} - a\lambda_1^2 - b\lambda_1\mu_1) - D_1^2(\lambda_3 - a_{11}\lambda_1) = 0,\\ D_2(\lambda_{11} - a\lambda_1^2 - b\lambda_1\mu_1) - D_1^2(\lambda_2 - b_{11}\lambda_1) = 0. \end{aligned} \tag{3.45}$$

Hence all second order derivatives of the functions λ and μ are found. This case is studied in a similar way to the case $\det(G) \neq 0$. Since our attention is focused on systems which have irreducible solutions with functional arbitrariness, further study will be devoted to the case $b = 0$.

For the case

$$b = 0 \tag{3.46}$$

equations (3.45) are two homogeneous quadratic forms with respect to λ_1 and μ_1. Because of the reduction theorem the coefficients of these forms have to be equal to zero:

$$aa'_{11} = 0, \quad ab'_{11} + b''_{11} = 0, \quad \frac{\partial a}{\partial \mu} = 0. \tag{3.47}$$

Equations (3.46) and (3.47) provide the involutiveness of system (3.36) with a one arbitrary function of a one argument. For classifying solutions of these systems it is necessary to study the following three cases: (a) $a'_{11} \neq 0$, (b) $a'_{11} = 0,\ b'_{11} \neq 0$, (c) $a'_{11} = 0,\ b'_{11} = 0$.

As mentioned in case (a) one can set $a_{11} = \lambda$, and the second equation of (3.47) leads to $b_{11} = c_1\lambda + c_2$ with some constants $c_1,\ c_2$. Taking the linear transformation of the independent variables $x'_1 = x_1 + c_2x_2,\ x'_2 = x_2,\ x'_3 = x_3 + c_1x_2$, one transforms system (3.36) to $b_{11} = 0$. The equation $a = 0$ yields $\lambda_{11} = 0$, $\partial b_{21}/\partial\lambda \neq 0$, and $a_{22} = \lambda + b_{21}(\partial b_{21}/\partial\lambda)^{-1}$. Since

$$\lambda_{11} = 0, \ \lambda_2 = 0, \ \lambda_3 = \lambda\lambda_1,$$

shifting the independent variables one obtains $\lambda = -x_1/x_3$. The general solution of (3.46) is

$$b_{22} = (a_{22} - \lambda)(\partial b_{21}/\partial\mu + b_{21}\psi), \tag{3.48}$$

with an arbitrary function $\psi(\mu)$. By the equivalence transformation $\lambda' = \lambda,\ \mu' = f(\mu)$ with the function $f(\mu)$ satisfying the equation $f'' - \psi f' = 0$, one can assume that $\psi = 0$ in (3.48). Changing the variables (x_1, x_2, x_3) to (λ, x_2, x_3), the remaining two equations in system (3.36) become

$$\begin{aligned} x_3\mu_3 + b_{21}(\partial b_{21}/\partial\lambda)^{-1}\mu_\lambda = 0,\\ x_3\mu_2 + b_{21}(\partial b_{21}/\partial\lambda)^{-1}(\partial b_{21}/\partial\mu)\mu_\lambda + b_{21} = 0. \end{aligned} \tag{3.49}$$

The general solution of the first equation is implicitly represented by the formulae

$$H(\mu, b_{21}x_3^{-1}, x_2) = 0$$

with an arbitrary function $H(\xi_1, \xi_2, \xi_3)$, where $\xi_1 = \mu, \xi_2 = b_{21}x_3^{-1}, \xi_3 = x_2$. Finding the derivatives μ_2 and μ_λ from this representation, and substituting them into the second equation of (3.49), one obtains

$$H_{\xi_3} - \xi_2 H_{\xi_1} = 0.$$

Hence, the general solution of system (3.49) is

$$\Phi(\mu - b_{21}x_2/x_3, b_{21}/x_3) = 0,$$

where Φ is an arbitrary function. Note that if the function Φ does not depend on the second argument, then the solution is invariant. Finally, it should be noted that the solution is a partially invariant solution with the defect $\delta = 1$.

Cases (b) and (c) are studied in a similar manner. Since $a'_{11} = 0$, using equivalence transformations one can set $a_{11} = 0$ in both cases. If $b'_{11} \neq 0$ (case (b)), system (3.36) is transformed to a system with $b_{11} = -\lambda$. Because of the second equation in (3.47) we have $a = 0$, and, hence, $\lambda_{11} = 0$. If $b'_{11} = 0$ (case (c)), system (3.36) is transformed to a system with $b_{11} = 0$. In this case one can assume that $\lambda = x_1$, and then $a = 0$. Thus, in both cases, $\lambda_{11} = 0$ and $a = 0$. Since $a = -b_{21}^{-1}\partial b_{21}/\partial\lambda = 0$, this means that $b_{21} = f(\mu)$. Using the transformation $\lambda' = \lambda,\ \mu' = \int f^{-1}(\mu)d\mu$, one can assume that $b_{21} = 1$. For irreducibility it is necessary to require $a_{22} \neq 0$. By choosing the function ϕ such that $\phi_\lambda = 1/a_{22}$, equation (3.46) can be used to find the coefficient b_{22}. Then, reducing the remaining two equations in (3.36) to a homogeneous linear system, one can find its solution.

Thus, the following theorem is valid.

Theorem 3.6. *Let the matrix A in (3.36) has real eigenvalues. Then systems of the form (3.36), having irreducible to invariant solutions with an arbitrary function, are equivalent to one of the following systems*

a) with the coefficients $(b_{21}(\partial b_{21}/\partial\lambda) \neq 0)$

$$a_{11} = -\lambda,\ b_{11} = 0,\ a_{22} = -\lambda - b_{21}(\frac{\partial b_{21}}{\partial\lambda})^{-1}, b_{22} = b_{21}\frac{\partial b_{21}}{\partial\mu}(\frac{\partial b_{21}}{\partial\lambda})^{-1},$$

with the general solution

$$\lambda = x_1/x_3, \quad \Phi(\mu - b_{21}x_2/x_3, b_{21}/x_3) = 0$$

b) with the coefficients

$$a_{11} = 0, \ b_{11} = -\lambda, \ a_{22} = 1/\phi_\lambda, \ b_{21} = 1, \ b_{22} = -\lambda + (\phi_\mu + \psi' e^{-\mu})/\phi_\lambda$$

with the general solution

$$\lambda = x_1/x_2, \ \Phi((x_3/x_2 + \phi)e^\mu + \psi, \ x_2 e^{-\mu}) = 0$$

c) with the coefficients

$$a_{11} = 0, \ b_{11} = 0, \ a_{22} = 1/\phi_\lambda, \quad b_{21} = 1, \quad b_{22} = \phi_\mu/\phi_\lambda$$

and the general solution

$$\lambda = x_1, \quad \Phi(\mu - x_2, \ x_3 + \phi) = 0$$

d) with the coefficients, satisfying conditions (3.43). The general solution is defined up to two arbitrary functions of a one argument.

Here $\Phi = \Phi(\xi_1, \xi_2)$, $\phi = \phi(\lambda, \mu)$, $\psi = \psi(\mu)$ are arbitrary functions and $\Phi_{\xi_2} \neq 0$. In case (d) system (3.36) is said to be written in terms of the Riemann invariants. Notice that the case where $b = 0$ is not included in the theorem.

4.2.4 Classification of plane isentropic double waves of gas flows

Double waves of plane isentropic potential flows were studied in [137]. The classification of double waves with the weaker condition of straight level lines were given in [161]. This section is devoted to classification of gas flows involving plane isentropic double waves, where the solutions are defined up to arbitrary functions. The requirements of the existence of functional arbitrariness lead to the study of a system of four quasilinear homogeneous equations of first order for two functions. Here this is demonstrated by obtaining such a system of equations.

The equations describing the motion of a two dimensional isentropic flow of a polytropic gas are

$$\frac{d\theta}{dt} + \kappa\theta div u = 0, \quad \frac{du_i}{dt} + \frac{\partial\theta}{\partial x_i}, \quad (i = 1, 2), \tag{3.50}$$

where $d/dt = \partial_t + u_1\partial_{x_1} + u_2\partial_{x_2}, \theta = c^2/\kappa, \theta_i = \partial\theta/\partial u_i, \psi_i = \theta_i^2 - \kappa\theta$, $(i = 1, 2)$, c is the sound speed, $\kappa = \gamma - 1$, γ is the polytropic exponent. The problem is to find the function[10] $\theta = \theta(u_1, u_2)$ for which a solution of (3.50) has functional arbitrariness.

---

[10] The functions $u_1(x_1, x_2, t), u_2(x_1, x_2, t)$ are assumed functionally independent.

If $\theta = const$, then it is simple to show that the general solution has two arbitrary functions of a one argument. This case is excluded from the further study. Hence, it is assumed $\theta_1^2 + \theta_2^2 \neq 0$, which because of the rotation of the coordinates allows us to assume that $\theta_1\theta_2\psi_1\psi_2 \neq 0$.

Substituting $\theta = \theta(u_1, u_2)$ into equations (3.50), one obtains an overdetermined system of equations. Partially prolonging this system by introducing the vortex $u_3 = \partial u_1/\partial x_2 - \partial u_2/\partial x_1$, one has the following system

$$\begin{aligned} S_i &= p_0^i + u_\alpha p_\alpha^i + \theta_\alpha p_i^\alpha = 0, \;\; (i = 1, 2), \\ S_3 &= p_0^3 + u_\alpha p_\alpha^3 + u_3(p_1^1 + p_2^2) = 0, \\ \Phi_1 = \psi_\alpha p_\alpha^\alpha + 2\theta_1\theta_2 p_1^2 &+ \theta_1\theta_2 u_3 = 0, \;\; \Phi_2 = p_2^1 - p_1^2 - u_3 = 0, \end{aligned} \tag{3.51}$$

where $x_0 = t$, $p_j^i = \partial u_i/\partial x_j$, $(i = 1, 2, 3;\; j = 0, 1, 2)$.

Since potential flows $(u_3 = 0)$ were classified earlier, further study is devoted to vortex flows $(u_3 \neq 0)$.

Before classifying double wave solutions, it is worth noticing that the equations

$$\begin{aligned} S_i &= p_0^i + u_\alpha p_\alpha^i + \theta_\alpha p_i^\alpha = 0, \;\; (i = 1, 2), \\ \Phi_1 &= \psi_\alpha p_\alpha^\alpha + \theta_1\theta_2 p_1^2 + \theta_1\theta_2 p_2^1 = 0 \end{aligned} \tag{3.52}$$

form a system of three quasilinear homogeneous partial differential equations of first order for the functions u_1 and u_2. If it is possible to obtain one more equation of the same type, one can use results of the previous theorem. For example, let the fourth equation be

$$c_3(p_2^1 - p_1^2) + c_4 p_1^1 + c_5 p_1^2 = 0 \tag{3.53}$$

with some coefficients $c_3 \neq 0$, c_4, c_5. In this case the system can be rewritten as

$$v_t = A v_{x_1}, \;\; v_{x_2} = B v_{x_1},$$

where the vector v is $v = (u_1, u_2)$. The equation $det(AB - BA) = 0$ becomes

$$\left(\psi_2 c_4^2 - 2\theta_1\theta_2 c_4 c_5 + \psi_1 c_5^2\right)\left((\theta_1^2 + \theta_2^2)c_3 + \theta_1(\theta_2 c_4 - \theta_2 c_5)\right) = 0.$$

If $\psi_2 c_4^2 - 2\theta_1\theta_2 c_4 c_5 + \psi_1 c_5^2 = 0$, the eigenvalues λ of the matrix B satisfy the equation

$$(\lambda + \xi)\left(\lambda + \xi - (\xi - \frac{\theta_1\theta_2}{\psi_2})(2 - \frac{c5}{c3})\right) = 0,$$

where ξ is a solution of the equation

$$\psi_2\xi^2 - 2\theta_1\theta_2\xi + \psi_1 = 0.$$

In the case $(\theta_1^2 + \theta_2^2)c_3 + \theta_1(\theta_2 c_4 - \theta_2 c_5) = 0$ the eigenvalues of the matrix B satisfy the equation

$$(\lambda\theta_1 - \theta_2)\left(\lambda\psi_2(\theta_2 c_4 - \theta_1 c_5) + c_4\theta_1(\theta_1^2 + +\kappa\theta) - c_5\theta_2\psi_1\right) = 0.$$

Thus the matrix B has two real eigenvalues, which is consistent with the previous theorem.

Further study is devoted to obtaining a fourth equation of the type (3.53). This equation is obtained by using the requirement of the functional arbitrariness of the solution of system (3.51). In fact, forming the combination

$$\Phi_3 = D_0\Phi_1 + (u_\alpha\theta_\alpha D_\alpha\Phi_1 + \theta_2(\theta_1^2 + \kappa\theta)D_1\Phi_2 + \theta_1\psi_2 D_2\Phi_2 \\ -\psi_1 D_1 S_1 - 2\theta_1\theta_2 D_1 S_2 - \psi_2 D_2 S_2 = 0,$$

one obtains

$$\kappa\theta(\theta_1 p_2^3 - \theta_2 p_1^3)b_{\alpha\beta}p_\alpha^\alpha p_\beta^\beta + u_3(a_\alpha p_\alpha^\alpha + u_3 a_3) = 0, \tag{3.54}$$

where D_i, $(i = 0, 1, 2)$ are the total derivatives with respect to x_i, and the coefficients b_{ik}, a_j, $(i, k = 1, 2;\ j = 1, 2, 3)$ depend on $\theta(u_1, u_2)$ and its derivatives:

$$b_{11} = -\psi_1^2 M/(4\theta_1\theta_2),\ b_{21} = b_{12} = (\psi_1\psi_2 - 2\theta_1^2\theta_2^2)M/(4\theta_1\theta_2),$$

$$b_{22} = \psi_2^2 M/(4\theta_1\theta_2),\ M = \psi_2(1+\theta_{11}) - 2\theta_1\theta_2\theta_{12} + \psi_1(1+\theta_{22}),$$

$$a_1 = \left((\theta_1^2 + \kappa\theta)M + (2(\kappa\theta)^2(1+\theta_{11}) - \psi_1(\theta_1^2 - \theta_2^2) - 2\theta_1^2\theta_2^2)\right)/(2\theta_1\theta_2),$$

$$a_2 = -\left((\theta_2^2 + \kappa\theta)M + 2(\kappa\theta^2(1+\theta_{22}) + \psi_2(\theta_1^2 - \theta_2^2) - 2\theta_1^2\theta_2^2)\right)/(2\theta_1\theta_2),$$

$$a_3 = M/4 + (\theta_1^2 + \theta_2^2 - 2\kappa\theta)/2.$$

Thus, any solution of system (3.51) satisfies the quasilinear first order equation (3.54). Since in the system (3.51), (3.54) the main derivatives are p_0^i, p_2^i, $(i = 1, 2, 3)$, the maximal arbitrariness of solutions with the given function $\theta = \theta(u_1, u_2)$ is equal to three functions of a one argument. This maximum is achieved if the system (3.51), (3.54) is involutive. For involutiveness it is necessary to require that:

$$\gamma = 2,\ \theta = \sum_{k=1}^{2}(u_k + c_k)2 + c_0, \tag{3.55}$$

where c_1, c_2 and c_0 are arbitrary constants. In this case equation (3.54) becomes

$$\theta(\theta_1 p_2^3 - \theta_2 p_1^3) + \theta^2 u_3(p_1^1 - p_2^2)/(\theta_1\theta_2) = 0,$$

The double wave (3.55) were obtained in [161] by using the assumption about straightness of level lines. Though the arbitrariness of the solution, pointed out there, is equal to two functions of one argument. Hence, the condition of the straightness of level lines narrows the functional arbitrariness.

If the system (3.51), (3.54) is not involutive, it has to be prolonged. Introducing the dependent variables $u_4 = p_1^1, u_5 = p_1^2$, the system (3.51), (3.54) is rewritten in ten quasilinear first order equations with the main derivatives p_0^i, p_2^i, $(i = 1, 2, ..., 5)$. In any k-th prolongation of this system the parametric derivatives are $\partial^k u_i / \partial x_1^k$, $(i = 3, 4, 5)$. Hence, all the Cartan characters except σ_1 are equal to zero, and there is the inequality $0 < \sigma_1 \leq 3$. A solution of (3.51), (3.54) is defined up to σ_1 arbitrary functions of a one argument.

Forming the combination[11]

$$D_0\Phi_3 + u_\alpha D_\alpha \Phi_3 - (\theta_2/\theta_1) D_1 S_3 - D_1 S_3 = 0,$$

and substituting into it the main derivatives, one obtains

$$\sum_{k=3}^{5} Q_k p_1^k + Q_6 = 0. \qquad (3.56)$$

where

$$\begin{aligned}Q_3 = {} & u_4\left(-\theta_1\theta_2(\psi_2 + 2\theta_1^2)M - \kappa\theta(\theta_1^2 + \theta_2^2)\psi_2\theta_{12} + 2\theta_1\theta_2\psi_2(\psi_1 + \psi_2)\right)/ \\ & (\theta_1^2\psi_2^2) + u_5\left(-(2\theta_1^2 + \theta_2^2)M - \kappa\theta(\theta_1^2 + \theta_2^2)(1 + \theta_{22})\right. \\ & \left.+2(\psi_2\theta_2^2 + \theta_1^2(\theta_2^2 + \kappa\theta))\right)/(\theta_1^2\psi_2) + \\ & u_3\left(2\kappa\theta\theta_1 M + \kappa\theta\theta_1\theta_2\psi_2\theta_{12} + \kappa\theta\theta_1(3\kappa\theta - \theta_2^2)(1 + \theta_{22}) - \right. \\ & \left.2\theta_1(\psi_2^2 + \theta_1^2(\theta_2^2 + \kappa\theta))\right)/(\theta_1\psi_2^2),\end{aligned}$$

$$\begin{aligned}Q_4 = {} & -u_4 2\psi_1 M/\psi_2^2 + u_5 2\theta_2 M/(\theta_1\psi_2) + u_3\left(-2\kappa\theta\theta_1\psi_2\theta_{12} + 2\theta_2\psi_1 M + \right. \\ & \left.2\kappa\theta\theta_2\psi_1(1 + \theta_{22}) + 2\theta_2(\theta_2^2\psi_2 - 2\theta_1^2\psi_1 - \theta_1^2\psi_2)\right)/(\theta_1\psi_2^2), \\ Q_5 = {} & -u_4 2\theta_2 M(2\theta_1^2 + \theta_2^2 - \kappa\theta)/(\theta_1\psi_2^2) - u_5 2M/\psi_2 + \\ & u_3\left(-2\kappa\theta\theta_2\psi_2\theta_{12} + 2\theta_1(\theta_2^2 + \kappa\theta)M + 2\kappa\theta\theta_1(\theta_2^2 + \kappa\theta)(1 + \theta_{22}) - \right. \\ & \left.2\theta_1(3\theta_2^2\psi_2 + 3\theta_1^2\theta_2^2 + \kappa\theta\theta_1^2)\right)/(\theta_1\psi_2^2).\end{aligned}$$

The expression for Q_6 is not essential for further study.

The case

$$Q_i = 0, \quad (i = 3, 4, 5), \qquad (3.57)$$

corresponds to the solution (3.55), where $\sigma_1 = 3$. Otherwise $Q_3^2 + Q_4^2 + Q_5^2 \neq 0$. Taking two prolongations of the system for the dependent variables $u_1, u_2, ..., u_5$, and excluding the main derivatives, one obtains the linear algebraic equations with respect to the derivatives p_{111}^i, $(i = 3, 4, 5)$:

$$\begin{gathered}\textstyle\sum_{k=3}^{5} Q_k p_{111}^k = f_1, \quad ((\theta_1^2 + \theta_2^2)Q_3 + \theta_1\theta_2 Q_4 - \theta_1^2 Q_5)p_{111}^3 = f_2, \\ \theta_2(-\psi_2 Q_3 + \theta_1^2 Q_5)p_{111}^3 + \theta_1\psi_1 Q_5 p_{111}^4 + \theta_1(-\psi_2 Q_4 + 2\theta_1\theta_2 Q_5)p_{111}^5 = f_3,\end{gathered} \qquad (3.58)$$

[11]This combination is obtained by excluding second order derivatives of the dependent variables from the prolongations of S_i, Φ_i, $(i = 1, 2, 3)$.

where the functions f_i, $(i = 1, 2, 3)$ depend on the derivatives of the functions $u_1, u_2, ..., u_5$ of order less than three. If the derivatives p^k_{111}, $(k = 3, 4, 5)$ can be found from system (3.58), then $\sigma_1 = 0$ and there are no arbitrary functions in the general solution. Hence, for solutions with functional arbitrariness, the determinant of the linear system (3.58) with respect to the derivatives p^k_{111}, $(k = 3, 4, 5)$ has to be equal to zero:

$$\left((\theta_1^2 + \theta_2^2)Q_3 + \theta_1\theta_2 Q_4 - \theta_1^2 Q_5\right)(\psi_2 Q_4 - 2\theta_1\theta_2 Q_4 Q_5 + \psi_1 Q_5) = 0, \tag{3.59}$$

Because of the representation of the functions Q_3, Q_4 and Q_5, equation (3.59) can be split into the product of three linear homogeneous forms of the type

$$c_3 u_3 + c_4 u_4 + c_5 u_5 = 0 \tag{3.60}$$

with some coefficients c_3, c_4, c_5 depending on the function $\theta(u_1, u_2)$ and its derivatives. For example, for the equation

$$(\theta_1^2 + \theta_2^2)Q_3 + \theta_1\theta_2 Q_4 - \theta_1^2 Q_5 = 0$$

these coefficients are

$$c_3 = \kappa\theta(\theta_2\theta_{12} - \theta_1(\theta_{22} - 1)), \quad c_4 = (-c_3(\theta_1^2 + \theta_2^2) + \theta_1^2 c_5)/(\theta_1\theta_2),$$
$$c_5 = \left(-(\theta_{11} - 1)\theta_2^2\psi_2 + \theta_1^2\psi_2(\theta_{22} - 1) + 2\theta_1\theta_2^2 c_3/(\kappa\theta)\right)/\theta_1.$$

Thus a fourth homogeneous quasilinear equation is obtained. Further detailed consideration of all cases leads to the following theorem.

Theorem 3.7. *There are only the following plane isentropic double waves having functional arbitrariness with the given function $\theta = \theta(u_1, u_2)$*
1) double waves reduced to invariant solutions,
2) double waves (3.55) with three arbitrary functions of a one argument,
3) potential double waves (3.32) with two arbitrary functions,
4) double waves reduced to the case $\theta(u_1)$.
In the last case the function $\theta(u_1, u_2)$ has to satisfy the equation

$$\theta'(\kappa\theta\psi_1\theta''' + \kappa\theta'(1+\theta'')(\psi_1 - \theta(2\theta'' - \kappa))) + \psi_1(1+\theta'')(\psi_1 - \kappa(1+\theta'')) = 0.$$

The coordinates of velocity are $u_1 = u_1(x_1, t)$, $u_2 = x_2 g_1(x_1, t) + g_2(x_1, t)$, where the functions $u_1(x_1, t)$, $g_1(x_1, t)$, $g_2(x_1, t)$ satisfy the involutive system

$$\frac{\partial u_1}{\partial t} + (u_1 + \theta')\frac{\partial u_1}{\partial x_1} = 0, \quad \frac{\partial g_i}{\partial t} + u_1\frac{\partial g_i}{\partial x_1} + g_1 g_i = 0, \quad (i = 1, 2),$$

$$\psi_1\frac{\partial u_1}{\partial x_1} = \kappa\theta g_1, \quad \frac{\partial g_1}{\partial x_1} = \frac{1 - (1 + \theta'')\kappa\theta}{\theta'\psi_1} g_1^2.$$

A solution of this system is defined up to a one function of a one argument. Notice also that the last solution corresponds to the class of solutions with a linear velocity profile with respect to a one spatial variable [159].

4.3 Unsteady space nonisentropic double waves of a gas

This section is devoted to double wave type solutions of unsteady nonisentropic gas flows in space. Particular solutions for unsteady nonisentropic double waves of a polytropic gas in space were studied in [86, 87, 180, 181, 182]. In this section it is assumed that the double waves are defined up to arbitrary functions. The classification of such double waves is given with respect to the state equation $\tau = \tau(p, S)$ of a gas.

Space flows of a gas are described by the equations

$$\frac{du}{dt} + \tau\nabla p = 0, \quad \frac{d\tau}{dt} - \tau\, div u = 0, \quad \frac{dS}{dt} = 0. \tag{3.61}$$

Here $\tau = \tau(p, S)$ is the state equation with $\tau_p \neq 0$, $\tau_S \neq 0$, $u = (u_1, u_2, u_3)$ is the velocity, p is the pressure, S is the entropy, $\tau = 1/\rho$, ρ is the density, $d/dt = \partial/\partial t + u_\alpha \partial/\partial x_\alpha$ (there is a summation with respect to a repeated Greek index from 1 to 3).

If p and S are functionally dependent on a solution of (3.61), then $p = p(S)$, $\tau = \tau(S)$, and system (3.61) can be rewritten in the form

$$\frac{\partial u_i}{\partial\lambda}\frac{d\lambda}{dt} + \phi_{x_i} = 0, \ \ (i = 1, 2, 3), \quad \frac{d\phi}{dt} = 0, \quad div\, u = 0, \tag{3.62}$$

where $\phi = \phi(S)$ is defined by the equation $\phi'(S) = \tau(S)p'(S) \neq 0$, and λ is some function which is functionally independent of ϕ. Choosing the variables ϕ and λ as the parameters of the double wave, from the first two equations of system (3.62) one obtains

$$\phi_{x_i}\frac{\partial u_j}{\partial\lambda} - \phi_{x_j}\frac{\partial u_i}{\partial\lambda} = 0, \ (i, j = 1, 2, 3;\ i \neq j). \tag{3.63}$$

Prolonging (3.63) by the operator $D/Dt = D_t + u_\alpha D_\alpha$, and substituting into it the derivatives

$$\frac{D}{Dt}\phi_{x_i} = D_i(\frac{D}{Dt}\phi) - \phi_{x_\alpha}(\frac{\partial u_\alpha}{\partial\lambda}\frac{\partial\lambda}{\partial x_i} + \frac{\partial u_\alpha}{\partial\phi}\frac{\partial\phi}{\partial x_i}) = -\phi_{x_\alpha}(\frac{\partial u_\alpha}{\partial\lambda}\frac{\partial\lambda}{\partial x_i} + \frac{\partial u_\alpha}{\partial\phi}\frac{\partial\phi}{\partial x_i}),$$

and $\phi_{x_i} = -\dfrac{\partial u_i}{\partial\lambda}\dfrac{d\lambda}{dt}$, one finds

$$\frac{\partial u_i}{\partial\lambda}\frac{\partial\lambda}{\partial x_j} - \frac{\partial u_j}{\partial\lambda}\frac{\partial\lambda}{\partial x_i} + \frac{d\lambda}{dt}\left(\frac{\partial^2 u_j}{\partial\lambda^2}\frac{\partial u_i}{\partial\lambda} - \frac{\partial^2 u_i}{\partial\lambda^2}\frac{\partial u_j}{\partial\lambda}\right) = 0, \tag{3.64}$$
$$(i, j = 1, 2, 3;\ i \neq j).$$

According to the Ovsiannikov theorem for solutions which are irreducible to invariant, the rank of the matrix, formed from the coefficients with respect to the derivatives $d\lambda/dt$, $\partial\lambda/\partial x_i$, $(i, j = 1, 2, 3; i \neq j)$ in the equations (3.64) and in the fifth equation of system (3.62), has to be less or equal to 2. This gives the equations $\partial u_i/\partial\lambda = 0,\ (i = 1, 2, 3)$, which contradict the condition for a nonisentropic flow. Thus, for irreducible double waves, the functions p and S are functionally independent.

Let us choose the pressure p and the entropy S as the parameters of the double wave, i.e., $u_i = u_i(p, S)$, $(i = 1, 2, 3)$. Introducing the new dependent variable $\phi = (div\ u)/\tau_p$, system (3.61) is reduced to the system

$$\frac{dp}{dt} - \tau\phi = 0,\quad S_0 = \frac{dS}{dt} = 0,\quad R_1 = u_{\alpha S}S_{x_\alpha} - H\phi = 0, \qquad (3.65)$$
$$\Phi_i = p_{x_i} + u_{ip}\phi = 0,\quad (i = 1, 2, 3),$$

where $H = \tau_p + u_{\alpha p}u_{\alpha p}$. It should be noted that $\phi \neq 0$, otherwise p is constant.

Differentiating (3.65), and forming their combinations, one obtains

$$D_i\Phi_j - D_j\Phi_i = u_{jp}\phi_{x_i} - u_{ip}\phi_{x_j} + (u_{jpS}S_{x_i} - u_{ipS}S_{x_j}) - \phi^2(u_{jpp}u_{ip} - u_{ipp}u_{jp}) = 0, \qquad (3.66)$$

$$R_2 = D(H\phi - u_{\alpha S}S_{x_\alpha})/Dt = H d\phi/dt - \phi(\tau u_{\alpha pS} + u_{\alpha p}u_{\beta p}u_{\beta S})S_{x_\alpha} + \phi^2(H^2 + \tau H_p) = 0, \qquad (3.67)$$

$$D(\Phi_i)/Dt = u_{ip}d\phi/dt + \tau\phi_{x_i} + \phi\zeta S_{x_i} - \phi^2(u_{ip}H - \tau u_{ipp}) = 0, \qquad (3.68)$$

where $i, j = 1, 2, 3;\ i \neq j$ and

$$\zeta = \tau_S + u_{\alpha p}u_{\alpha S},\ \xi = \tau_S + 2u_{\alpha p}u_{\alpha S},\ D/Dt = D_t + u_\alpha D_\alpha.$$

Excluding the derivatives ϕ_{x_i}, ϕ_{x_j} from (3.66) by using equations (3.68), one has

$$(\zeta u_{ip} - \tau u_{ips})S_{x_j} - (\zeta u_{jp} - \tau u_{jps})S_{x_i} = 0,\quad (i, j = 1, 2, 3;\ i \neq j). \qquad (3.69)$$

Further study is separated into the two cases: $H \neq 0$ and $H = 0$.

4.3.1 The case $H \neq 0$

Finding ϕ from the third equation of system (3.65), and substituting it into the other equations of this system, these equations together with (3.69) form a homogeneous quasilinear system of seven equations with respect to p and S. According to the Ovsiannikov theorem the equations of the system have to be dependent for the existence of solutions irreducible to invariant. This gives the conditions

$$\tau u_{ipS} - \zeta u_{ip} = 0,\quad (i = 1, 2, 3). \qquad (3.70)$$

Assume that $u_{ip} = 0, \ (i = 1, 2, 3)$. In this case

$$p = h(t), \quad \phi = h'/\tau, \, S_t + u_\alpha(S)S_{x_\alpha} = 0, \tag{3.71}$$

where the functions $u_i(S)$ are arbitrary, and the function $h(t)$ satisfies the equation

$$h'' = -(h')^2 \left(\ln(|\tau_p|)\right)_p . \tag{3.72}$$

Since p and S are functionally independent, from the last expression one obtains
$\left(\ln(|\tau_p|\right)_{pS} = 0$, which implies

$$\tau = A_1(S)g(p) + A_2(S), \tag{3.73}$$

with arbitrary functions $A_1(S)$, $A_2(S)$ and $g(p)$. Substituting τ into (3.72), and integrating it with respect to t, one has $g(h(t)) = c_1 t + c_2$, where $c_1 \neq 0$ and c_2 are arbitrary constants. Without loss of generality one can set $c_2 = 0$.

Thus, for the state equation (3.73) there are irreducible double waves, where $u_i(S), \ (i = 1, 2, 3)$ are arbitrary functions of the entropy S, the pressure p is defined from the equation $g(p) = c_1 t$, and the entropy S satisfies the system of two partial differential equations

$$\frac{dS}{dt} = 0, \quad u'_\alpha S_{x_\alpha} = c_1 A_1/(c_1 t A_1 + A_2). \tag{3.74}$$

System (3.74) is involutive and its solution has a one arbitrary function of two arguments. For example, for $A_2 = 0$ its solution is

$$tu_1(S) - x_1 + \psi\,(tu_2(S) - x_2, tu_3(S) - x_3) = 0,$$

where $\psi(\eta_1, \eta_2)$ is an arbitrary function, and $\eta_1 = tu_2(S) - x_2$, $\eta_2 = tu_3(S) - x_3$.

Now let us assume that $u_{\alpha p} u_{\alpha p} \neq 0$. For the sake of distinctness it is assumed that $u_{1p} \neq 0$. Equations (3.70) imply the existence of functions $F_i = F_i(p), \ (i = 0, 2, 3)$ such that

$$\tau = u_{1p}(F_0 - u_1 - F_2 u_2 - F_3 u_3), \quad u_{ip} = F_i u_{1p}, \quad (i = 2, 3). \tag{3.75}$$

Excluding the derivatives $d\phi/dt$ from equations (3.68), with the help of equation (3.67), one obtains

$$\Psi_i = \tau H \phi_{x_i} + \phi(H\zeta S_{x_i} + u_{ip}\xi u_{\alpha p} S_{x_\alpha}) - \phi^2 c_i = 0, \tag{3.76}$$

where

$$c_i = 2u_{ip}H^2 + \tau H_p u_{ip} - \tau H u_{ipp}, \quad (i = 1, 2, 3).$$

The combination

$$\tau\phi\xi u_{\alpha p} D_\alpha R_1 - u_{\alpha S}(D\Psi_\alpha/Dt - \tau D_\alpha R_2) + \phi(\xi u_{\alpha p} u_{\alpha S} + \zeta H) u_{\beta p} D_\beta S_0 = 0$$

gives one more equation of first order

$$R_3 = H(u_{\alpha p}S_{x_\alpha})[\tau\xi(u_{\beta SS}S_{x_\beta}) + \phi(2H\xi u_{\alpha p}u_{\alpha S} - 2\xi\tau u_{app}u_{\alpha S} - H\tau\xi_S)] + \phi^2 b = 0. \tag{3.77}$$

The expression for the function $b = b(p, S)$ is very cumbersome, and it is not important for the further study.

Notice that if $u_{\alpha p}S_{x_\alpha} = 0$, an irreducible double wave only exists in the case where $u_i = u_i(S), \quad (i = 2, 3)$. In fact, differentiating the last equation along the trajectory, and using the equation $\frac{dS}{dt} = 0$, one obtains $D(u_{\alpha p}S_{x_\alpha})/Dt = \tau\phi u_{\alpha pp}S_{x_\alpha} = 0$. Because of the Ovsiannikov theorem for irreducible double waves one has

$$\det\begin{pmatrix} u_{1p} & u_{2p} & u_{3p} \\ u_{1S} & u_{2S} & u_{3S} \\ u_{1pp} & u_{2pp} & u_{3pp} \end{pmatrix} = u_{1p}\det\begin{pmatrix} 1 & F_2 & F_3 \\ u_{1S} & u_{2S} & u_{3S} \\ 0 & F_2' & F_3' \end{pmatrix} = 0.$$

Integrating this equation with respect to S, it becomes

$$(F_2'F_3 - F_3'F_2)u_1 + F_3'u_2 - F_2'u_3 = g(p) \tag{3.78}$$

with an arbitrary function $g(p)$. Differentiating the last equation with respect to p, one gets

$$(F_2''F_3 - F_3''F_2)u_1 + F_3''u_2 - F_2''u_3 = g'(p). \tag{3.79}$$

The equations (3.78) and (3.79) give

$$\begin{aligned} (F_2'F_3'' - F_3'F_2'')(u_3 - F_3u_1) &= F_3'g'(p) - F_3''g(p), \\ (F_2'F_3'' - F_3'F_2'')(u_2 - F_2u_1) &= F_2'g'(p) - F_2''g(p). \end{aligned} \tag{3.80}$$

If $(F_2'F_3'' - F_3'F_2'') \neq 0$, then from the last equations and (3.78) one finds that $u_i = u_i(p), \quad (i = 1, 2, 3)$, but because of the equation $R_1 = 0$, this contradicts to the condition $H\phi \neq 0$. Hence, $(F_2'F_3'' - F_3'F_2'') = 0$. If $(F_2')^2 + (F_3')^2 \neq 0$, for example, $F_2' \neq 0$, then by a rotation of the coordinates and a Galilean transformation this solution can be transformed to a solution with $u_3 = 0$. This means that such a double wave is reduced to a two–dimensional double wave[12]. Thus, for double waves irreducible to invariant, one has to assume $F_2' = F_3' = 0$. By the rotation of the coordinates one can account that $u_i = u_i(S), \quad (i = 2, 3)$. From the equation $u_{\alpha p}S_{x_\alpha} = 0$, one has $S_{x_1} = 0$, and from the relation $D_1R_1 = 0$ one finds $\phi_{x_1} = H_p u_{1p}\phi^2/H$. Substituting S_{x_1} and ϕ_{x_1} into (3.76) (i = 1), one obtains

$$2u_{1p}H - \tau u_{1pp} = 0. \tag{3.81}$$

[12]Classification of two–dimensional double waves is studied in [109].

Since the first equation of (3.75) is $\tau = u_{1p}(F_0 - u_1)$, after substituting τ into H, equation (3.81) becomes

$$2u_{1p}F_0' + u_{1pp}(F_0 - u_1) = 0.$$

The system (3.65), (3.67), (3.76) together with the equation $u_{\alpha p}S_{x_\alpha} = 0$ is involutive and its general solution has a one arbitrary function of a one argument.

Now let us consider the case $u_{\alpha p}S_{x_\alpha} \neq 0$.

At first it is shown that system (3.65) has no solutions with arbitrary functions in the case $\xi \neq 0$. In fact, let

$$r = rank \begin{pmatrix} u_{1p} & u_{2p} & u_{3p} \\ u_{1S} & u_{2S} & u_{3S} \end{pmatrix} = 1,$$

then $u_2 = u_2(u_1),\ u_3 = u_3(u_1)$. Hence, $u_2'(u_1) = F_2(p)$ and $u_3'(u_1) = F_3(p)$. If $(u_2'')^2 + (u_3'')^2 \neq 0$, then $u_i = u_i(p),\ (i = 1, 2, 3)$. Using equation (3.70) this leads to the contradiction $u_{1p} = 0$. In the case $u_2'' = u_3'' = 0$ the double wave is reduced by the rotation of the coordinates to a two-dimensional flow. Therefore, $r = 2$. Let us introduce the new dependent variable[13] $\lambda = u_{\alpha p}S_{x_\alpha}$. Since $\xi \neq 0$, the equation

$$R_{3+i} = -\tau D_i R_2 + D\Psi_i/Dt - \phi\xi u_{ip}u_{\alpha p}D_\alpha S_0 - \phi\zeta H D_i S_0 = 0$$

implies

$$L_i = \lambda_{x_i} - a_{i\alpha}S_{x_\alpha} - c = 0,\ (i = 1, 2, 3)$$

with some functions $c = c(p, S, \lambda, \phi), a_{ij} = a_{ij}(p, S),\ (i, j = 1, 2, 3)$. Moreover, there is also the equation

$$L_0 = \frac{D}{Dt}(\lambda - u_{\alpha p}S_{x_\alpha}) = \frac{d\lambda}{dt} - \tau\phi u_{\alpha pp}S_{x_\alpha} + \tau_p\phi\lambda = 0.$$

Hence, the dependent variables p, S, ϕ, λ satisfy the overdetermined system of equations (3.65), (3.67), (3.76) and $L_i = 0,\ (i = 0, 1, 2, 3)$. The parametric derivatives of this system are defined by the equations

$$u_{\alpha p}S_{x_\alpha} - \lambda = 0,\ \ u_{\alpha S}S_{x_\alpha} - H\phi = 0.$$

Let us consider the following prolongations

$$\begin{gathered} D_i R_1 = u_{\alpha S}S_{x_\alpha x_i} + \cdots = 0,\ \ D_i R_3 = \tau\lambda H\xi u_{\alpha SS}S_{x_\alpha x_i} + \cdots = 0, \\ DL_i/Dt - D_i L_0 + a_{i\alpha}D_\alpha S_0 = \tau\phi u_{\alpha pp}S_{x_\alpha x_i} + \cdots = 0, \\ D_i(u_{\alpha p}S_{x_\alpha} - \lambda) = u_{\alpha p}S_{x_\alpha x_i} + \cdots = 0,\quad (i = 1, 2, 3). \end{gathered}$$

Here, only the coefficients with respect to the highest order derivatives are given. Since $H\lambda\phi\xi \neq 0$, and the parametric derivatives are in the set of the

[13]With respect to the consistency theory the space of the dependent variables is partially prolonged.

derivatives $S_{x_ix_j}$, $(i, j = 1, 2, 3)$, for solutions with functional arbitrariness the rank of the matrix formed from these coefficients has to be smaller than the number of the derivatives $S_{x_ix_j}$, $(i, j = 1, 2, 3)$. This condition gives the inequality

$$r_0 = rank \begin{pmatrix} u_{1p} & u_{2p} & u_{3p} \\ u_{1s} & u_{2S} & u_{3S} \\ u_{1pp} & u_{2pp} & u_{3pp} \\ u_{1SS} & u_{2SS} & u_{3SS} \end{pmatrix} \leq 2.$$

As in the case $u_{\alpha p} S_{x_\alpha} = 0$ the equation

$$\det \begin{pmatrix} u_{1p} & u_{2p} & u_{3p} \\ u_{1S} & u_{2S} & u_{3S} \\ u_{1pp} & u_{2pp} & u_{3pp} \end{pmatrix} = 0$$

leads to the double wave with $u_2(S)$, $u_3(S)$. Since $r_0 \leq 2$, one finds that $u_2' u_3'' - u_2'' u_3' = 0$, which means a reduction to a two-dimensional flow.

Therefore, for irreducible double waves with functional arbitrariness in the case $u_{\alpha p} S_{x_\alpha} \neq 0$, one obtains $\xi = 0$. In this case from (3.70) and $\xi = 0$ one gets $\tau = f_0 u_{1p}^2$ with some function $f_0(p)$. Taking the combinations

$$D_{x_i} \Psi_j - D_{x_j} \Psi_i = 0, \quad (i = 1, 2, 3; i \neq j),$$

one finds the first order differential equations

$$S_{x_i} a_j - S_{x_j} a_i = -\phi(u_{ip} d_j - u_{jp} d_i), \quad (i, j = 1, 2, 3; i \neq j), \tag{3.82}$$

where

$$a_1 = \tau H u_{1p}[(H_p/H)_S - 2(u_{1pp}/u_{1p})_S + 2(H/\tau)_S], \quad d_1 = 0,$$

$$a_i = F_i a_1, d_i = F_i d_1 + F_i'(u_{1p}(2H^2 + \tau H_p) - 2\tau H u_{1pp}) - \tau H u_{1p} F_i'',$$
$$(i = 2, 3).$$

Solving with respect to ϕ the equation $R_1 = 0$, and substituting ϕ into the other equations of (3.65) and (3.82), one obtains a system of eight homogeneous quasilinear differential equations. The Ovsiannikov theorem implies that the equations

$$H a_1 (F_i S_{x_1} - S_{x_i}) + u_{1p} d_i u_{\alpha S} S_{x_\alpha} = 0, \quad (i = 2, 3)$$

have to be linearly dependent. This leads to $a_1 = 0$ and $d_i = 0$, $(i = 2, 3)$, i.e.,

$$(H_p/H)_S - 2(u_{1pp}/u_{1p})_S + 2(H/\tau)_S = 0, \tag{3.83}$$

$$F_i'(u_{1p}(2H^2 + \tau H_p) - 2\tau H u_{1pp}) - \tau H u_{1p} F_i'' = 0, \ (i = 2, 3). \tag{3.84}$$

Because of (3.83), from the equations

$$D\Psi_i/Dt - \tau D_i R_2 + \phi H u_{\alpha p} u_{\alpha S} D_i S_0 = 0, \quad (i = 1, 2, 3)$$

one obtains

$$2\tau^2 H u_{1ppp} - 2\tau u_{1pp}(2H^2 + \tau H_p) + u_{1p} H(4H^2 - \tau H_p + \tau\tau_{pp}) = 0. \quad (3.85)$$

The system of equations, consisting of the equation

$$\xi = \tau_S + 2u_{\alpha p} u_{\alpha S} = 0 \quad (3.86)$$

and (3.83), (3.84),(3.85) guarantee the involutiveness of the overdetermined system (3.65), (3.67), (3.76) with the arbitrariness of two functions of a single argument.

Let us study the consistency of the system (3.83)-(3.86) for the functions $u_1(p, S)$, $u_2(p, S)$, $\tau(p, S)$. Since $\tau H u_{1p} \neq 0$ and equations (3.84) are homogeneous linear algebraic equations with respect to $(u_{1p}(2H^2 + \tau H_p) - 2\tau H u_{1pp})$ and $\tau H u_{1p}$, this gives $F_2' F_3'' - F_3' F_2'' = 0$. Hence, without loss of generality one can assume that $F_3 = 0$ or $u_3 = u_3(S)$, so one only needs to study two cases: $F_2' \neq 0$ and $F_2' = 0$.

Let $F_2' \neq 0$. From (3.84) one finds

$$H_p/H = 2u_{1pp}/u_{1p} - 2H/\tau + F_2''/F_2'. \quad (3.87)$$

In this case equation (3.83) is satisfied identically. Since $\tau = f_0 u_{1p}^2$, from equation (3.86) one obtains

$$u_{2S} = -(f_0 u_{1pS} + u_{1S})/F_2. \quad (3.88)$$

Because $u_{2p} = F_2 u_{1p}$, equating the mixed derivatives, and integrating it with respect to S, one finds

$$u_{1pp} = -(f_0 F_2)^{-1}(u_{1p}(F_2 f_0' - f_0 F_2' + F_2(1 - F_2^2)) - u_1 F_2' + F_4), \quad (3.89)$$

where $F_4 = F_4(p)$ is an arbitrary function. Differentiating[14] (3.89) with respect to p, one finds u_{1ppp}. Substituting the derivatives u_{1pp} and u_{1ppp} into (3.85) and (3.87), one finds $F_4 = cF_2'$ and the function $u_1 = u_1(p, S)$ has to satisfy the equation of the type

$$a(p)u_{1p} + (u_1 - c) = 0,$$

where c is constant. Without loss of generality one can assume that $c = 0$. From the last equation one finds that the function $u_1(p, S)$ has the representation $u_1 = h_1(p)A(S)$.

[14] Since all calculations are cumbersome, only the method of calculation is discussed here.

The relation $u_2 = F_{2p}u_{1p}$ and equation (3.88) show that $u_2 = h_2(p)A(S)$. This, and the equation $\tau = f_0u_{1p}^2$, lead to the following representation of the state equation $\tau = g(p)A^2(S)$. Here the function $A = A(S)$ is arbitrary, and the functions $g = g(p)$ and $h_i = h_i(p)$, $(i = 1, 2)$ satisfy two ordinary differential equations. One of them is obtained from equation (3.86):

$$g + h_1h_1' + h_2h_2' = 0, \tag{3.90}$$

which can also be rewritten in the form $\tau + u_\alpha u_{\alpha p} = 0$. Another equation follows from (3.89). The function $u_3 = u_3(S)$ is arbitrary.

By virtue of (3.90) the pressure does not depend on time $p = p(x_1, x_2, x_3)$. Notice also that, instead of the function $h_2(p)$, one can search for the function $F_2(p)$. In this case $h_2' = h_1'F_2$, and the equations relating the functions $h_1(p)$, $F_2(p)$ and $g(p)$ are[15]

$$\begin{gathered} F_2F_2''((h_1')^2(1 + F_2^2) + g') = 2(F_2')^2h_1'(g'h_1 - h_1'g) \\ +F_2F_2'gg'' + 2((F_2')^2g + (h_1')^2F_2F_2'(1 + F_2^2))((h_1')^2(1 + F_2^2) + g'), \end{gathered} \tag{3.91}$$

$$F_2gh_1'' = F_2(1 + F_2^2)(h_1')^3 - F_2'h_1(h_1')^2 + h_1'(F_2g' - gF_2'). \tag{3.92}$$

Remark 3.2. *In [180, 181, 182] for a polytropic gas*[16] *another representation of the solution found here is given. Namely, if a solution has the form* $u_1^2 = \psi(p)A^2(S)$, *then for the functions* $F_2(p)$ *and* $\psi(p)$ *there is the following system of ordinary differential equations*

$$\begin{aligned} g(2\psi\psi'' - (\psi')^2)/(2\psi\psi') = g' + (1 + F_2^2)(\psi')^2/(4\psi) - \\ (g + \psi'/2)F_2'/F_2, \end{aligned} \tag{3.93}$$

$$\begin{aligned} g(g' + (1 + F_2^2)(\psi')^2/(4\psi)) = (1 + F_2^2)^2(\psi')^2/(8\psi^2) + \\ (g'(1 + F_2^2) + gF_2F_2')\psi'/(2\psi) + \psi'g'F_2'/F_2 + \\ g(g'' + 2g'F_2'/F_2). \end{aligned} \tag{3.94}$$

Remark 3.3. *Because of (3.84), equation* $\tau + u_\alpha u_{\alpha p} = 0$ *guarantees the identical satisfaction of equation (3.85) for steady space double waves [109].*

Now assume that $F_2' = 0$. Without loss of generality one can set $F_2 = 0$ or $u_2 = u_2(S)$. Substituting $\tau = f_0u_{1p}^2$ into the first equation of (3.70) and integrating it with respect to S, gives

$$f_0u_{1p}^2 + u_1 = g,$$

[15] Intermediate calculations are omitted.

[16] The function $u_3(S)$ obtained in [180, 181, 182] is related with $A(S)$: $u_3 = cA(S)$, where c is constant. This is because of the special representation of the double wave studied there.

where $g = g(p)$ is arbitrary. The general solution of the last equation is $u = h_1(p)A(S) + \psi(p)$, where $f_0 = -h_1/h_1'$, and $A(S)$ and $\psi(S)$ are arbitrary functions of their arguments. From the analysis of the remaining two equations (3.83), (3.85) one obtains $\psi = c_1 h_1 + c_2$ with constant c_1, c_2. Without loss of generality one can set $c_1 = 0, c_2 = 0$, i.e.,

$$u_1 = h_1(p)A(S), \quad \tau = -h_1 h_1' A^2(S). \tag{3.95}$$

In this case equations (3.83), (3.85) are satisfied identically. The functions $u_2(S)$ and $u_3(S)$ are arbitrary. Notice that the pressure also does not depend on t in this case.

Thus, the study of the case $H \neq 0$ is complete.

4.3.2 The case $H = 0$

Let $H = 0$. Solving the first equation of (3.65) with respect to ϕ, and substituting it into others, one obtains homogeneous quasilinear equations. Applying the Ovsiannikov theorem to these equations together with the equations (3.67), (3.69), one finds that for irreducible double waves

$$rank \begin{pmatrix} u_{1S} & u_{2S} & u_{3S} \\ a_1 & a_2 & a_3 \\ b_2 & -b_1 & 0 \\ b_3 & 0 & -b_1 \\ 0 & b_3 & -b_2 \end{pmatrix} \leq 2,$$

where $b_i = \tau u_{ips} - \zeta u_{ip}$, $a_i = b_i - \xi u_{ip}$, $(i = 1, 2, 3)$. Notice that this implies

$$b_\alpha a_\alpha = 0, \qquad u_{\alpha S} b_\alpha = 0. \tag{3.96}$$

For further study let us make the change[17] of the independent variables x_1, x_2, x_3, t to p, S, x_3, t, i.e.,

$$x_1 = P(p, S, x_3, t), \quad x_2 = Q(p, S, x_3, t).$$

Finding all derivatives of the functions p and S through the derivatives of the functions P and Q, and substituting them into (3.61), they become

$$\begin{gathered} \tau Q_S - u_{1p}(BP_S - AQ_S) = 0, \quad \tau P_S + u_{2p}(BP_S - AQ_S) = 0, \\ \tau(Q_{x_3} P_S - P_{x_3} Q_S) = u_{3p}(BP_S - AQ_S) = 0, \\ BP_p - AQ_p = 0, \quad (u_{2S} - u_{3S} Q_{x_3})P_p - (u_{1S} - u_{3S} P_{x_3})Q_p = 0, \end{gathered} \tag{3.97}$$

where

$$A = u_1 - u_3 P_{x_3} - P_t, \quad B = u_2 - u_3 Q_{x_3} - Q_t$$

[17]Without loss of generality one can assume that $p_{x_1} S_{x_2} - p_{x_2} S_{x_1} \neq 0$.

and

$$P_pQ_S - P_SQ_p \neq 0. \tag{3.98}$$

Since $P_S^2 + Q_S^2 \neq 0$, one has $(BP_S - AQ_S) \neq 0$. The third equation, with the help of first and second, can be transformed to

$$u_{3p} + u_{1p}P_{x_3} + u_{2p}Q_{x_3} = 0. \tag{3.99}$$

Forming a combination of the first equation multiplied by u_{2p} and second multiplied by u_{1p}, one has

$$u_{1p}P_S + u_{2p}Q_S = 0. \tag{3.100}$$

Using this equation, the first and the second equations become

$$u_{ip}(\psi - u_{1p}P_t - u_{2p}Q_t) = 0, \ \ (i = 1, 2).$$

where $\psi = \tau + u_{\alpha p}u_{\alpha p}$. Because of $u_{\alpha p}u_{\alpha p} \neq 0$, the last equations are reduced to the equation

$$u_{1p}P_t + u_{2p}Q_t = \psi. \tag{3.101}$$

Integrating equations (3.99) and (3.101) with respect to x_3 and t, one obtains

$$u_{1p}P + u_{2p}Q + x_3u_{3p} - t\psi = \chi, \tag{3.102}$$

where $\chi = \chi(p, S)$. Differentiating (3.102) with respect to p, and using equation (3.100), it becomes

$$u_{1pS}P + u_{2pS}Q + u_{3pS}x_3 - t\psi_S = \chi_S, \tag{3.103}$$

Equations (3.102), (3.103) show that if $\Delta_3 \equiv u_{1pS}u_{2p} - u_{2pS}u_{1p} \neq 0$, then one can define P and Q:

$$P = \left(x_3\Delta_1 + t\begin{vmatrix} \psi & \psi_S \\ u_{2p} & u_{2ps} \end{vmatrix} + \begin{vmatrix} \chi & \chi_S \\ u_{2p} & u_{2ps} \end{vmatrix}\right)/\Delta_3 \tag{3.104}$$

$$Q = \left(x_3\Delta_2 - t\begin{vmatrix} \psi & \psi_S \\ u_{1p} & u_{1ps} \end{vmatrix} - \begin{vmatrix} \chi & \chi_S \\ u_{1p} & u_{1ps} \end{vmatrix}\right)/\Delta_3. \tag{3.105}$$

System (3.97) with the help of (3.96) is reduced to the equation

$$(u_2 - u_3Q_{x_3} - Q_t)P_p - (u_1 - u_3P_{x_3} - P_t)Q_p = 0. \tag{3.106}$$

Substituting the representations of P and Q (3.104), (3.105) into (3.106), a linear equation with respect to x_3 and t is obtained. Splitting it with respect to x_3 and t, one finds three equations. The free term is a linear second order hyperbolic equation for the function $\chi = \chi(p, S)$ of the form

$$\tau\chi_{pS} - \zeta\chi_p + q_1\chi_S + q_2\chi = 0, \tag{3.107}$$

where the functions $q_1(p, S)$ and $q_2(p, S)$ depend on the $\tau(p, S)$, $u_i(p, S)$, $(i = 1, 2, 3)$ and their derivatives. The coefficients corresponding to x_3 and t give

$$\Delta_\alpha b_{\alpha p} = 0, \ (u_{\alpha p} b_\alpha)\Delta_3 + b_1 b_{2p} - b_2 b_{1p} = 0. \qquad (3.108)$$

Thus, if $\Delta_3 \neq 0$, there are flows with straight level lines, and the functions $\tau(p, S)$ and $u_i(p, S)$, $(i = 1, 2, 3)$ satisfy the overdetermined system of equations consisting of five differential equations: (3.96), (3.108) and $H = 0$. Analysis of this overdetermined system in the general case is difficult. If one assumes that $\psi = 0$, which corresponds to steady flows, then for the state equation of the form $\tau = g(p)A^2(S)$ it is shown in [111] that this system is only consistent if a gas is polytropic with the exponent $\gamma = 2$. The general solution of the gas dynamics equations in this case is defined up to a one arbitrary function of a one argument.

Let $u_{ips}u_{jp} - u_{jps}u_{ip} = 0$, $(i, j = 1, 2, 3; i \neq j)$. Without loss of generality it can be assumed that $u_{1p} \neq 0$. Hence,

$$u_{ip} = F_i u_{1p}, \ (i = 2, 3) \qquad (3.109)$$

with some functions $F_i = F_i(p)$, $(i = 2, 3)$. Equations (3.102), (3.103) give $\psi = F_0 u_{1p}$, $\chi = F_4 u_{1p}$ and

$$P = -F_2 Q - x_3 F_3 + t F_0 + F_4, \qquad (3.110)$$

where $F_0 = F_0(p)$, $F_4 = F_4(p)$ are some functions of integrations. Notice that equations (3.96) are satisfied, $\tau = -(u_1 + u_2 F_2 + u_3 F_3 - F_0)u_{1p}$, and equation (3.110) implies

$$x_1 + x_2 F_2 + x_3 F_3 + t F_0 = F_4.$$

The problem of finding a solution of (3.97) is reduced to solving the system of two linear equations with respect to the dependent function $Q(p, S, x_3, t)$:

$$\begin{aligned} \omega Q_p + \beta(u_3 Q_{x_3} - u_2 + Q_t) &= 0, \\ \omega_S Q_p + \beta(u_{3S} Q_{x_3} - u_{2S}) &= 0. \end{aligned} \qquad (3.111)$$

Here

$$\omega = u_1 + u_2 F_2 + u_3 F_3 - F_0, \ \beta = F_4' - F_2' Q - x_3 F_3' + t F_0'.$$

The reversibility condition (3.98) gives $\beta Q_S \neq 0$. Since $\tau = -\omega u_{1p}$, one has $\omega \neq 0$ and $\xi = -\omega^2 (u_{1p}/\omega)_S$. Notice also that differentiating the equation $\tau + \omega u_{1p} = 0$ with respect to p, and using the condition $H = 0$, one obtains

$$F_2' u_2 + F_3' u_3 - F_0' = -\omega u_{1pp}/u_{1p}. \qquad (3.112)$$

Excluding Q_p from (3.111), and because of $\beta \neq 0$, one obtains

$$(u_3/\omega)_S Q_{x_3} + (1/\omega)_S Q_t = (u_2/\omega)_S. \quad (3.113)$$

Analysis shows that $\omega_S \neq 0$. In fact, assume that $\omega_S = 0$. In this case $u_{3S} \neq 0$, because otherwise from (3.113) and $\omega_S = 0$ one finds $u_{2S} = 0$ and $u_{1S} = 0$, which contradicts to the condition $\tau_S \neq 0$. The general solution of (3.113) is

$$Q = x_3(u_{2S}/u_{3S}) + G,$$

where $G = G(t, p, S)$. Substituting this into the first equation of (3.111), and splitting it with respect to x_3, one has

$$\omega(u_{2S}/u_{3S})_p - (F_2'(u_{2S}/u_{3S}) + F_3')(u_3(u_{2S}/u_{3S}) + G_t - u_2) = 0, \quad (3.114)$$

$$\omega G_p + (F_4' - F_2' G + tF_0')(u_3(u_{2S}/u_{3S}) + G_t - u_2) = 0. \quad (3.115)$$

If $F_2'u_{2S} + F_3'u_{3S} = 0$, then from equation (3.114) one finds $(u_{2S}/u_{3S})_p = 0$, which by virtue of $\omega_S = 0$ leads to the contradiction $\tau_S = 0$. Hence, $F_2'u_{2S} + F_3'u_{3S} \neq 0$. Since $\omega_S = 0$, differentiating (3.112) with respect to S, one finds

$$F_2'u_{2S} + F_3'u_{3S} = -\omega(u_{1pp}/u_{1p})_S.$$

From another point of view, because $\omega_S = u_{1S} + u_{2S}F_2 + u_{3S}F_3$, differentiating the equation $\omega_S = 0$ with respect to p, and using (3.109), one finds $F_2'u_{2S} + F_3'u_{3S} = -u_{1pS}(1 + F_2^2 + F_3^2)$. Hence,

$$\tau(u_{1pS}/u_{1p})_p = -u_{1p}u_{1pS}(1 + F_2^2 + F_3^2).$$

Substituting $(1 + F_2^2 + F_3^2) = -\tau_p/u_{1p}^2$, one obtains $\left(u_{1pS}/(\tau u_{1p})\right)_p = 0$ or $\left(u_{1pS}/(\omega u_{1p}^2)\right)_p = 0$. Since $\omega_S = 0$, the general solution of the last equation is

$$u_{1p} = 1/(\omega g + \lambda) \quad (3.116)$$

with some functions $g = g(S)$ and $\lambda = \lambda(p)$. Hence, $\tau = -1/(g + \lambda/\omega)$ and

$$1 + F_2^2 + F_3^2 = \lambda\omega_p - \lambda'\omega.$$

Because $\tau_S\tau_p \neq 0$, the functions $g(S)$ and $\lambda(p)$ satisfy the inequality $g'(\lambda/\omega)_p \neq 0$.

Differentiating (3.112) with respect to p, one has

$$F_2''u_2 + F_3''u_3 - F_0'' = -(\omega(u_{1pp}/u_{1p}))_p - u_{1p}(F_2F_2' + F_3F_3'). \quad (3.117)$$

Since $(F_2')^2 + (F_3')^2 \neq 0$, let us assume, for example, that $F_3' \neq 0$. If $(F_2'/F_3')' \neq 0$, then one can find u_2 and u_3 from equations (3.112), (3.117).

Differentiating them with respect to p, and substituting into equations (3.109), one obtains polynomial equations with respect to $g(S)$ with coefficients which only depend on p. Since $g'(S) \neq 0$, these coefficients have to be equal to zero. Analysis of these equations leads to a contradiction. A similar contradiction is obtained in the case $(F_2'/F_3')' = 0$.

Therefore it is only necessary to study the case $\omega_S \neq 0$.

The general solution of (3.113) is

$$Q = t\frac{(u_2/\omega)_S}{(1/\omega)_S} + G(p, S, \lambda),$$

where $\lambda = x_3 - t\frac{(u_3/\omega)_S}{(1/\omega)_S}$. Substituting the representation of Q into (3.115), and splitting it with respect to t, one has

$$(2\omega_S u_{1p} + \tau_S)(k_2 - k_3 G_\lambda) = 0, \tag{3.118}$$

$$G_p - G_\lambda(F_4' - F_2'G - F_3'\lambda)u_{3S}/\omega_S = (F_4' - F_2'G - F_3'\lambda)u_{2S}/\omega_S, \tag{3.119}$$

where

$$k_i = F_i\omega_S - u_{iS}(1 + F_2^2 + F_3^2), \ (i = 2, 3).$$

Let $2\omega_S u_{1p} + \tau_S = 0$. Further analysis of the equations $\tau + \omega u_{1p} = 0$ and $H = 0$ leads to the following: $F_0 = 0$, and the coordinates of the velocity can be represented as follows: either

$$u_i = A(S)h_i(p), \ \ (i = 1, 2, 3), \tag{3.120}$$

or

$$u_i = A(S)h_i(p), \ \ (i = 1, 2), \ \ u_3 = g_3(S), \ \ (h_3 = 0), \tag{3.121}$$

or

$$u_1 = A(S)h_1(p), \ \ \ u_i = g_i(S), \ (i = 2, 3), \ \ (h_2 = 0, \ h_3 = 0),), \tag{3.122}$$

where the $g_i(S)$ are arbitrary functions, and

$$g' + h_\alpha' h_\alpha' = 0, \ \ \ g + h_\alpha h_\alpha' = 0. \tag{3.123}$$

The last equation means that the pressure p does not depend on t. Notice that the function $G(p, S, \lambda)$ only satisfies the one equation (3.119).

Let $2\omega_S u_{1p} + \tau_S \neq 0$. If in this case $k_3 = 0$, then $k_2 = 0$, i.e.,

$$\begin{aligned} F_2\omega_S - u_{2S}(1 + F_2^2 + F_3^2) = 0, \\ F_3\omega_S - u_{3S}(1 + F_2^2 + F_3^2) = 0. \end{aligned} \tag{3.124}$$

Considering these equations as algebraic equations with respect to ω_S and $(1 + F_2^2 + F_3^2)$, one obtains that discriminant

$$F_2 u_{3S} - F_3 u_{2S} = 0. \tag{3.125}$$

Differentiating this equation with respect to p, one has

$$F_2' u_{3S} - F_3' u_{2S} = 0. \tag{3.126}$$

If $F_2 F_3' - F_2' F_3 \neq 0$, then $u_{2S} = 0$ and $u_{3S} = 0$, but equations (3.124) lead to $F_2 = F_3 = 0$. This corresponds to a reducible double wave. If $F_2 F_3' - F_2' F_3 = 0$, then without loss of generality one can assume $F_3 = 0$. Equations (3.124) lead to $u_{3S} = 0$, hence, this double wave is also reducible. Therefore $k_3 \neq 0$. Integrating (3.118), one has $G = \lambda(k_2/k_3) + \Phi(p, S)$. Substituting this representation of the function $G(p, S, \lambda)$ into equation (3.119), and splitting it with respect to λ, one obtains

$$\omega_S(k_3 k_{2p} - k_2 k_{3p}) + (k_2 F_2' + k_3 F_3')(k_2 u_{3S} + k_3 u_{2S}) = 0, \tag{3.127}$$

$$\omega_S k_3 \Phi_p + (F_4' - F_2' \Phi)(k_2 u_{2S} + k_3 u_{3S}) = 0. \tag{3.128}$$

Notice that the equation $\tau + \omega u_{1p} = 0$ together with its derivative with respect to p, after using the condition $H = 0$, gives the linear system of equations for the functions u_2 and u_3:

$$\begin{aligned} u_2 F_2 + u_3 F_3 &= F_0 - u_1 - \tau/u_{1p}, \\ u_2 F_2' + u_3 F_3' &= F_0' + \tau u_{1pp}/u_{1p}. \end{aligned} \tag{3.129}$$

If $F_2 F_3' - F_2' F_3 \neq 0$, one can find u_2 and u_3 from system (3.129)

$$\begin{aligned} u_2 &= [F_3'(F_0 - u_1 - \tau/u_{1p}) - F_3(F_0' + \tau u_{1pp}/u_{1p})]/(F_2 F_3' - F_2' F_3), \\ u_3 &= [-F_2'(F_0 - u_1 - \tau/u_{1p}) + F_2(F_0' + \tau u_{1pp}/u_{1p})]/(F_2 F_3' - F_2' F_3). \end{aligned} \tag{3.130}$$

Direct calculations show that in this case the equation

$$F_2(u_{2p} - F_2 u_{1p}) + F_3(u_{3p} - F_3 u_{1p}) = 0$$

is identically satisfied. This means that the equations (3.109)

$$u_{2p} = F_2 u_{1p}, \quad u_{3p} = F_3 u_{1p}$$

are dependent. Therefore equations (3.127) and (3.109) form two equations for two functions $\tau(p, S)$ and $u_1(p, S)$. The remaining equation (3.128) is an equation for the function $G(p, S, \lambda)$.

If $F_2F_3' - F_2'F_3 = 0$, then without loss of generality one can assume that $F_3 = 0$. Notice that one can also assume that $F_2 = 0$. In fact, let $F_2 \neq 0$. One can find u_2 and u_{1p} from system (3.129). Substituting them into (3.127), one obtains that $F_2'u_{3S} = 0$. Because in this case $k_3 = (1 + F_2^2)u_{3S} \neq 0$, one finds that $F_2' = 0$, which can be transformed to the case $F_2 = 0$. Thus one can assume that $F_2 = F_3 = 0$. The functions $\tau(p, S)$ and $u_1(p, S)$ only have to satisfy the two equations:

$$\tau = (F_0 - u_1)u_{1p}, \qquad u_{1pp}(F_0 - u_1) + F_0'u_{1p} = 0 \tag{3.131}$$

The functions $u_2(S)$ and $u_3(S)$ are arbitrary, and the function $G(p, S, \lambda)$ satisfies equation (3.128). Because $H = 0$ one has $\tau_p = -u_{1p}^2$. Thus, excluding u_1 from equations (3.131), one finds that the state equation satisfies the equation

$$2\alpha F_0'\tau_p\sqrt{-\tau_p} + \tau\tau_{pp} = 0,$$

where $\alpha^2 = 1$.

To review all calculations, one finds that space nonisentropic nonisobaric unsteady flows of an ideal gas of the double wave type, irreducible to invariant solutions with arbitrary functions are only the following types (in order of their derivation).

1. Double waves with one arbitrary function of two arguments for a gas with the state equation (3.73). In this case the functions $u_i = u_i(S)$, $(i = 1, 2, 3)$ are arbitrary, the pressure is defined by the equation $g(p) = c_1t$, and the entropy satisfies the involutive system of two differential equations (3.74).

2. Double waves with one arbitrary function of a single argument, where $u_i = u_i(S)$, $(i = 2, 3)$ are arbitrary, $u_1 = u_1(p, S)$ and $\tau(p, S)$ are defined by the equations

$$\tau = u_{1p}(F_0 - u_1), \quad u_{1pp}(F_0 - u_1) + 2u_{1p}F_0' = 0 \tag{3.132}$$

with an arbitrary function $F_0(p)$. Excluding $u_1(p, S)$ from (3.132), one finds that such double waves exist for state equations satisfying the condition

$$\begin{aligned}\tau\tau_{pp}[-\alpha(F_0')^3 + 3\alpha\tau_pF_0' + (\tau_p - (F_0')^2)\sqrt{(F')^2 - 4\tau_p}]+\\ \tau\tau_pF_0''[\alpha(F_0') - 2\alpha\tau_p + F_0'\sqrt{(F_0')^2 - 4\tau_p}]+\\ 2F_0'\tau_p^2[\alpha(F_0')^2 - 4\alpha\tau_p + F_0'\sqrt{(F_0')^2 - 4\tau_p}] = 0,\end{aligned}$$

where $\alpha^2 = 1$. The functions $p = p(x_1, t)$ and $S = S(x_2, x_3, t)$ are defined by the involutive overdetermined system (3.65), (3.67), (3.76). Notice that in this case it is assumed $H = -u_{1p}F_0' \neq 0$

3. Double waves with two arbitrary functions of a single argument, with the state equation $\tau = g(p)A^2(S)$, in which $u_1 = h_1(p)A(S)$, and the other

velocity coordinates u_2, u_3 have the form: either $u_2 = h_2(p)A(S)$, $u_3 = u_3(S)$, or $u_2 = u_2(S)$, $u_3 = u_3(S)$. In the first case the functions $h_1(p), h_2(p)$ and $g(p)$ satisfy the system of two ordinary differential equations (4.3.1), (3.92) (or (3.93), (3.94)), $h_1h_1'' + h_2h_2'' \neq 0$, the function $u_3(S)$ is arbitrary. In the second case the functions $u_2 = u_2(S)$, $u_3 = u_3(S)$ are arbitrary, and $h_1(p)$ and $g(p)$ are related by the condition $g + h_1h_1' = 0$, $h_1h_1'' \neq 0$. For these double waves the pressure is steady.

Such solutions for a polytropic gas were considered in [180, 181, 182], but there the functions $u_2(S)$ and $u_3(S)$ are related to $A(S)$ by a linear dependence. In the presented study they are arbitrary.

4. Double waves with straight level lines and with two arbitrary functions of a single argument. The arbitrary functions are arbitrary functions of the general solution of the equation (3.107). The functions $\tau(p, S), u_i(p, S)$, $(i = 1, 2, 3)$ satisfy the overdetermined system of five differential equations: (3.96), (3.108) and $H = 0$. A complete analysis of this system for the general case of the state equation is difficult. But in the particular case of a polytropic gas with the exponent $\gamma = 2$, and with the additional condition $\tau + u_\alpha u_{\alpha p} = 0$ which corresponds to a steady flow, this system is compatible and has solutions with a one function of a single argument.

5. Double waves with a one arbitrary function of two arguments. The arbitrary function is an arbitrary function of the general solution of equation (3.119). The pressure in these flows is also steady. The functions $g(p), h_i(p)$, $(i = 1, 2, 3)$ are related by (3.123), where $\tau = g(p)A^2(S)$ and $u_1 = h_1(p)$ $A(S)$. The remaining velocity coordinates are either $u_i = h_i(p)A(S)$, $(i = 2, 3)$, $h_2'h_3 - h_2h_3' \neq 0$, or $u_2 = h_2(p)A(S)$, and $u_3 = u_3(S)$ is arbitrary, or $u_2 = u_2(S)$ and $u_3 = u_3(S)$ are arbitrary.

6. Double waves with a one arbitrary function of a single argument, which is an arbitrary function of the general solution of equation (3.128). The function $\tau(p, S)$ and $u_i(p, S)$, $(i = 1, 2, 3)$ are defined as follows. Either $u_2 = u_2(S)$, $u_3 = u_3(S)$ are arbitrary, $u_1 = F_0 - \alpha\tau/\sqrt{-\tau}_p$, and the state equation $\tau = \tau(p, S)$ satisfies the equation

$$2\alpha F_0'\tau_p\sqrt{-\tau}_p + \tau\tau_{pp} = 0,$$

where $\alpha^2 = 1$ and F_4 is an arbitrary function; or u_2 and u_3 are defined by equations (3.130), and $\tau = \tau(p, S)$ and $u_1 = u_1(p, S)$ are found from the following two equations: one is (3.127) and another is one of equations (3.109).

Thus, there is the following theorem.

Theorem 3.8. *There are only these six types of space nonisentropic nonisobaric unsteady flows of an ideal gas of the double wave type, which are irreducible to invariant solutions with arbitrary functions.*

5. Double waves in a rigid plastic body

Applications of multiple waves in multi–dimensional gas dynamics admit the possibility that the degenerate hodograph method can also be applied to the theory of plasticity, where so far only simple waves have been applied. In multi–dimensional plasticity theory there are only a few known results. For example, there is one class of double waves constructed in [136]. This section is devoted to double waves of the Prandtl–Reuss equations[18]

$$\frac{\partial v_\alpha}{\partial x_a} = 0, \quad \rho \frac{\partial v_i}{\partial t} = \frac{\partial \sigma}{\partial x_i} + \frac{\partial S_{i\alpha}}{\partial x_\alpha}, \ (i = 1, 2, 3), \tag{3.133}$$
$$\frac{\partial v_i}{\partial x_j} + \frac{\partial v_j}{\partial x_i} = 2\Psi S_{ij} \ (i = 1, 2, 3),$$
$$S_{\alpha\beta} S_{\alpha\beta} = 2k^2.$$

Here (S_{ij}) is the deviator of the strength tensor $(\sum_\alpha S_\alpha = 0)$, $v = (v_1, v_2, v_3)$ is the velocity vector, $S_i = S_{ii}$, $(i = 1, 2, 3)$, Ψ is the coefficient in the associated flow rule. Without loss of generality one can assume $\rho = 1$.

5.1 Unsteady plane waves

Plane deformations of an ideal rigid–plastic body with the von Mises yield criterion are described by the equations[19]

$$\frac{\partial v_\alpha}{\partial x_a} = 0, \quad \rho \frac{\partial v_i}{\partial t} = \frac{\partial \sigma_{i\alpha}}{\partial x_\alpha}, \quad (i = 1, 2), \tag{3.134}$$
$$\frac{\sigma_{11} - \sigma_{22}}{2\sigma_{12}} = \frac{\partial v_1/\partial x_1 - \partial v_2/\partial x_2}{\partial v_1/\partial x_2 + \partial v_2/\partial x_1},$$
$$(\sigma_{11} - \sigma_{22})^2 + 4\sigma_{12}^2 = 4k^2,$$

where σ_{ij} are the components of the stress tensor, $v = (v_1, v_2)$ is the two-dimensional velocity vector, k is the yield strength in pure shear.

5.1.1 Double waves

Substituting the relations

$$\sigma_{11} = \sigma - k \sin 2\theta, \quad \sigma_{22} = \sigma + k \sin 2\theta, \quad \sigma_{12} = k \cos 2\theta \tag{3.135}$$

into equations (3.134), they are reduced to the homogeneous system of four quasilinear differential equations for the unknown functions $\sigma(t, x_1, x_2)$,

[18] Irreducible double waves with functional arbitrariness of the Prandtl–Reuss equations are studied in [110, 112].

[19] Summation from 1 to 2 with respect to a repeat Greek index is assumed in this subsection.

$\theta(t, x_1, x_2)$, $v_i(t, x_1, x_2)$, $(i = 1, 2)$:

$$\frac{\partial v_1}{\partial t} = \frac{\partial \sigma}{\partial x_1} - 2k(\cos 2\theta \frac{\partial \theta}{\partial x_1} + \sin 2\theta \frac{\partial \theta}{\partial x_2}), \tag{3.136}$$

$$\frac{\partial v_2}{\partial t} = \frac{\partial \sigma}{\partial x_2} - 2k(\sin 2\theta \frac{\partial \theta}{\partial x_1} - \cos 2\theta \frac{\partial \theta}{\partial x_2}),$$

$$\frac{\partial v_1}{\partial x_1} + \frac{\partial v_2}{\partial x_2} = 0, \quad \frac{\partial v_2}{\partial x_1} + \frac{\partial v_1}{\partial x_2} - 2\cot 2\theta \frac{\partial v_2}{\partial x_2} = 0.$$

The trivial case $\theta = const$ is excluded from further consideration.

If in a double wave the coordinates of the velocity (v_1, v_2) are functionally independent, then one can choose them as the parameters of the wave. Otherwise one has to study the case $v_2 = \Phi(v_1)$.

Let us consider the first case, where the coordinates v_1 and v_2 of the velocity are functionally independent:

$$\sigma = \sigma(v_1, v_2), \quad \theta = \theta(v_1, v_2). \tag{3.137}$$

Substituting (3.137) into (3.136), one obtains the overdetermined system of four differential equations

$$v_t + G_1 v_y = 0, \quad v_x + G_2 v_y = 0, \tag{3.138}$$

where $x = x_1$, $y = x_2$, $v = (v_1, v_2)^t$,

$$G_1 = \begin{pmatrix} a_1 & a_2 - 2a_1\cot(2\theta) \\ -a_2 & -a_1 \end{pmatrix}, \quad G_2 = \begin{pmatrix} 0 & 1 \\ 1 & -2\cot(2\theta) \end{pmatrix},$$

$$a_1 = \sigma_2 + 2k(\theta_1 \sin 2\theta - \theta_2 \cos 2\theta), \quad a_2 = \sigma_1 + 2k(\theta_1 \cos 2\theta + \theta_2 \sin 2\theta),$$

$$\sigma_i = \frac{\partial \sigma}{\partial v_i}, \quad \theta_i = \frac{\partial \theta}{\partial v_i}, \quad (i = 1, 2).$$

Further study depends on the value of $r = rank(G_1G_2 - G_2G_1)$. Cumbersome calculations show that in the case $\det(G_1G_2 - G_2G_1) \neq 0$ there are no irreducible double waves. Hence, $\det(G_1G_2 - G_2G_1) = 0$, which leads to the equation

$$a_2 = a_1 \frac{\alpha + \cos 2\theta}{\sin 2\theta}, \quad (\alpha = \pm 1). \tag{3.139}$$

If $r = 0$, then $a_i = 0$, $(i = 1, 2)$, and this corresponds to a steady motion, which is an invariant solution. Therefore, for irreducible double waves $r = 1$.

Assuming that the Jacobian $\partial(v_1, v_2)/\partial(x, y) = 0$, one obtains a contradiction to the assumed functional independence of v_1 and v_2. It is obtained by changing the independent variables (t, x, y) to (v_1, v_2, x) or to (v_1, v_2, y). Hence, the Jacobian $\partial(v_1, v_2)/\partial(x, y) \neq 0$. By the inverse function theorem one can define the functions

$$x = P(v_1, v_2, t), \quad y = Q(v_1, v_2, t) \tag{3.140}$$

where the independent variables are (v_1, v_2, t). Finding the derivatives of the functions v_i, $(i = 1, 2)$ through the derivatives of the functions P and Q, system (3.138) becomes

$$\begin{aligned} -P_t Q_2 + P_2 Q_t - a_1 P_2 + P_1(a_2 - 2a_1 \cot 2\theta) = 0, \\ -P_1 Q_t + Q_1 P_t + a_2 P_2 - a_1 P_1 = 0, \\ Q_2 + P_1 = 0, \quad Q_1 + P_2 + 2\cot 2\theta\, P_1 = 0, \end{aligned} \tag{3.141}$$

where $P_i = \partial P/\partial v_i$, $Q_i = \partial Q/\partial v_i$, $(i = 1, 2)$, $P_t = \partial P/\partial t$, $Q_t = \partial Q/\partial t$, and

$$P_1 Q_2 - P_2 Q_1 \neq 0. \tag{3.142}$$

Since system (3.141) is a linear and homogeneous system of algebraic equations with respect to the variables P_i, Q_i, $(i = 1, 2)$, and because of (3.142), its determinant is equal to zero:

$$\Delta = (P_t)^2 - (Q_t)^2 - 2P_t Q_t \cot 2\theta - ((a_2)^2 - (a_1)^2 - 2a_1 a_2 \cot(2\theta)) = 0.$$

Since (3.139), this gives

$$P_t = \lambda Q_t, \tag{3.143}$$

where

$$\lambda \equiv (\beta + \cos 2\theta)/\sin 2\theta, \quad \beta = \pm 1.$$

Integrating (3.143) with respect to t, one obtains

$$P = \lambda Q + \chi, \tag{3.144}$$

where $\chi = \chi(v_1, v_2)$ is an arbitrary function of the integration. Substituting (3.144) into the last two equations of system (3.141), one finds that

$$\begin{aligned} Q_2 + \lambda Q_1 + (\lambda_1 Q + \chi_1) = 0, \\ (\lambda - 2\cot 2\theta) Q_2 + Q_1 + (\lambda_2 Q + \chi_2) = 0, \end{aligned} \tag{3.145}$$

where $\lambda_i = \partial\lambda/\partial v_i$, $\chi_i = \partial\chi/\partial v_i$, $(i = 1, 2)$. Since λ satisfies the equation

$$\lambda^2 - 2\lambda \cot(2\theta) - 1 = 0,$$

it is possible to eliminate Q_2 and Q_1 from (3.145):

$$(\lambda_1 - \lambda\lambda_2) Q + \chi_1 - \lambda\chi_2 = 0.$$

If $\lambda_1 - \lambda\lambda_2 \neq 0$, then $Q_t = 0$ and $P_t = 0$. This corresponds to a steady solution, which is excluded from the study. Hence,

$$\lambda_1 - \lambda\lambda_2 = 0, \quad \chi_1 - \lambda\chi_2 = 0. \tag{3.146}$$

Substituting $\lambda = (\beta + \cos 2\theta)/\sin 2\theta$ into the first equation, one finds that

$$\theta_1 - \lambda\theta_2 = 0.$$

Using this in equation (3.139), one obtains

$$\sigma_1 - \lambda\sigma_2 = 0.$$

Thus, $\sigma = \sigma(\lambda)$, $\chi = \chi(\lambda)$. From the first equation of (3.145) one can find the derivative Q_2. Substituting Q_2 into the second equation of (3.141), one obtains

$$(\lambda_2 Q + \chi_2)(\lambda \sin 2\theta - \alpha - \cos 2\theta) = 0.$$

This gives $\alpha = \beta$. Hence, the first equation of system (3.141) becomes

$$(a_2\lambda + a_1)Q_1 + (a_2\lambda + Q_t)(\lambda_2 Q + \chi_2) = 0. \tag{3.147}$$

Therefore, the function $Q(v_1, v_2, t)$ satisfies the system of two quasilinear differential equations: namely the first equation of system (3.145), and equation (3.147). Reducing this system to a linear homogeneous system and taking the Poisson bracket, one has

$$Q_t(cQ + d) = 0,$$

where $c = \frac{\partial}{\partial v_1}((1 + \lambda^2)\sigma' - 2\beta k)^{-1}$ and $d = \frac{\partial}{\partial v_1}(((1 + \lambda^2)\sigma' - 2\beta k)^{-1}\chi')$. Since $Q_t \neq 0$, this equation leads to $c = 0,\ d = 0$, and after integrating them with respect to v_1 one obtains

$$(1 + \lambda^2)\sigma' - 2\beta k = \varphi(v_2), \quad \chi' = \psi(v_2)$$

with the arbitrary functions $\varphi(v_2),\ \psi(v_2)$.

If $((\sigma'(1+\lambda^2))')^2 + (\chi'')^2 \neq 0$ then, because of the inverse function theorem, one finds $\lambda = \lambda(v_2)$. Since of (3.146) it gives $\lambda = const$, which corresponds to the excluded trivial case $\theta = const$. Hence, $(\sigma'(1 + \lambda^2))' = 0$ and $\chi'' = 0$, or

$$\chi = c_2\lambda - c_1, \tag{3.148}$$

$$\sigma + 2\beta k\theta + c_3\theta = c_4, \tag{3.149}$$

with some arbitrary constants c_i, $(i = 1, 2, 3, 4)$. Substituting (3.148) into (3.144), and by virtue of (3.140), one has $x + c_1 = (y + c_2)\lambda$ or $\lambda = (x + c_1)/(y + c_2)$. Without loss of generality one can assume that $c_1 = 0$, $c_2 = 0$:

$$\lambda = x/y. \tag{3.150}$$

From (3.149), (3.150) and because of $\lambda = \lambda(\theta)$, the stress state in the double wave is steady, i.e., the stress tensor does not depend on time. In this case

$$v_i = tH_i(x, y) + g_i(x, y), \quad (i = 1, 2). \tag{3.151}$$

Note that

$$\tan(\theta) = \alpha(y/x)^{\alpha}, \ (\alpha = \pm 1). \tag{3.152}$$

Because of equations (3.136), (3.149), (3.150) the functions H_i are defined by the equations

$$H_i = c_3 \lambda_{x_i}/(1+\lambda^2), \ \ (i = 1, 2). \tag{3.153}$$

The functions $g_i(x, y)$ satisfy the system of differential equations

$$\begin{gathered} \partial g_1/\partial x + \partial g_2/\partial y = 0, \\ \partial g_2/\partial x + \partial g_1/\partial y + ((1-\lambda^2)/\lambda)\partial g_2/\partial y = 0. \end{gathered} \tag{3.154}$$

The general solution of system (3.154) is

$$\begin{gathered} g_1 = (\Phi_1 + \Phi_2 + \lambda(\lambda^2+1)\Phi_1')/(1+\lambda^2)^{1/2}, \\ g_2 = ((1+\lambda^2)\Phi_1' - \lambda(\Phi_1 + \Phi_2))/(1+\lambda^2)^{1/2} \end{gathered} \tag{3.155}$$

with arbitrary functions $\Phi_1 = \Phi_1(\lambda)$ and $\Phi_2 = \Phi_2(r)$, where $r = \sqrt{x^2+y^2}$. The velocity found from (3.151), (3.155) in the polar coordinate system

$$x = r\cos\varphi, \ \ y = r\sin\varphi, \ \ r = \sqrt{x^2+y^2}).$$

has the following form

$$v_r = -\Phi_1(\varphi), \ \ v_\varphi = c_3 t/r + \Phi_1(\varphi) + \Phi_2(r). \tag{3.156}$$

Remark 3.4. *The constructed solution belongs to the class of solutions of equations (3.134) where*

$$\sigma = \sigma(x, y), \ \theta = \theta(x, y), \ v_i = tH_i(x, y) + g_i(x, y), \ (i = 1, 2). \tag{3.157}$$

Substituting this representation of a solution into (3.134) this system is split. For the functions σ, θ, H_1, H_2 one obtains the closed system of differential equations

$$\begin{gathered} \frac{\partial R_1}{\partial x} + \tan\theta \frac{\partial R_1}{\partial y} = H_1 + H_2 \tan\theta, \\ \frac{\partial R_2}{\partial x} - \cot\theta \frac{\partial R_2}{\partial y} = H_1 - H_2 \cot\theta, \\ \frac{\partial H_1}{\partial x} + \frac{\partial H_2}{\partial y} = 0, \\ \frac{\partial H_2}{\partial x} + \frac{\partial H_1}{\partial y} - 2\cot 2\theta \frac{\partial H_2}{\partial y} = 0. \end{gathered} \tag{3.158}$$

For the functions $g_i(x, y)$, $(i = 1, 2)$ there is the system

$$\frac{\partial g_1}{\partial x} + \frac{\partial g_2}{\partial y} = 0, \quad \frac{\partial g_2}{\partial x} + \frac{\partial g_1}{\partial y} - 2\cot 2\theta \frac{\partial g_2}{\partial y} = 0.$$

Here

$$R_i = \sigma + (-1)^i 2k\theta, \quad (i = 1, 2),$$

and $\theta(x, y)$ is considered as a known function, which is a solution of system (3.158).

The basis of the Lie algebra admitted by (3.158) consists of the generators

$$X_i = \partial_{x_i}, \quad X_3 = \partial_{R_1} + \partial_{R_2}, \quad X_4 = H_\alpha \partial_{H_\alpha} - x_\alpha \partial_{x_\alpha},$$
$$X_5 = H_2 \partial_{H_1} - H_1 \partial_{H_2} + x_2 \partial_{x_1} - x_1 \partial_{x_2} + 2k(\partial_{R_1} - \partial_{R_2}),$$
$$X_{5+i} = x_i(\partial_{R_1} + \partial_{R_2}) + \partial_{H_i} (i = 1, 2).$$

The optimal system of one-dimensional subalgebras is composed of the generators

$$X_3, \; X_6, \; X_4 + \gamma X_5 + \beta X_3, \; X_5 + \beta X_3, \; X_1 + \beta X_6 + \gamma X_7,$$

where γ, β are arbitrary constants.

Let us give a solution invariant with respect to the generator X_4. In the polar coordinate system (r, φ) it is

$$H_\varphi = c_3/r, H_r = (-2k\sin(2(\theta - \varphi)) + c_1 \sin(c_2 - 2\varphi))/r,$$
$$2k\cos(2(\theta - \varphi)) = c_3 + c_1 \cos(c_2 - 2\varphi),$$
$$\sigma = c_1(\sin(c_2 - 2\varphi) - k\sin(2(\theta - \varphi)))/2 + c_4.$$

with constant c_i, $(i = 1, 2, 3, 4)$. This solution can be adjoint with the double wave constructed above [111].

5.2 Steady three-dimensional double waves

Equations (3.133) are nonhomogeneous. However from (3.133) for $i = j = 1$ one can find[20] $\Psi = \frac{1}{S_1}\frac{\partial v_1}{\partial x_1}$, and excluding it from the remaining equations of (3.133), there is the closed homogeneous system of quasilinear equations with respect to the nine dependent variables:

$$\frac{\partial \sigma}{\partial x_i} + \frac{\partial S_{i\alpha}}{\partial x_\alpha} = 0, \quad (i = 1, 2, 3), \quad S_{\alpha\beta} S_{\alpha\beta} = 2k^2, \tag{3.159}$$

$$S_1\left(\frac{\partial v_i}{\partial x_j} + \frac{\partial v_j}{\partial x_i}\right) - 2S_{ij}\frac{\partial v_1}{\partial x_1} = 0, \; (i, j = 1, 2, 3), \tag{3.160}$$

[20] Without loss of generality it is assumed $S_1 \neq 0$.

Equation $div(\mathbf{v}) = 0$ follows from (3.159) and the equation $S_1 + S_2 + S_3 = 0$.

Because of rotation there are only two possibilities for the double waves of system (3.159): either the functions v_1 and v_2 are functionally independent, or $v_i = v_i(v_1),\ (i = 2, 3)$.

5.2.1 Functionally independent v_1 and v_2

Let the Jacobian $\dfrac{\partial(v_1, v_2)}{\partial(x_1, x_2)} \neq 0$. The coordinates v_1 and v_2 can be chosen as the parameters of the double wave. All other functions σ, S_{ij} and v_3 are functions of (v_1, v_2)

$$v_3 = v_3(v_1, v_2),\ \ \sigma = \sigma(v_1, v_2),\ \ S_{ij} = S_{ij}(v_1, v_2),\ \ (i, j = 1, 2, 3). \quad (3.161)$$

Substituting this representation into (3.159), one obtains the homogeneous system of quasilinear equations of first order $Gf = 0$, with $f = (v_{1,3}, v_{2,3}, v_{1,1}, v_{2,1}, v_{1,2}, v_{2,2})^t$, and the matrix

$$G = \begin{pmatrix} 0 & 0 & -S_2 & 0 & 0 & S_1 \\ 0 & 0 & -2S_{12} & S_1 & S_1 & 0 \\ S_1 & 0 & -2S_{13} + S_1 v_{3,1} & S_1 v_{3,2} & 0 & 0 \\ 0 & S_1 & -2S_{23} & 0 & S_1 v_{3,1} & S_1 v_{3,2} \\ S_1 v_{3,1} & S_1 v_{3,2} & -S_3 & 0 & 0 & 0 \\ S_{13,1} & S_{13,2} & \sigma_1 + S_{1,1} & \sigma_2 + S_{1,2} & S_{12,1} & S_{12,2} \\ S_{23,1} & S_{23,2} & S_{12,1} & S_{12,2} & \sigma_1 + S_{2,1} & \sigma_2 + S_{2,2} \\ \sigma_1 + S_{3,1} & \sigma_2 + S_{3,2} & S_{13,1} & S_{13,2} & S_{23,1} & S_{23,2} \end{pmatrix}.$$

Here

$$v_{i,j} = \frac{\partial v_i}{\partial x_j},\ \ v_{3,i} = \frac{\partial v_3}{\partial v_i},\ \ \sigma_i = \frac{\partial \sigma}{\partial v_i},\ \ S_{kj,i} = \frac{\partial S_{kj}}{\partial v_i},\ \ v_{3,in} = \frac{\partial^2 v_3}{\partial v_i \partial v_n},$$
$$(i, n = 1, 2;\ k, j = 1, 2, 3).$$

Since $S_1 \neq 0$, the matrix G satisfies the condition $rank(G) \geq 4$. This shows that the first four equations of the system $Gf = 0$ are independent and can be rewritten in the form:

$$\frac{\partial \mathbf{u}}{\partial x_2} = G_2 \frac{\partial \mathbf{u}}{\partial x_1}, \quad (3.162)$$

$$\frac{\partial \mathbf{u}}{\partial x_3} = G_1 \frac{\partial \mathbf{u}}{\partial x_1}. \quad (3.163)$$

Here $\mathbf{u} = (v_1, v_2)^t$, $\xi = (2S_{23} - 2S_{12}v_{3,1} - S_2 v_{3,2})/S_1$,

$$G_1 = \begin{pmatrix} 2S_{13}/S_1 - v_{3,1} & -v_{3,2} \\ \xi & v_{3,1} \end{pmatrix} \ \ G_2 = \begin{pmatrix} 2S_{12}/S_1 & -1 \\ S_2/S_1 & 0 \end{pmatrix}.$$

According to the Ovsiannikov theorem, for any irreducible double wave $rank(G) \leq 4$. Therefore, when studying irreducible double waves one has to require $rank(G) = 4$.

Let A_{ij} be the square 5×5 matrix, formed from the first four rows and i-th row ($5 \leq i \leq 8$), and the first, second, fourth, sixth and j-th columns of the matrix G ($j = 3$ or $j = 5$). The determinants of these matrices $a_{ij} = \det A_{ij}$ are

$$a_{ij} = 0, \ (5 \leq i \leq 8). \tag{3.164}$$

The equations $a_{65} = 0$ and $a_{75} = 0$ serve to find the function $\sigma = \sigma(v_1, v_2)$:

$$\begin{aligned} \sigma_1 &= -S_{23,1}v_{3,2} + S_{23,2}v_{3,1} + S_{12,2} - S_{2,1}, \\ \sigma_2 &= -S_{13,1}v_{3,2} + S_{13,2}v_{3,1} + S_{12,1} - S_{1,2}. \end{aligned}$$

For double waves with functional arbitrariness one needs to study the matrix

$$(G_1G_2 - G_2G_1) = \frac{2}{S_1}\begin{pmatrix} Z_1 & Z_2 \\ (S_2Z_2 + 2S_{12}Z_1)/S_1 & -Z_1 \end{pmatrix},$$

where

$$Z_1 = S_{23} - S_{12}v_{3,1} - S_2v_{3,2}, \quad Z_2 = -S_{13} + S_1v_{3,1} + S_{12}v_{3,2}.$$

For the existence of solutions with functional arbitrariness it is necessary that $r = rank(G_1G_2 - G_2G_1) < 1$.

The cases $r = 0$ and $r = 1$ are considered sequentially.

1°. Let $r = 0$, then $G_1G_2 - G_2G_1 = 0$. Then this means that $Z_1 = 0$, $Z_2 = 0$ or

$$S_{23} = S_{12}v_{3,1} + S_2v_{3,2}, \quad S_{13} = S_1v_{3,1} + S_{12}v_{3,2}. \tag{3.165}$$

Differentiating equations (3.162) with respect to x_3 and subtracting equations (3.163) differentiated with respect to x_2, and taking into account (3.165), one finds

$$(D_3G_2 + G_2(D_1G_1) - D_2G_1 - G_1(D_1G_2))\frac{\partial \mathbf{u}}{\partial x_1} = 0. \tag{3.166}$$

Substituting the derivatives $\frac{\partial \mathbf{u}}{\partial x_2}$ and $\frac{\partial \mathbf{u}}{\partial x_3}$ into (3.167), one obtains the two quadratic homogeneous forms with respect to the derivatives $\frac{\partial \mathbf{u}}{\partial x_1}$:

$$g(S_{12}v_{3,22} + S_1v_{3,12}) = 0, \quad g(S_1v_{3,11} - S_2v_{3,22}) = 0, \tag{3.167}$$

where

$$g = S_2(\frac{\partial v_1}{\partial x_1})^2 - 2S_{12}\frac{\partial v_1}{\partial x_1}\frac{\partial v_2}{\partial x_1} + S_1(\frac{\partial v_2}{\partial x_1})^2.$$

According to the Ovsiannikov, theorem for irreducible solutions $g \neq 0$. Hence equations (3.167) give

$$S_{12}v_{3,22} + S_1v_{3,12} = 0, \quad S_1v_{3,11} - S_2v_{3,22} = 0. \tag{3.168}$$

Notice that if $v_{3,22} = 0$, then $v_{3,11} = 0$ and $v_{3\,12} = 0$, which means that $v_3 = C_1 v_1 + C_2 v_2 + C_3$ with some constant C_i, $(i = 1, 2, 3)$. Using rotation and Galilean transformations, this solution can be reduced to a plane deformation. Hence one needs to consider $v_{3,22} \neq 0$. In this case, from equations (3.168), one finds S_{12} and S_2. Equations (3.168) provide necessary and sufficient conditions for involutiveness of the overdetermined system (3.162), (3.163). It should be also noted that the double waves in this case have straight level lines.

Thus one needs to study the equations (3.164), (3.164), (3.165), (3.168) and the von Mises yield condition

$$S_{\alpha\beta} S_{\alpha\beta} = 2k^2. \tag{3.169}$$

A complete analysis of these equations has not been performed, but it is known that this system is compatible: one of its solution is given in [136], though the way to study the compatibility of this system is clear. At first, from these equations, one can find all components S_{ij} of the deviator of the stress tensor through the function $v_3(v_1, v_2)$ and its derivatives. From a further analysis one can find all derivatives of order five of the function $v_3(v_1, v_2)$, and by comparing the mixed derivatives of the function $v_3(v_1, v_2)$ of the next order one can obtain the involutive conditions for this system.

Remark 3.5. *Solutions with double–functional arbitrariness (representation in Riemann invariants) were constructed in [136]:*

$$v_1 = \rho \cos\phi, \; v_2 = \rho \sin\phi, \; v_3 = Aw,$$

$$\rho = A\, ch\, w, \;\; w = R^1 + R^2, \;\; \phi = R^1 - R^2,$$

$$S_1 = k(\cos^2\phi(th^2 w + 1) - 1), \; S_2 = k(\sin^2\phi(th^2 w + 1) - 1),$$

$$S_{12} = \frac{k}{2}\sin^2 2\phi(th^2 w + 1), \; S_{13} = k\frac{\cos\phi}{ch\, w} th\, w, \; S_{23} = k\frac{\sin\phi}{ch\, w} th\, w,$$

where R^1, R^2 are the wave parameters. They are also Riemann invariants.

2°. Now let $r = 1$. Hence one has

$$S_1 Z_1^2 + 2S_{12} Z_1 Z_2 + S_2 Z_2^2 = 0, \tag{3.170}$$

$$Z_1^2 + Z_2^2 \neq 0. \tag{3.171}$$

From these equations one finds that $S_{12}^2 - S_1 S_2 \geq 0$ and $Z_1 = hZ_2$, where $h = h(v_1, v_2)$ is a solution of the quadratic equation

$$S_1 h^2 + 2S_{12} h + S_2 = 0.$$

The equation $Z_1 - hZ_2 = 0$, after substituting for Z_1 and Z_2, becomes

$$S_{23} + hS_{13} + S_{12}(hv_{3,2} - v_{3,1}) + S_1 h v_{3,1} + S_2 v_{3,2} = 0. \tag{3.172}$$

Since the Jacobian $\frac{\partial(v_1,v_2)}{\partial(x_1,x_2)} \neq 0$, one can make a transition to the variables (v_1, v_2, x_3):

$$x_1 = P(v_1, v_2, x_3),\ x_2 = Q(v_1, v_2, x_3). \tag{3.173}$$

The quasilinear system (3.162), (3.163) for the functions $v_i(x_1, x_2, x_3)$, $(i = 1, 2)$ is transformed into the system for the functions $P(v_1, v_2, x_3)$, $Q(v_1, v_2, x_3)$, which after some algebraic combinations becomes

$$\begin{gathered} S_1P_1 - S_2Q_2 = 0, \\ S_1P_2 + S_1Q_1 + 2S_{12}Q_2 = 0, \\ S_1(Q_3 + v_{3,2})Q_1 + (2S_{12}(Q_3 + v_{3,2}) + S_1(P_3 + v_{3,1}) - 2Z_2)Q_2 = 0, \\ S_1(P_3 + v_{3,1})Q_1 - (S_2(Q_3 + v_{3,2}) + 2Z_1)Q_2 = 0. \end{gathered} \tag{3.174}$$

Here $P_i = \frac{\partial P}{\partial v_i}$, $Q_i = \frac{\partial Q}{\partial v_i}$, $(i = 1, 2)$, $P_3 = \frac{\partial P}{\partial x_3}$, $Q_3 = \frac{\partial Q}{\partial x_3}$, and

$$P_1Q_2 - P_2Q_1 \neq 0. \tag{3.175}$$

System (3.174) is a linear and homogeneous algebraic system of equations with respect to P_i, Q_i, $(i = 1, 2)$. By virtue of the inequality (3.175) its determinant has to be equal to zero, i.e.,

$$\begin{gathered} S_1(P_3 + v_{3,1})^2 + 2S_{12}(Q_3 + v_{3,2})(P_3 + v_{3,1}) + S_2(Q_3 + v_{3,2})^2 - \\ 2Z_2(P_3 + v_{3,1}) + 2Z_1(Q_3 + v_{3,2}) = 0, \end{gathered}$$

or

$$S_1(P_3+v_{3,1}) = Z_2-S_{12}(Q_3+v_{3,2})+\gamma((S_1h+S_{12})(Q_3+v_{3,2})-Z_2). \tag{3.176}$$

where $\gamma = \pm 1$. Notice that in the case $\gamma = -1$ from (3.174) one can find the derivatives P_1, P_2 and Q_1, which lead to a contradiction of condition (3.175). Hence, $\gamma = 1$. Integrating (3.176) with respect to x_3, one has

$$P = hQ - x_3a + \chi, \tag{3.177}$$

where $a = v_{3,1} - hv_{3,2}$, $\chi = \chi(v_1, v_2)$ is an arbitrary function, and $a_i = \frac{\partial a}{\partial v_i}$, $\chi_i = \frac{\partial \chi}{\partial v_i}$, $h_i = \frac{\partial h}{\partial v_i}$, $(i = 1, 2)$. Substituting (3.177) into (3.174), and taking into account (3.172), it is reduced to

$$Q(h_1 - hh_2) - x_3(a_1 - ha_2) + \chi_1 - h\chi_2 = 0, \tag{3.178}$$

$$S_1Q_1 + (hS_1 + 2S_{12})Q_2 + S_1(h_2Q - x_3a_2 + \chi_2) = 0, \tag{3.179}$$

$$2Q_2(S_1v_{3,1} + S_{12}v_{3,2} - S_{13}) + S_1(h_2Q - x_3a_2 + \chi_2)(Q_3 + v_{3,2}) = 0. \tag{3.180}$$

If $h_1 - hh_2 \neq 0$, then from (3.178) one defines Q. Substituting found Q into equations (3.179), (3.180), and splitting them with respect to x_3 one obtains four equations for the functions $h = h(v_1, v_2)$, $v_3 = v_3(v_1, v_2)$ and $\chi = \chi(v_1, v_2)$. Analysis of the compatibility of this system[21] leads to a solution which is reducible to invariant. Hence, $h_1 - hh_2 = 0$ and therefore

$$a_1 - ha_2 = 0, \quad \chi_1 - h\chi_2 = 0. \tag{3.181}$$

A complete analysis of this case is also very cumbersome. Here only the particular case where $a = 0$ and $h_2 \neq 0$ is given. In this case $v_3 = v_3(h)$, $\chi = \chi(h)$,

$$S_{23} = -hS_{13} - h_2 v_3'(h^2 S_1 + S_2), \; S_2 = -h^2 S_1 - 2hS_{12}. \tag{3.182}$$

It should also be noted that for irreducible double waves in this case $Q_3 \neq 0$. In fact, let $Q_3 = 0$, then $P_3 = 0$, and this is a reducible double wave: it is reduced to plane deformation.

By virtue of the equation $h_1 - hh_2 = 0$ there exists a function $\lambda = \lambda(v_1, v_2)$, which is functionally independent of $h(v_1, v_2)$ and

$$\lambda_1 = -1(1 + h^2)^{-1/2}, \; \lambda_2 = h(1 + h^2)^{-1/2}. \tag{3.183}$$

Let us change the variables (v_1, v_2) with (h, λ). The equations $a_{55} = 0$, $a_{85} = 0$, $a_{i3} = 0$, $(i = 5, 6, 7, 8)$, and the von Mises yield condition are reduced to the equations

$$S_{12} = (1 - h^2)S_1/(2h), \quad (1 + h^2)^2 S_1^2 + 4h^2(1 + h^2)S_{13}^2 = 4h^2 k^2, \tag{3.184}$$

$$h(1 + h^2)\frac{\partial S_1}{\partial h} + (h^2 - 1)S_1 = 0, \quad (1 + h^2)\frac{\partial S_{13}}{\partial h} + hS_{13} = 0. \tag{3.185}$$

Equations (3.179) and (3.180) are reduced to the equations

$$(1 + h^2)Q_h + hQ = -h\chi'. \tag{3.186}$$

$$\sqrt{1 + h^2}((1+h^2)h_2 v_3' S_1 - 2hS_{13})Q_\lambda + (Q + \chi')h_2((1+h^2)S_1 Q_3 + 2hS_{13}) = 0. \tag{3.187}$$

Integrating (3.185) and using the von Mises yielding condition one finds

$$S_1 = 2hA(1 + h^2)^{-1}, \; S_{13} = B(1 + h^2)^{-1/2}, \tag{3.188}$$

where the functions $A = A(\lambda)$ and $B = B(\lambda)$ are related by the equation $B^2 + A^2 = k^2$.

Substituting the obtained expressions for the components of the deviator of the stress tensor (S_{ij}) into equations (3.164), one has $\sigma = \sigma(h)$ and

$$\sigma' = -v_3' B' - A' h_2^{-1}(1 + h^2)^{-1/2} - 2A(1 + h^2)^{-1}. \tag{3.189}$$

[21] Because this analysis is cumbersome it is omitted here.

Equation (3.186) can be also integrated to give

$$Q(h, \lambda, x_3) = (1 + h^2)^{-1/2}(g(h) + \phi(\lambda, x_3)). \qquad (3.190)$$

Here $\phi(\lambda, x_3)$ is an arbitrary function of integration such that $\phi_{x_3} \neq 0$. The function $g = g(h)$ is related to the function $\chi(h)$ through the equation $g' = -h(1+h^2)^{-1/2}\chi'$. In this case $P_1Q_2 - P_2Q_1 = h_2(Q + \chi')\phi_\lambda \neq 0$. Equation (3.187) is reduced to the equation

$$f\left(\phi + g + \chi'\sqrt{1+h^2}\right)^{-1}\phi_\lambda + A\phi_{x_3} + B = 0. \qquad (3.191)$$

with $f = Av_3'(1+h^2) - Bh_2^{-1}\sqrt{1+h^2}$. Differentiating (3.191) with respect to h, and using the condition $\phi_\lambda \neq 0$, one obtains $\frac{\partial}{\partial h}\left(f\left(\phi + g + \chi'\sqrt{1+h^2}\right)^{-1}\right) = 0$. Differentiating this relation with respect to x_3, and using the condition $\phi_{x_3} \neq 0$, one has $f_h = 0$, and, hence, $\left(g + \chi'\sqrt{1+h^2}\right)' = 0$. This gives

$$\chi = c_3 h + c_4, \quad g = -c_3\sqrt{1+h^2},$$

where c_3, c_4 are constants, which are not essential: one can assume that $c_3 = 0$ and $c_4 = 0$. This means that

$$\chi = 0, \quad g = 0. \qquad (3.192)$$

Let us analyze the equation $f_h = 0$. To find the derivative $\partial h_2/\partial h$ one can use the relations

$$\frac{\partial}{\partial v_2} + h\frac{\partial}{\partial v_1} = h_2(1+h^2)\frac{\partial}{\partial h}, \quad h\frac{\partial}{\partial v_2} - \frac{\partial}{\partial v_1} = \sqrt{1+h^2}\frac{\partial}{\partial \lambda}.$$

Hence,

$$\frac{\partial h_2}{\partial h} = \frac{h_{22}}{h_2} + \frac{hh_2}{1+h^2}, \quad \frac{\partial h_2}{\partial \lambda} = -\frac{h_2^2}{\sqrt{1+h^2}}.$$

Then the equation $f_h = 0$ becomes

$$Bh_{22}\sqrt{1+h^2} + Ah_2^3\left(v_3''(1+h^2) + 2hv_3'\right) = 0. \qquad (3.193)$$

Because of (3.193), after differentiating (3.189) with respect to h, one finds

$$\sigma'' + \frac{2h}{1+h^2}\sigma' = 0.$$

Since the general solution of this equation is

$$\sigma = K\arctan(h) + c_5,$$

equation (3.189) is reduced to the equation

$$v_3' B'(1+h^2) + A' h_2^{-1}\sqrt{1+h^2} = -2A - K. \tag{3.194}$$

Differentiating it with respect to λ one obtains

$$(1+h^2)v_3'(B')^3 k_0^2 = A^3\left(A''(2A+K) - (3A')^2\right),$$

and, differentiating this with respect to h, one has

$$\left((1+h^2)v_3'\right)' B' = 0.$$

From this equation one obtains the following results. If $B' = 0$, then $A' = 0$, and therefore, $K = -2A$. In this case if $B = 0$, then from (3.193) one obtains

$$v_3 = b\arctan(h) + c_6.$$

If $B' \neq 0$, then

$$v_3 = b\arctan(h) + c_6$$

and $h_{22} = 0$. Here $b,\ c_5,\ c_6$ are constant.

For further study one needs to consider the following three cases: a) $B = 0$, b) $B \neq 0$, $B' = 0$, and c) $B' \neq 0$.

Let us assume that $B = 0$. Then $A = \pm k$, and $K = -2A$. In this case $f = bA$ and the general solution of equation (3.191) is

$$\lambda\phi = bx_3 + \Phi(\phi) \tag{3.195}$$

with an arbitrary function $\Phi(\phi)$.

Thus, the general solution in the case under consideration has two arbitrary functions of a single argument: one arbitrary function is in the definition $h = h(v_1, v_2)$ and another is in the definition $\Phi = \Phi(\phi)$.

Let us analyze this solution in the original space of the independent variables (x_1, x_2, x_3).

From (3.173), (3.177), (3.190), (3.192) and (3.195) one obtains $h(v_1, v_2) = x_1/x_2$ and $\lambda(v_1, v_2) = \beta(bx_3 + \Phi(R))/R$, where $\beta = sign(x_2), R = \sqrt{x_1^2 + x_2^2}$. These expressions of the functions $h = h(v_1, v_2)$ and $\lambda = \lambda(v_1, v_2)$ serve for the representation of the components of the velocity v_1, v_2 as functions of the independent variables x_1, x_2, x_3. Differentiating $h = h(v_1, v_2)$ and $\lambda = \lambda(v_1, v_2)$ with respect to x_3, and using (3.183), one obtains

$$hv_{1,x_3} + v_{2,x_3} = 0, \quad -v_{1,x_3} + hv_{2,x_3} = bx_2^{-1}.$$

From the last equations one obtains

$$\begin{aligned} v_1 &= -bx_2x_3R^{-2} + g_1(x_1, x_2),\\ v_2 &= bx_1x_3R^{-2} + g_2(x_1, x_2). \end{aligned} \tag{3.196}$$

If, as in the case of plane deformation, we introduce the angle θ through the formula

$$h = (\cos 2\theta + \gamma)/\sin 2\theta, \tag{3.197}$$

then from (3.184), (3.188), and (3.194) one finds

$$\begin{gathered} S_{13} = S_{23} = 0, \quad S_1 = -S_2 = -k\sin 2\theta, \\ S_{12} = k\cos 2\theta, \quad v_3 = -b\theta + c_5, \\ \sigma = -2\gamma k\theta + c_5, \; v_3 = -b\theta + c_6, \; \tan(\theta) = \gamma(x_2/x_1)^{\gamma}. \end{gathered} \tag{3.198}$$

where $A = -\gamma k$, $\gamma = \pm 1$. Substituting (3.196)-(3.198) into equations (3.162), (3.163), they are reduced to the equations

$$\frac{\partial g_1}{\partial x_2} + 2ctg(2\theta)\frac{\partial g_1}{\partial x_1} + \frac{\partial g_2}{\partial x_1} = 0, \quad \frac{\partial g_2}{\partial x_2} + \frac{\partial g_1}{\partial x_1} = 0.$$

The general solution of the last equations is

$$\begin{gathered} g_1 = (\phi_1 + \phi_2 + h(1+h^2)\phi_1'/(1+h^2)^{1/2}, \\ g_2 = ((1+h^2)\phi_1' - h(\phi_1+\phi_2)/(1+h^2)^{1/2}, \end{gathered} \tag{3.199}$$

with some arbitrary functions $\phi_1 = \phi_1(h)$ and $\phi_2 = \phi_2(R)$.

Let $B \neq 0$, $B' = 0$. Integrating (3.193) with respect to v_2, one obtains

$$h_2 = \frac{1}{\eta + \mu}, \tag{3.200}$$

where $\eta(h) = AB^{-1}\left(v_3'\sqrt{1+h^2} + \int h(1+h^2)^{-1/2}v_3'\,dh\right)$ and $\mu = \mu(v_1)$ is an arbitrary function. The form of this function is obtained from the study of the compatibility condition of two differential equations for the function $h = h(v_1, v_2)$: the first equation is $h_1 - hh_2 = 0$ and second is (3.200). This system of differential equations is consistent if, and only if, $\mu = -v_1$. Since

$$h\frac{\partial}{\partial v_2} - \frac{\partial}{\partial v_1} = \sqrt{1+h^2}\frac{\partial}{\partial \lambda},$$

direct calculations show that $f_\lambda = -B$, or the function[22] $f(\lambda)$ is $f = -B\lambda$. In this case the general solution of equation (3.191) is

$$\lambda = \phi\Phi(x_3 + AB^{-1}\phi),$$

where $\Phi = \Phi(\zeta)$ is an arbitrary function.

Therefore, in case b) an irreducible double wave has two arbitrary functions of a single argument: $v_3 = v_3(h)$ and $\Phi = \Phi(\zeta)$. The construction of the

[22]The function $\lambda(v_1, v_2)$ is defined up to an arbitrary constant.

solution in the space of the independent variables (x_1, x_2, x_3) can be done as follows. At first one has to choose the functions $v_3(h)$ and $\Phi(\zeta)$ and then one defines the function $h = h(v_1, v_2)$ from the equation

$$v_2 + hv_1 = \int \eta(h)\, dh.$$

After that one defines the function

$$\lambda(v_1, v_2) = \sqrt{1+h^2}\left(-v_1 + AB^{-1}\int \frac{v_3' h\, dh}{\sqrt{1+h^2}}\right)$$

and, at last, one finds the coordinates of the velocity $v_1(x_1, x_2, x_3)$, $v_2(x_1, x_2, x_3)$ using the equations

$$h(v_1, v_2) = x_1/x_2, \quad \lambda(v_1, v_2) = \beta R \Phi\left(x_3 + \beta A B^{-1} R\right),$$

where $R = \sqrt{x_1^2 + x_2^2}$, $\beta = sign(x_2)$. The stress state is defined by the relations

$$S_1 = -S_2 = \frac{2Ah}{1+h^2}, \quad S_{12} = \frac{A(1-h^2)}{1+h^2},$$

$$S_{13} = \frac{B}{\sqrt{1+h^2}}, \quad S_{23} = -hS_{13}, \quad \sigma = -2A\arctan(h) + c_5.$$

Let $B \neq 0$, $B' \neq 0$. Since $h_{22} = 0$ in this case, without loss of generality

$$h = -\frac{v_2}{v_1}, \quad \lambda = -\sqrt{v_1^2 + v_2^2}.$$

By virtue of $v_3 = b\arctan(h) + c_6$ one obtains $f = Ab - B\lambda$, and equation (3.194) becomes

$$B'b + A'\lambda = -2A - K.$$

where $B = \alpha(k^2 - A^2)^{1/2}$, $\alpha = \pm 1$. Equation (3.191) is reduced to the equation

$$(Ab - \lambda B)\phi_\lambda + \phi(A\phi_{x_3} + B) = 0.$$

The general solution of this equation is

$$\Phi\left(\phi\zeta(\lambda),\ x_3 - \phi\zeta(\lambda)\int A(Ab - \lambda B)^{-1}\zeta^{-1}(\lambda)\, d\lambda\right) = 0$$

with some arbitrary function Φ.

5.2.2 The case $v_i = v_i(v_1)$, $(i = 2, 3)$

Assume now that $v_i = v_i(v_1)$, $(i = 2, 3)$. Substituting $v_i = v_i(v_1)$, $(i = 2, 3)$ into (3.160), one obtains

$$\begin{aligned}
-S_2 w_1 + S_1 v_2' w_2 = 0,\\
-2S_{12} w_1 + S_1(w_2 + v_2' w_1) = 0,\\
-2S_{13} w_1 + S_1(w_3 + v_3' w_1) = 0,\\
-2S_{23} w_1 + S_1(v_2' w_3 + v_3' w_2) = 0,\\
(S_1 + S_2) w_1 + S_1 v_3' w_3 = 0,
\end{aligned} \tag{3.201}$$

where for conciseness we use the notation $w_i = \partial v_1/\partial x_i$, $(i = 1, 2, 3)$. Since system (3.201) is a linear and homogeneous system with respect to w_i, $(i = 1, 2, 3)$, and it is assumed that $\sum_i w_i^2 \neq 0$, one has

$$\begin{aligned}
S_1(v_2')^2 - 2S_{12}v_2' + S_2 = 0,\\
S_1(v_3')^2 - 2S_{13}v_3' - S_1 - S_2 = 0,\\
S_{13}v_2' + S_{12}v_3' - S_1 v_2' v_3' - S_{23} = 0.
\end{aligned}$$

Since for $v_2' = 0$ the solution is reduced to a plane deformation, one needs to consider the case $v_2' \neq 0$. From the last equations one finds S_{12}, S_2, S_{23}. The result of substituting them into the von Mises yield condition is to give a squared equation for the component S_{13}. From this equation one obtains $| S_1(h+1)h^{-1/2}/(2k) | < 1$, where $h = (v_2')^2 + (v_3')^2$. Let us introduce an angle θ such that $\sin\theta = S_1(h+1)h^{-1/2}/(2k)$.

If $S_1 = S_1(v_1)$ or $\theta = \theta(v_1)$), then $\sigma = \sigma(v_1)$ and the solution is reduced to a simple wave. Thus, one can choose S_1 and v_1 as the parameters of the double wave.

Substituting $\sigma = \sigma(S_1, v_1)$ into equations (3.159), and taking into account the two first independent equations in (3.201), one obtains

$$b_{i\alpha}\partial S_1/\partial x_\alpha + b_{i4}w_1 = 0, \quad (i = 1, 2, 3), \tag{3.202}$$

where the forms of the coefficients b_{ij}, $(i = 1, 2, 3;\ j = 1, 2, 3, 4)$ are very cumbersome and so are not given here. All further calculations are performed by on computer. Here only the method of the of the calculations and the final results are presented.

For irreducible double waves one has to have in (3.202) no more than two independent equations. This means that the rank of the matrix $B = (b_{ij})$ is not greater than two. If one denotes with B_j a square matrix constructed from the matrix B without the j-th column, then $\det(B_j) = 0$, $(j = 1, 2, 3, 4)$. From the equation $\det(B_4) = 0$ one finds

$$\frac{\partial\sigma}{\partial S_1}\left((\frac{\partial\sigma}{\partial S_1})^2 - \frac{(h+1)^2}{4h\cos^2\theta} \right) = 0.$$

This equation means that either $\sigma = \sigma(v_1)$ or $\sigma = \pm S_1(h+1)h^{-1/2}/(2\cos\theta) + \phi(v_1)$. In both cases the result of substituting the expression for σ into $\det(B_3) = 0$ is a polynomial with respect to $tg\theta$ with coefficients which only depend on v_1. Since θ and v_1 are functionally independent, these coefficients must be equal to zero. However, among these expressions there are contradictory equations, for example, $h + 1 = 0$.

Hence, in the case $v_i = v_i(v_1)$, $(i = 2, 3)$ there are no irreducible double waves.

6. Triple waves of isentropic potential gas flows

The system of equations defining solutions of the triple wave type of the three-dimensional isentropic potential gas dynamics equations were obtained in [156]. Before [157] the question about the existence of non similar triple waves with $\gamma \neq 1$ was open. A set of exact solutions of the triple wave type for $1 < \gamma < 2$ was constructed in [157]. These solutions depend on three arbitrary functions of a single argument. Some applications of these solutions are also given in the cited article. However, a complete analysis of consistency of the equations describing the triple waves has not been performed. This section is devoted to solving this problem.

A three-dimensional isentropic potential motion of a gas is described by the equations

$$\frac{d\mathbf{u}}{dt} + \nabla\theta = 0, \quad \frac{d\theta}{dt} - \kappa\theta div\, \mathbf{u} = 0, rot\, \mathbf{u} = 0, \qquad (3.203)$$
$$rot\, \mathbf{u} = 0.$$

where $\kappa = \gamma - 1 > 0, \theta = c^2/\kappa, \mathbf{u} = (u_1, u_2, u_3)$ the vector of velocity, c is the sound speed, γ is the polytropic exponent of a gas $d/dt = \partial/\partial t + u_\alpha \partial/\partial x_\alpha$. For triple wave type flows there are two possibilities: the coordinates of the velocity vector u_1, u_2, u_3 in some domain D of the space $R^4(x_1, x_2, x_3, t)$ are either functionally independent or functionally dependent, for example, $u_3 = \phi(u_1, u_2)$).

In the first case one can set $\theta = \theta(u_1, u_2, u_3)$. Substituting $\theta = \theta(u_1, u_2, u_3)$ into equations (3.203) and using the condition of potentiality, one obtains the overdetermined system of quasilinear equations

$$(S_1, S_2, S_3)^t = E\frac{\partial \mathbf{u}}{\partial t} + G_\alpha \frac{\partial \mathbf{u}}{\partial x_\alpha} = 0, \qquad (3.204)$$

where

$$f_1 = \frac{\partial u_1}{\partial x_2} - \frac{\partial u_2}{\partial x_1} = 0, \;\; f_2 = \frac{\partial u_2}{\partial x_3} - \frac{\partial u_3}{\partial x_2} = 0,$$

$$f_3 = \frac{\partial u_1}{\partial x_3} - \frac{\partial u_3}{\partial x_1} = 0, \;\; f_4 = \Psi_\alpha \frac{\partial u_\alpha}{\partial x_\alpha} + 2\theta_1\theta_2 \frac{\partial u_2}{\partial x_1} + 2\theta_2\theta_3 \frac{\partial u_3}{\partial x_2} + 2\theta_1\theta_3 \frac{\partial u_3}{\partial x_1} = 0.$$

Here $\theta_i = \partial\theta/\partial u_i$, $\Psi_i = \theta_i^2 - \kappa\theta$, E is a 3×3 unit matrix, $G_i = \Delta_i E$, $\Delta_i = u_i + \theta_i$, $(i = 1, 2, 3)$, $x_0 = t$, $p_{ij} = \frac{\partial^2 \mathbf{u}}{\partial x_i \partial x_j}$, $(i, j = 0, 1, 2, 3)$, D_i are the total derivatives with respect to x_i. Without loss of generality it is assumed that $\Psi_3 \neq 0$, which is possible because of a rotation.

A compatibility analysis of system (3.204) shows that the maximal arbitrariness of the general solution of this system is equal to two functions of two arguments.

For further study it is necessary to consider the problem of the algebraic dependence of the second order derivatives. This is considered in a matrix form. The matrix composed of the coefficients of the prolonged system (3.204) with respect to x_i, $(i = 0, 1, 2, 3)$ is

$$
\begin{array}{ccccccccccc}
p_{11} & p_{12} & p_{13} & p_{22} & p_{23} & p_{33} & p_{01} & p_{02} & p_{03} & p_{00} & \\
G_1 & G_2 & G_3 & 0 & 0 & 0 & E & 0 & 0 & 0 & D_1S \\
0 & G_1 & 0 & G_2 & G_3 & 0 & 0 & E & 0 & 0 & D_2S \\
0 & 0 & G_1 & 0 & G_2 & G_3 & 0 & 0 & E & 0 & D_3S \\
0 & 0 & 0 & 0 & 0 & 0 & G_1 & G_2 & G_3 & E & D_0S \\
\Phi_1 & \Phi_2 & \Phi_3 & 0 & 0 & 0 & 0 & 0 & 0 & 0 & D_1\Phi \\
0 & \Phi_1 & 0 & \Phi_2 & \Phi_3 & 0 & 0 & 0 & 0 & 0 & D_2\Phi \\
0 & 0 & \Phi_1 & 0 & \Phi_2 & \Phi_3 & 0 & 0 & 0 & 0 & D_3\Phi \\
 & & & & & & & & & & \\
0 & 0 & 0 & 0 & 0 & \Phi_1 & 0 & \Phi_2 & \Phi_3 & 0 & D_0\Phi \\
H_1 & H_2 & H_3 & 0 & 0 & 0 & 0 & 0 & 0 & 0 & D_1f_1 \\
0 & H_1 & 0 & H_2 & H_3 & 0 & 0 & 0 & 0 & 0 & D_2f_1 \\
0 & 0 & H_1 & 0 & H_2 & H_3 & 0 & 0 & 0 & 0 & D_3f_1 \\
0 & 0 & 0 & 0 & 0 & 0 & H_1 & H_2 & H_3 & 0 & D_0f_1
\end{array}
$$

Here the first row means derivatives with respect to which the coefficients are given below, $\Phi_i = \frac{\partial \Phi}{\partial(\partial \mathbf{u}/\partial x_i)}$, $H_i = \frac{\partial f1}{\partial(\partial \mathbf{u}/\partial x_i)}$, $M_i = H_i\Phi_3^{-1}$, $\Phi = (f_2, f_3, f_4)^t$. The last column of each row presents the equation from which these coefficients are obtained.

Since $\det \Phi_3 = \Psi_3 \neq 0$, the part of independent equations consists of the first 7 rows. Moreover from these equations one can find the derivatives p_i, p_{0i}, $(i = 1, 2, 3)$. Excluding these derivatives from the last five equations, one has

$$
\begin{array}{cccc}
p_{11} & p_{12} & p_{22} & \\
0 & 0 & 0 & D_0\Phi + \Delta_\alpha D_\alpha \Phi - \Phi_\alpha D_\alpha S \\
H_1 & H_2 & 0 & D_1 f_1 \\
0 & H_1 & H_2 & D_2 f_1 \\
M_1\Phi_1 & M_1\Phi_2 + M_2\Phi_1 & M_2\Phi_2 & -D_3 f_1 + M_\alpha D_\alpha \Phi \\
0 & 0 & 0 & D_0 f_1 + \Delta_\alpha D_\alpha f_1 - H_\alpha D_\alpha S.
\end{array}
\tag{3.205}
$$

From the representation of the matrices M_i, Φ_i it follows that

$$M_i\Phi_j + M_j\Phi_i = 0, \quad (i, j = 1, 2) \tag{3.206}$$

Hence, a solution of (3.204) also satisfies the following equations of first order

$$D_0\Phi + \Delta_\alpha D_\alpha \Phi - \Phi_\alpha D_\alpha S = 0, \tag{3.207}$$

$$D_3 f_1 - M_\alpha D_\alpha \Phi = 0, \tag{3.208}$$

$$D_0 f_1 + \Delta_\alpha D_\alpha f_1 - H_\alpha D_\alpha S = 0. \tag{3.209}$$

These equations have to be appended to the original system.

Among the relations (3.207)-(3.209) there is only a nonzero relation because of system (3.204), and this is the third equation in (3.207)

$$D_0 f_4 + \Delta_\alpha D_\alpha f_4 - \Psi_\alpha D_\alpha S_\alpha - 2\theta_1\theta_2 D_1 S_2 - 2\theta_1\theta_3 D_1 S_3 - 2\theta_2\theta_3 D_2 S_3 = 0. \tag{3.210}$$

Substituting into this equation the derivatives $\partial \mathbf{u}/\partial x_3$ and $\partial u_1/\partial x_2$, found from system (3.204), one obtains a quadratic form with respect to the derivatives $p^i_j = \partial u_i/\partial x_j$, $(i, j = 1, 2, 3;\ j < i)$

$$\begin{aligned} f_5 = c_1(p^1_1)^2 + 2c_2 p^1_1 p^2_1 + 2c_3 p^1_1 p^3_1 + 2c_4 p^1_1 p^2_2 + 2c_5 p^1_1 p^3_2 + \\ + c_6(p^2_1)^2 + 2c_7 p^2_1 p^3_1 + 2c_8 p^2_1 p^2_2 + 2c_9 p^2_1 p^3_2 + \\ + c_{10}(p^3_1)^2 + 2c_{11} p^3_1 p^2_2 + 2c_{12} p^3_1 p^3_2 + \\ c_{13}(p^2_2)^2 + 2c_{14} p^2_2 p^3_2 + \\ c_{15}(p^3_2)^2 = 0, \end{aligned} \tag{3.211}$$

where the coefficients $c_i (i = 1, 2, \ldots, 15)$ are expressed through $\theta, \theta_i, \theta_{ij}$. For example, $c_{15} = \Psi_3 M_{23} (M_{ij} = \Psi_i(1 + \theta_{jj}) - 2\theta_i\theta_j\theta_{ij} + +\Psi_j(1 + \theta_{ii}))$. The expressions for the other coefficients are very cumbersome. If for any rotation of the coordinates the value $M_{23} = 0$, the following relations are necessary

$$\begin{aligned} \Psi_3\theta_{12} + \theta_1\theta_2(1 + \theta_{33}) - \theta_1\theta_3\theta_{23} - \theta_2\theta_3\theta_{13} = 0, \\ \Psi_1\theta_{23} + \theta_2\theta_3(1 + \theta_{11}) - \theta_1\theta_2\theta_{13} - \theta_1\theta_3\theta_{12} = 0, \\ \Psi_2\theta_{13} + \theta_1\theta_3(1 + \theta_{22}) - \theta_2\theta_3\theta_{12} - \theta_1\theta_2\theta_{23} = 0, \\ M_{12} = 0, M_{13} = 0, M_{23} = 0. \end{aligned} \tag{3.212}$$

The last system is linear and homogeneous with respect to θ_{ij}, $(1+\theta_{ii})$, $(i, j = 1, 2, 3;\ i \neq j)$, and its determinant is equal to $2(\kappa\theta)^4(\theta_1^2 + \theta_2^2 + \theta_3^2 - \kappa\theta)^2$. Hence, if $(\theta_1^2+\theta_2^2+\theta_3^2-\kappa\theta)^2 \neq 0$, then $\theta_{ij} = 0,\ \theta_{ii} = -1,\ (i, j = 1, 2, 3;\ i \neq j)$, and one finds

$$\theta = C_0 - \sum_{i=1}^{3} (u_i + C_i)^2/2, \tag{3.213}$$

where C_i, $(i = 1, 2, 3)$ are constant. In this case $f_5 = 0$, system (3.204) is involutive and its general solution has two arbitrary functions of two arguments.

However, it should be mentioned that after the Galilei transformation $\bar{x}_i = x_i + C_i t$ the representation (3.213) corresponds to the Bernoulli integral, and the gas motion corresponds to the general potential steady space flows.

If $\theta_1^2 + \theta_2^2 + \theta_3^2 - \kappa\theta = 0$, from equations (3.212) one has that $\gamma + 2 = 0$, but this contradicts to the condition $\kappa > 0$.

Thus, without loss of generality one can set $M_{23} \neq 0$.

Repeating the calculations, as done when obtaining equations (3.205)-(3.207), where the value f_1 is replaced by $f_1' = (f_1, f_5)^t$, one concludes that the equations (3.205), (3.207), (3.209) conserve their form (this means that instead f_1 we take $(f_1, f_5)^t$. Hence, there is only the new equation with respect to the equations (3.207)-(3.209)

$$f_6 = D_0 f_5 + \Delta_\alpha D_\alpha f_5 - (\partial f_5/\partial p_1^\alpha) D_1 S_\alpha - (\partial f_5/\partial p_2^\alpha) D_2 S_\alpha = 0. \quad (3.214)$$

Excluding the derivatives $\partial\mathbf{u}/\partial t$ in (3.214), one has

$$2f_6 = p_\alpha^\beta (a_{1\alpha} D_1 \Delta_\beta + a_{2\alpha} D_2 \Delta_\beta) = 0, \quad (3.215)$$

where $a_{1i} = \partial f_5/\partial p_1^i$, $a_{2i} = \partial f_5/\partial p_2^i$, $(i = 1, 2, 3)$. Here, substituting the other main derivatives of system (3.204), (3.210) is very cumbersome [23].

After all substitutions, equation (3.214) becomes $f_6 = Ag$, with the function A, which is defined through $\theta(u_1, u_2, u_3)$ and its derivatives

$$A = -\Psi_1\theta_{23}^2 - \Psi_2\theta_{13}^2 - \Psi_3\theta_{12}^2 + 2\theta_1\theta_2\theta_{13}\theta_{23} + 2\theta_1\theta_3\theta_{12}\theta_{23} + 2\theta_2\theta_3\theta_{12}\theta_{13} -$$
$$2\theta_1\theta_2\theta_{12}(1+\theta_{33}) - 2\theta_1\theta_3\theta_{13}(1+\theta_{22}) - 2\theta_2\theta_3\theta_{23}(1+\theta_{11}) +$$
$$\Psi_1(1+\theta_{22})(1+\theta_{33}) + \Psi_2(1+\theta_{11})(1+\theta_{33}) + \Psi_3(1+\theta_{11})(1+\theta_{22}) = 0.$$

The function g is an homogeneous cubic form with respect to the derivatives p_j^i, $(i, j = 1, 2, 3;\ j < i)$. Here the expression of g is also very cumbersome.

From another point of view, from the equations (3.204), (3.210) and the condition $M_{23} \neq 0$, one finds that $g = c_{15}\Delta$, where $\Delta = \dfrac{\partial(u_1, u_2, u_3)}{\partial(x_1, x_2, x_3)}$. Hence, the relation $g = 0$ leads to $\Delta = 0$. But in this case, from equation (3.204), one obtains a contradiction to the functional dependence of the functions (u_1, u_2, u_3). Therefore one gets

$$A = 0. \quad (3.216)$$

Remark 3.6. *For consistency of the overdetermined system (3.204), (3.210) it is necessary and sufficient*[24] *to find matrices Λ_i, Λ_{3i}, $(i = 1, 2, 3)$ such that*

$$D_t \begin{pmatrix} \Phi \\ f_1' \end{pmatrix} + \Lambda_\alpha D_\alpha \Phi - \begin{pmatrix} \Phi_a \\ f_1 \end{pmatrix} D_\alpha S = 0, \quad \Lambda_{3a} D_\alpha \begin{pmatrix} \Phi \\ f_1' \end{pmatrix} = 0.$$

[23] The calculations were done by computer program developed in [53] using the system of analytical calculations REFAL.

[24] See development of these conditions in the next chapter.

For the case discussed, these matrices are

$$\Lambda_i = \Delta_i \begin{pmatrix} E & 0 \\ 0 & E_2 \end{pmatrix}, \quad \Lambda_{3i} = (M_i', -\delta_{3i} E_2), (i = 1, 2, 3),$$

where E_2 is a unit 2×2 matrix, δ_{ij} is the Kronecker symbol. The general solution of this system is defined up to one function of two arguments.

Thus, equation (3.216) is necessary and sufficient for involutiveness of the overdetermined system (3.204), (3.210), and the general solution has a one arbitrary function of two arguments.

The second representation of a triple wave type solution is $u_3 = \phi(u_1, u_2)$. Substituting the expression of u_3 into system of equations (3.203), one has

$$(S_1, S_2, S_3)^t = E\frac{\partial V}{\partial t} + G_\alpha \frac{\partial V}{\partial x_\alpha} = 0, \tag{3.217}$$

where

$$f = (f_1, f_2, f_3)^t = \frac{\partial V}{\partial x_3} - \phi_\alpha \frac{\partial V}{\partial x_\alpha} = 0, \; f_4 = \frac{\partial u_1}{\partial x_2} - \frac{\partial u_2}{\partial x_1} = 0.$$

Here the sum with respect to a repeated Greek index is taken from 1 to 2, $V = (u_1, u_2, \theta)'$, $\phi_i = \partial\phi/\partial u_i$, and

$$G_1 = (u_1 + \phi\phi_1)E + \begin{pmatrix} 0 & 0 & 1 \\ 0 & 0 & 0 \\ \kappa\theta(1 + \phi_1^2) & 2\kappa\theta\phi_1\phi_2 & \end{pmatrix},$$

$$G_2 = (u_2 + \phi\phi_2)E + \begin{pmatrix} 0 & 0 & 1 \\ 0 & 0 & 0 \\ 0 & \kappa\theta(1 + \phi_1^2) & 0 \end{pmatrix}.$$

Performing similar calculations, as in the case of the representation $\theta = \theta(u_1, u_2, u_3)$, one obtains one more equation

$$\begin{aligned} f_5 = {} & (\partial\theta/\partial x_2)^2\phi_{22} + (\partial\theta/\partial x_1)^2\phi_{11} + 2(\partial\theta/\partial x_1)(\partial\theta/\partial x_2)\phi_{12} + \\ & \kappa\theta\Psi(\tfrac{\partial\theta}{\partial x_i}, \tfrac{\partial u_i}{\partial x_j}, \phi, \phi_i, \phi_{ij}), \end{aligned} \tag{3.218}$$

where the function f_5 is a homogeneous quadratic form with respect to the derivatives $\frac{\partial\theta}{\partial x_i}, \frac{\partial u_i}{\partial x_j}$, $(i, j = 1, 2)$, and Ψ is a linear function with respect to $\frac{\partial\theta}{\partial x_i}$ $(i = 1, 2)$.

The case where $\sum_{i,j} \phi_{ij}^2 = 0$ is reduced to the general solution of a plane flow. Hence, one needs to assume $\sum_{i,j} \phi_{ij}^2 \neq 0$. Using the rotation of coordinate axes (x_1, x_2) one can set $\phi_{22} \neq 0$.

Analysis of the first prolongation of the system (3.217), (3.218) shows that for the consistency of this system it is necessary to satisfy one more equation

$$f_6 = D_3 f_5 - \sum_{a=1}^{3}(a_{1a} D_1 f_\alpha + a_{2a} D_2 f_\alpha) - \phi_1 D_1 f_5 - \phi_2 D_2 f_5 = 0,$$

where

$$a_{i\alpha} = \partial f_5/\partial(\partial u_i/\partial x_\alpha), a_{i3} = \partial f_5/\partial(\partial\theta/\partial x_i), (i, \alpha = 1, 2)$$

Substituting the main derivatives of the system (3.217), (3.218) into this equation, one finds $f_6 = (\phi_{12}^2 - \phi_{11}\phi_{22})g = 0$, for some function g.

Since it is a triple wave, the Jacobian $\Delta = \frac{\partial(u_1,u_2,\theta)}{\partial(x_1,x_2,t)} \neq 0$, otherwise u_1, u_2, θ are functionally dependent. Moreover, there is the equality $g = \phi_{22}\Delta$, which means that $g \neq 0$, and, hence,

$$\phi_{12}^2 - \phi_{11}\phi_{22} = 0. \qquad (3.219)$$

Therefore the conditions (3.219) are necessary and sufficient for the existence of a triple wave type solution in the case $u_3 = \phi(u_1, u_2)$. As in the previous case, the equation $f_6 = 0$ provides the involutiveness of the overdetermined system (3.217), (3.218), with a one arbitrary function of two arguments.

Theorem 3.9. *Equations (3.214) (or (3.219)) are necessary and sufficient for the existence of irreducible triple waves of potential flows of a polytropic gas. The equations of a triple wave type (3.204), (3.211) (or (3.217), (3.218)) are involutive with a one arbitrary function of two arguments.*

Remark 3.7. *Conditions (3.216) and (3.219) were obtained in [156] assuming that $\Delta \neq 0$. Here it was shown that this assumption is necessary and sufficient for the existence of triple waves. Except for condition (3.216) (or (3.219)) in [156], this was obtained for two more equations for the distribution function by a hodograph transformation in equations (3.204) (or in (3.217))). By using computer symbolic calculations, it is proven that after this transformation the equation $f_6 = 0$ is satisfied identically. Thus one can conclude that the system of the two equations for the distribution function obtained in [156] with a given function $\theta = \theta(u_1, u_2, u_3)$ (or $u_3 = \phi(u_1, u_2)$) is involutive, and its general solution has one arbitrary function of two arguments.*

Chapter 4

METHOD OF DIFFERENTIAL CONSTRAINTS

Practically all methods for finding exact particular solutions of partial differential equations require the analysis of the compatibility of overdetermined systems. One method differs from another in the way in which overdetermined systems are obtained. For example, functionally invariant solutions have to satisfy first order differential equations, which are bilinear equations with respect to first order derivatives; solutions with degenerate hodographs are picked out by relations between the dependent variables; in group analysis additional relations are obtained from the requirement that the solution should be invariant or partially invariant, as are relations between invariants.

Since the Cartan theorem asserts that any compatible system of partial differential equations after a finite number of prolongations becomes an involutive system, the main tool of compatibility theory is the analysis of involutiveness. If an analytic system of partial differential equations is an involutive system, the Cartan-Khäler theorem solves the problem of the existence of a solution.

The theory of compatibility is followed by the basic definitions of the method of differential constraints, formulated by N.N.Yanenko [177]. The method of differential constraints has mostly been applied to systems of quasilinear partial differential equations with two independent variables. The first problem arising in applications of the method of differential constraints is the involutiveness problem of an original system of partial differential equations with differential constraints. Since the Cartan-Khäler theorem only proves the existence of a solution for analytic systems, the problem of the existence of a solution for nonanalytic involutive systems arises. This problem is solved by using the notion of characteristics for an overdetermined system of partial differential equations. Characteristic curves also play the main role in defining a class of solutions generalizing simple waves. Finding a solution for this class is reduced to integrating a system of ordinary differential equations along characteristics. The generalized simple waves have properties similar to simple waves. For example, the solution of the Goursat problem can be given in terms of generalized simple waves. The general study of generalized simple waves is followed by a section devoted to deriving this class of solutions for gas dynamic equations.

The second part of the chapter considers applications of the method of differential constraints to systems of quasilinear equations with more than two independent variables. Following the general study are examples of differential constraints for multi-dimensional gas dynamic equations. As was mentioned, invariant solutions can also be described by differential constraints. The following section studies the relations between the method of differential constraints and Lie-Bäcklund groups of transformations. The chapter ends with the construction of one class of solutions for systems of quasiliniear equations with more than two independent variables.

1. Theory of compatibility

This section gives the necessary knowledge of involutive systems. Because this theory is a special subject of mathematical analysis, the statements are given without proofs[1].

There are two approaches for studying compatibility. These approaches are related to the works of E.Cartan and C.H.Riquier.

The Cartan approach is based on the calculus of exterior differential forms. The problem of the compatibility of a system of partial differential equations is then reduced to the problem of the compatibility of a system of exterior differential forms. Cartan studied the formal algebraic properties of systems of exterior forms. For their description he introduced special integer numbers, named characters. With the help of the characters he formulated a criterion for a given system of partial differential equations to be involutive.

The Riquier approach has a different theory of establishing the involution[2]. The main advantage of this approach being that there is no necessity to reduce a system of partial differential equations that is being studied to exterior differential forms. Calculations in the Riquier approach are shorter than in the Cartan approach. The main operations of the study of compatibility in the Riquier approach are prolongations of a system of partial differential equations and the study of the ranks of some matrices. In this section the Riquier approach is discussed.

Let a system of q-th order differential equations (S) be defined by the equations

$$(S) \qquad \Phi^i(x, u, p) = 0, \quad (i = 1, 2, ..., s). \tag{4.1}$$

[1]Detailed theory of involutive systems can be found in [24, 44, 93, 138]. Short history of the theory can be found in [138].

[2]This method can be found in [93] and [138].

Here $x = (x_1, x_2, \ldots, x_n)$ are the independent variables, $u = (u^1, u^2, \ldots, u^m)$ are the dependent variables, $p = (p^j_\alpha)$ is the set of the derivatives $p^j_\alpha = \dfrac{\partial^{|\alpha|}u^j}{\partial x^\alpha}$, $(j = 1, 2, \ldots, m;\ |\alpha| \le q)$, $\alpha = (\alpha_1, \alpha_2, \ldots, \alpha_n)$, $|\alpha| = \alpha_1 + \alpha_2 + \ldots + \alpha_n$. All constructions are considered in some neighborhood of the point $X_o = (x_o, u_o, p_o) \in (S)$. First the algebraic properties of a symbol of the system (S) are studied. The symbol G_q of the system (S) at the point X_o is defined as the vector space of vectors with the coordinates (ξ^j_α), $(j = 1, 2, ..., m;\ |\alpha| = q)$, where the coordinates ξ^j_α satisfy the algebraic equations

$$\sum_{j=1}^{m}\sum_{|\alpha|=q} \xi^j_\alpha \frac{\partial \Phi^i}{\partial p^j_\alpha}(X_o) = 0, \quad (i = 1, 2, ..., s).$$

The subspace of the symbol G_q composed by the vectors with

$$\xi^j_{\beta,l} = 0, \ (|\beta| = q - 1;\ l = 1, 2, ..., k;\ j = 1, 2, ..., m)$$

is denoted by $(G_q)^k$, $(k = 1, 2, \ldots, n-1)$. Here $\beta, l = (\beta_1, \beta_2, ..., \beta_{l-1}, \beta_l + 1, \beta_{l+1}, ..., \beta_n)$, $(G_q)^0 = G_q$, and $(G_q)^n = \{0\}$. Let the dimensions of the vector spaces $(G_q)^k$ be τ_k. For example,

$$\tau_o = m\begin{pmatrix} n+q-1 \\ q \end{pmatrix} - rank\left(\frac{\partial \Phi^i}{\partial p^j_\alpha}(X_o)\right), \quad \tau_n = 0.$$

The number

$$\sum_{k=0}^{n-1} \tau_k$$

is called the Cartan number. With the help of the numbers τ_k, $(k = 0, 1, ..., n)$ the Cartan characters are defined by the formulae

$$\sigma_{k+1} = \tau_k - \tau_{k+1}, \ (k = 1, ..., n - 1).$$

Note that $\tau_o = \sum_{k=1}^{n} \sigma_k$ and the Cartan number can be expressed through the Cartan characters

$$\sum_{k=0}^{n-1} \tau_k = \sum_{k=1}^{n} k\sigma_k.$$

Let G_{q+1} be the symbol of the prolonged system (DS):

$$(DS) \qquad D_l\Phi^i(x, u, p) = 0, \ (l = 1, 2, \ldots, n;\ i = 1, 2, ..., s). \qquad (4.2)$$

Here the operator D_l is the total derivative with respect to x_l

$$D_l = \frac{\partial}{\partial x_l} + \sum_{|\alpha|}\sum_{j=1}^{m} p^j_{\alpha,l}\frac{\partial}{\partial p^j_\alpha}.$$

Definition 4.1. *The system of differential equations composed by the system* (S) *and* (DS) *is called a first prolongation of the system* (S).

The theory of compatibility is a local theory, i.e., all properties are considered in some neighborhood of a point X_o, and all manifolds and functions are assumed to be the necessary number of times continuously differentiable. Moreover, the Cartan theorem works only for analytical functions.

Note that Cartan characters depend on the order of the independent variables $(x_1, x_2, x_3, ..., x_n)$: so any change of the order can change the Cartan characters. There is the estimate

$$\dim(G_{q+1}) \leq \sum_{k=1}^{n} k\sigma_k.$$

Definition 4.2. *A coordinate system of the independent variables in which there is the equality*

$$\dim(G_{q+1}) = \sum_{k=1}^{n} k\sigma_k$$

is called a quasiregular coordinate system.

Definition 4.3. *If there exists a quasiregular coordinate system, then a symbol* G_q *is called an involutive symbol.*

After studying the algebraic properties of the system (S) one has to analyze the differential structure of the manifold defined by the equations (DS). From the system (DS) one can find $N = \dim(G_{q+1})$ derivatives of the highest $q+1$ order. These derivatives are called the main derivatives of the system (DS) of order $q+1$.

Definition 4.4. *If a system* (S) *with an involutive symbol posses the property that after substituting the main derivatives of the prolonged system* (DS) *of order* $q+1$, *the remaining equations of the system* (DS) *are identities because of the system* (S), *then system* (S) *is called involutive.*

Theorem 4.1. *(Cartan). Any analytic system of partial differential equations after a finite number of prolongations becomes either involutive or incompatible.*

Theorem 4.2. *(Cartan–Khäler). If a system* (S) *of order* q *is involutive and analytic, there exists one and only one analytic solution of the Cauchy problem with given* σ_k *functions of* k *arguments* $(k = 1, 2, \ldots, n-1)$.

Remark 4.1. *The property of analyticity of an involutive system is not a necessary condition for the existence of a solution. There are theorems of existence of involutive systems of the class* C^1 *[107].*

Remark 4.2. *Any study of compatibility requires a large amount of symbolic calculations. These calculations consist of consecutive algebraic operations: prolongation of a system, substitution of some expressions (transition onto manifold), and the determination of ranks of matrices (for obtaining the Cartan characters). Because these operations are very labor intensive, it is necessary to use a computer for symbolic calculations. One of the first applications of computers for studying the compatibility of systems of partial differential equations was carried out by V.A.Shurygin and N.N.Yanenko [155]. The Cartan method was programmed in [5], and later in [168]. Another algorithm was realized in [52] and in [53, 54]. This short review does not pretend to be comprehensive, as it only reflects the attempts of the author and his colleagues.*

2. Method formulation

The method of differential constraints is one of the methods for constructing particular exact solutions of partial differential equations. The method is based on the following idea[3].

Consider a system of differential equations

$$S_i(x, u, p) = 0, \quad (i = 1, 2, \ldots, s). \tag{4.3}$$

Assume that a solution of system (4.3) satisfies the additional differential equations

$$\Phi_k(x, u, p) = 0, \quad (k = 1, 2, \ldots, q). \tag{4.4}$$

Definition 4.5. *Differential equations (4.4) are called differential constraints.*

The system (4.3), (4.4) is an overdetermined system. The method of differential constraints requires the overdetermined system (4.3), (4.4) to be compatible. The form of the differential constraints (the functions Φ_k) and the part of the equations of the initial system (the functions S_i) may not be known a priori. They may be determined by solving the inverse problem of the compatibility theory of the systems (4.3), (4.4). The inverse problem defines the differential constraints such that the augmented system has a solution with the required flexibility.

An application of the method of differential constraints for finding solutions of the system (S) involves two stages. The first stage is to find the set of differential constraints (4.4) under which the overdetermined system is compatible. The system $(S\Phi)$ may be further augmented by new relations in the process of

[3]The idea of the method was proposed by N.N.Yanenko [177] in 1964. A survey of the method can be found in [160]. In particular, the method of differential constraints is discussed in [147].

compatibility analysis (reducing the system $(S\Phi)$ to an involutive one). The second stage of the method is the construction of solutions of the involutive overdetermined system. Since a solution of the system (S) has to satisfy the differential constraints (Φ), it allows an easier construction of a particular solution of the given system (4.3).

Definition 4.6. *A solution of system (4.3) satisfying (4.4) is called a solution characterized by the differential constraints (4.4).*

The requirements for the compatibility of system (4.3), (4.4) is very general and includes (almost) all known methods of determining solutions of partial differential equations: group–invariant solutions, nonclassical and weak symmetries, partially invariant solutions, separation of variables, as well as many others.

Increasing the number of requirements on the differential constraints narrows the generality of the method and makes it more suitable for finding exact particular solutions. V.P.Shapeev [152, 152] suggested that requiring the involutiveness of the augmented system (4.3), (4.4) would be one of the possible ways forward in this direction.

Definition 4.7. *System (4.3) has the D–property, or the differential constraints (4.4) are admitted by system (4.3), if the overdetermined system (4.3), (4.4) is involutive.*

Classification of differential constraints and solutions characterized by them is carried out with respect to the functional arbitrariness of solutions of the overdetermined system (4.3), (4.4), and the order of the highest derivatives included in the differential constraints (4.4). With this refinement the method of differential constraints becomes a practical tool for obtaining exact particular solutions.

However the requirement of involutiveness for the augmented system involves a large arbitrariness in choice of differential constraints. Further restrictions were proposed in [80, 81]: the conditions of the involutiveness for the augmented system must coincide with the determining equations of the one–parameter Lie group admitted by system (4.3).

3. Quasilinear systems with two independent variables

In this section the method of differential constraints is applied to the quasilinear system of partial differential equations

$$\frac{\partial u}{\partial t} + Q\frac{\partial u}{\partial x} - f = 0. \tag{4.5}$$

Here $Q = Q(x, t, u)$ is an $m \times m$ matrix, $f = f(x, t, u)$ is a vector, E_r is an $r \times r$ unit matrix, $x_1 = x,\ x_2 = t$. For the sake of simplicity, only solutions characterized by first order differential constraints

$$\Phi_k(x, t, u, u_x) = 0, \quad (k = 1, 2, \dots, q). \tag{4.6}$$

are studied[4]. The natural requirement

$$rank\left(\frac{\partial\Phi_k}{\partial u_x}\right) = q.$$

is assumed.

3.1 Involutive conditions

Without loss of generality the system of differential equations and the differential constraints can be rewritten in the more suitable form

$$S \equiv Lu_t + ALu_x - Lf = 0, \tag{4.7}$$

$$\Phi = B_1Lu_x + \Psi = 0. \tag{4.8}$$

Here $L = L(x,t,u)$ is a nonsingular $m \times m$ matrix, $A = LQL^{-1}$, the function $\Psi = \Psi(x,t,u,y)$ depends on x, t, u and $y = B_2Lu_x$, B_1 and B_2 are rectangular $q \times m$ and $(q-1) \times m$ matrices with the elements

$$(B_1)_{ij} = \delta_{ij}, \ (1 \le i \le q, \ 1 \le j \le m),$$
$$(B_2)_{kj} = \delta_{q+k,j}, \ (1 \le k \le m-q, \ 1 \le j \le m),$$

δ_{ij} is the Kronecker's symbol. The matrices B_1 and B_2 have the following properties:

$$\begin{gathered} B_1B_1' = E_q, \ B_2B_2' = E_{m-q}, \ B_1'B_1 + B_2'B_2 = E_m, \\ B_1B_2' = 0, \ B_2B_1' = 0. \end{gathered} \tag{4.9}$$

For a hyperbolic system (4.5) the matrix A can be chosen to be diagonal. If the matrix A is a diagonal matrix, then the matrices B_jAB_j are also diagonal and $B_iAB_j = 0, \ (i, j = 1, 2; \ i \neq j)$.

Remark 4.3. *In applications it is convenient to fix the matrix L and taking the matrices B_1 and B_2 in the form*

$$(B_1)_{ij} = \delta_{i\sigma(j)}, \ (1 \le i \le q, \ 1 \le j \le m),$$
$$(B_2)_{kj} = \delta_{q+k,\sigma(j)}, \ (1 \le k \le m-q, \ 1 \le j \le m),$$

where σ is some permutation. In this case these matrices also satisfy (4.9).

There is the following theorem [107].

[4] Differential constraints of higher order of the system (S) can be reduced to differential constraints of first order for the prolonged system. This means that solutions characterized by differential constraints of higher order can be are studied in similar fashion.

Theorem 4.3. *The overdetermined system (4.7), (4.8) is involutive if and only if*

$$\left(D_t\Phi + ZAB_1'D_x\Phi - ZD_xS\right)_{|(S\Phi)} = 0, \tag{4.10}$$

$$ZA - ZAB_1'Z = 0, \tag{4.11}$$

where $Z = B_1 + \Psi_y B_2$ *and the manifold* $(S\Phi)$ *is defined by the formulae*

$$(S\Phi) \equiv \{(x, u, p)|\ S(x, u, p) = 0,\ \ \Phi(x, u, p) = 0\}.$$

Proof.

Let us prove the necessity of the conditions (4.11). The symbol G_1 consists of the set of vectors $(\xi_{1,0}, \xi_{0,1}) \in R^{2m}$, which satisfy the equations

$$L\xi_{0,1} + AL\xi_{1,0} = 0,$$
$$ZL\xi_{1,0} = 0.$$

Since the matrix L is nonsingular, the dimensions are $\tau_0 = \dim(G_1) = m - q$, $\tau_1 = \dim\left((G_1)^{(1)}\right) = 0$. Hence, the sum $\sum \tau_i$ is minimal. This means that if the symbol G_1 is involutive, the coordinate system (x, t) is quasi-regular. The space G_2 is a subspace of the vector space of vectors with the coordinates $(\xi_{2,0}, \xi_{1,1}, \xi_{0,2}) \in R^{3m}$. This subspace is defined by the equations

$$L\xi_{0,2} + AL\xi_{1,1} = 0,$$
$$L\xi_{1,1} + AL\xi_{2,0} = 0,$$
$$ZL\xi_{1,1} = 0,\ \ ZL\xi_{0,2} = 0.$$

Thus,

$$\dim(G_2) = 3m - rank\ H,$$

where the matrix H has a block structure and consists of the blocks

$$H_{ij} = \begin{cases} AL, & (j = i,\ i \le 2), \\ L, & (j = i + 1,\ i \le 2), \\ ZL, & (i = 3,\ j = 1;\ i = 4,\ j = 2), \end{cases}$$

and 0 in other cases. Here i is the number of a block-row and j is the number of a block-column. For the involutive symbol G_1

$$\dim(G_2) = \tau_0 + \tau_1 = m - q,$$

which gives

$$rank\ H = 2m + q.$$

Since the rank of the matrix composed of the first 3 block-rows ($i \le 3$) is equal to $2m + q$, there is linear dependence of the last block-row ($i = 4$) from the previous block-rows

$$\sum_{\alpha=1}^{3} M_\alpha H_{\alpha j} = H_{4j}, \ (j = 1, 2, 3),$$

where M_α, $(\alpha = 1, 2, 3)$ are coefficients of the linear dependence. The last relations can be rewritten as follows

$$M_1 = Z, \ M_2 = 0, \ ZA + M_3 Z = 0.$$

Multiplying the last equation by B_1' from the right hand side, one obtains

$$M_3 = -ZAB_1'.$$

This corresponds to equation (4.11). The linear dependence between second derivatives leads to the relations

$$D_t\Phi + ZAB_1' D_x\Phi - ZD_x S = 0,$$

which only contain derivatives of first order. Since the system (4.7), (4.8) is involutive, these equations must be satisfied according to system (4.7), (4.8). This proves that relations (4.10) are valid.

The inverse statement of the theorem is proved similarly.•

Remark 4.4. *In applications equations (4.11) are checked first, although they are contained in (4.10).*

Remark 4.5. *After substituting Z, equations (4.11) are rewritten*

$$B_1AB_2' - \Psi_y B_2 AB_1' + \Psi_y B_2 AB_2' - B_1 AB_1' \Psi_y = 0.$$

If the matrix A is a diagonal matrix with the diagonal elements λ_i $(i = 1, 2, ..., m)$, then $B_1AB_2' = 0$, $B_2AB_1' = 0$, the matrices B_1AB_1', B_2AB_2' are diagonal and equations (4.11) become

$$(\lambda_i - \lambda_j)\frac{\partial \Psi_i}{\partial y_j} = 0, \ (i = 1, 2, ..., q; \ j = 1, 2, ..., m - q).$$

This means that the Ψ_i can only depend on y_j such that $(\lambda_i - \lambda_j) = 0$. In particular, for a strictly hyperbolic system $(\lambda_i - \lambda_j) \neq 0$, $(i \neq j)$, and equations (4.11) are reduced to [67]

$$\Psi_y = 0.$$

Thus, for strictly hyperbolic quasilinear systems, the differential constraints (4.8) are quasilinear[5]*.*

[5] Similar results were also obtained in [45].

3.2 Theorems of Existence

If system (4.7), (4.8) is analytic, then its involutiveness provides the uniqueness and existence of the Cauchy problem. There are also weaker requirements on the smoothness of system (4.7), (4.8), which are sufficient for the uniqueness and existence.

Further in this subsection it is assumed that

$$L \in C^1(D),\ A \in C^1(D),\ f \in C^1(D),\ \Psi \in C^1(D) \tag{4.12}$$

in some open domain $D \subset R^m \times R^2$.

Lemma 4.1. *If system (4.7), (4.8) satisfies (4.10), (4.11), then there exist matrices $\Theta_1, \Theta_2 \in C(D)$ such that*

$$D_t\Phi + ZAB_1'D_x\Phi - ZD_xS + \Theta_1 S + \Theta_2\Phi \equiv 0.$$

Proof.

Let us calculate the expression

$$\begin{gathered}\Omega \equiv D_t\Phi + ZAB_1'D_x\Phi - ZD_xS = \\ = Z\tfrac{\partial L}{\partial u}\langle p_{0,1}, p_{1,0}\rangle + Z\tfrac{\partial L}{\partial t}p_{1,0} + \tfrac{\partial\Psi}{\partial u}p_{0,1} + \tfrac{\partial\Psi}{\partial t} + ZAB_1'\tfrac{\partial\Psi}{\partial u}p_{1,0} \\ -Z\tfrac{\partial A}{\partial u}\langle p_{1,0}, Lp_{1,0}\rangle - Z\tfrac{\partial A}{\partial x}Lp_{1,0} + ZAB_1'\tfrac{\partial\Psi}{\partial x} - Z\tfrac{\partial L}{\partial u}\langle p_{1,0}, p_{0,1}\rangle \\ +Z\tfrac{\partial(Lf)}{\partial u}p_{1,0} - Z\tfrac{\partial L}{\partial x}p_{0,1} + Z\tfrac{\partial(Lf)}{\partial x}.\end{gathered}$$

Here $p_{1,0} = \frac{\partial u}{\partial x}$, $p_{0,1} = \frac{\partial u}{\partial t}$, the vector $\frac{\partial L}{\partial u}\langle\xi, \eta\rangle \in R^m$ is the vector with the coordinates

$$\left(\frac{\partial L}{\partial u}\langle\xi, \eta\rangle\right)_i = \frac{\partial L_{i\alpha}}{\partial u^\beta}\xi_\beta\eta_\alpha,$$

where ξ and η are arbitrary vectors. Substituting into Ω the values

$$\begin{gathered}p_{1,0} = L^{-1}B_1'\Phi + L^{-1}(B_2'y - B_1'\Psi), \\ p_{0,1} = L^{-1}S - L^{-1}AB_1'\Phi + L^{-1}A(B_1'\Psi - B_2'y) + f,\end{gathered}$$

found from the equations

$$\begin{gathered}S = Lp_{0,1} + ALp_{1,0} - Lf, \\ \Phi = B_1Lp_{1,0} + \Psi,\ y = B_2Lp_{1,0},\end{gathered} \tag{4.13}$$

one obtains

$$\Omega = \Omega_1(y, u, x, t) + \Omega_2(S, \Phi, y, u, x, t),$$

where

$$\begin{gathered}\Omega_2 = Z\left\{\frac{\partial L}{\partial u}\langle\xi, \eta\rangle + \frac{\partial L}{\partial t}\eta + AB_1'\frac{\partial\Psi}{\partial u}\eta - \frac{\partial A}{\partial u}\langle\eta, L\eta\rangle\right. \\ \left.-Z\frac{\partial L}{\partial u}\langle\eta, \xi\rangle + \frac{\partial(Lf)}{\partial u}\eta - \frac{\partial L}{\partial x}\xi\right\} + \frac{\partial\Psi}{\partial u}\xi,\end{gathered}$$

$$\xi = L^{-1}S - L^{-1}AB_1'\Phi, \ \eta = L^{-1}B_1'\Phi.$$

From the form of Ω_2 it follows that $\Omega_{2|(S\Phi)} = 0$. Since Ω_1 does not depend on S and Φ, and because of (4.10) there is $\Omega_1 = \Omega_{|(S\Phi)} = 0$. Hence, Ω_1 is identically equal to zero, and therefore

$$\Omega = \Omega_2.$$

Substituting the expressions for S and Φ in (4.13) where necessary, the expression Ω_2 can be rewritten in a linear homogeneous form with respect to S and Φ.•

As was mentioned, for hyperbolic systems the matrix A can be chosen to be diagonal. In this case the matrix L consists of the left eigenvectors of the matrix Q.

Theorem 4.4. *Let equations (4.10), (4.11) be satisfied for a hyperbolic system (4.7) with (4.12). There exists a unique solution $u(x,t) \in C^1$ of the Cauchy problem of system (4.7), (4.8) with the initial data $u(x,0) = \varphi(x) \in C^1$ satisfying the differential constraints (4.8) at $t = 0$.*

Proof.

Let $\varphi(x)$ satisfy the differential constraints (4.8) at $t = 0$

$$B_1L(\varphi(x), x, 0)\varphi'(x) + \Psi\left(B_2L(\varphi(x), x, 0)\varphi'(x), \varphi(x), x, 0\right) = 0.$$

There exists a unique solution[6] $u(x,t) \in C^1$ of the Cauchy problem of the hyperbolic system (4.7) with the initial values $u(x,0) = \varphi(x) \in C^1$. This solution has the property that each coordinate of the vector–function $\mathcal{P}(x,t) = L(u(x,t), x, t)u_x(x,t)$ is continuously differentiable along its characteristic curves. In fact

$$\begin{gathered}D_xS \equiv \mathcal{P}_t + A\mathcal{P}_x - \tfrac{\partial(Lf)}{\partial x} - \tfrac{\partial(Lf)}{\partial u}L^{-1}\mathcal{P} + \tfrac{\partial A}{\partial u}\langle L^{-1}\mathcal{P}, \mathcal{P}\rangle \\ +\tfrac{\partial A}{\partial x}\mathcal{P} - \tfrac{\partial L}{\partial u}\langle f - L^{-1}A\mathcal{P}, P\rangle - \tfrac{\partial L}{\partial t}L^{-1}\mathcal{P} + \tfrac{\partial L}{\partial u}\langle L^{-1}\mathcal{P}, f - L^{-1}A\mathcal{P}\rangle \\ +\tfrac{\partial L}{\partial x}(f - L^{-1}A\mathcal{P}) = 0\end{gathered} \tag{4.14}$$

Here the value $\mathcal{P}_t + A\mathcal{P}_x$ is a vector, the i-th coordinate of which is a derivative of the function $\mathcal{P}_i$ along the characteristic curve $\frac{dx}{dt} = \lambda_i$:

$$(\mathcal{P}_t + A\mathcal{P}_x)_i = \frac{d_i\mathcal{P}}{dt}.$$

Let us prove that each j–th ($j = 1, 2, ..., q$) coordinate of the function

$$\begin{gathered}\widetilde{\Phi}(x,t) = \Phi\left(u_x(x,t), u(x,t), x, t\right) = B_1L(u(x,t), x, t)u_x(x,t) \\ + \Psi\left(B_2L(u(x,t), x, t)u_x(x,t), u(x,t), x, t\right)\end{gathered}$$

[6]See, for example, [147].

is continuously differentiable along the characteristic curve $\frac{dx_j}{dt} = \lambda_j$, $(j = 1, 2, ..., q)$, where $\lambda_j = (B_1AB_1')_{jj}$ is an eigenvalue of the matrix B_1AB_1'.

Assume that $\lambda_{k_1} = \lambda_{k_2} = ... = \lambda_{k_s} = \lambda_j$. Since Φ_j does not depend on the components $\mathcal{P}_i$ except i equal to $k_1, k_2, \ldots, k_s$, which are continuously differentiable along the characteristic curve $\frac{dx}{dt} = \lambda_j$, the function $\widetilde{\Phi}_j(x,t)$ is also continuously differentiable along this characteristic curve. The derivative $\frac{d_j\widetilde{\Phi}_j}{dt} = \frac{\partial\widetilde{\Phi}_j}{\partial t} + \lambda_j\frac{\partial\widetilde{\Phi}_j}{\partial x}$ is obtained as follows

$$\frac{d_j\widetilde{\Phi}_j}{dt} = \lim_{\Delta t\to 0}\frac{\widetilde{\Phi}_j(x_j(t+\Delta t),t+\Delta t)-\widetilde{\Phi}_j(x_j(t),t)}{\Delta t} =$$
$$= \lim_{\Delta t\to 0}\left\{\sum_{\mu=1}^{s}\frac{\partial\Phi_j}{\partial\mathcal{P}_{k_\mu}}\left(\frac{\mathcal{P}_{k_\mu}(x_j(t+\Delta t),t+\Delta t)-\mathcal{P}_{k_\mu}(x_j(t),t)}{\Delta t}\right)+\right.$$
$$\left.\frac{\partial\Phi_j}{\partial u}\left(\frac{u(x_j(t+\Delta t),t+\Delta t)-u(x_j(t),t)}{\Delta t}\right)+\frac{\partial\Phi_j}{\partial x}\left(\frac{x_j(t+\Delta t)-x_j(t)}{\Delta t}\right)+\frac{\partial\Phi_j}{\partial t}\right\} =$$
$$= \sum_{\mu=1}^{s}\frac{\partial\Phi_j}{\partial\mathcal{P}_{k_\mu}}\frac{d_j\mathcal{P}_{k_\mu}}{dt} + \lambda_j\left(\frac{\partial\Phi_j}{\partial u}p_{1,0} + \frac{\partial\Phi_j}{\partial x}\right) + \frac{\partial\Phi_j}{\partial u}p_{0,1} + \frac{\partial\Phi_j}{\partial t}.$$

Composing the vector $\frac{d\widetilde{\Phi}}{dt}$ with the coordinates $\frac{d_j\widetilde{\Phi}_j}{dt}$, one obtains

$$\frac{d\widetilde{\Phi}}{dt} = \frac{\partial\Phi}{\partial\mathcal{P}}(\mathcal{P}_t + A\mathcal{P}_x) + B_1AB_1'\left(\frac{\partial\Phi}{\partial u}p_{1,0} + \frac{\partial\Phi}{\partial x}\right) + \frac{\partial\Phi}{\partial u}p_{0,1} + \frac{\partial\Phi}{\partial t}.$$

Using (4.14), the derivative $\frac{d\widetilde{\Phi}}{dt}$ is rewritten

$$\frac{d\widetilde{\Phi}}{dt} = \frac{\partial\Phi}{\partial\mathcal{P}}\{\tfrac{\partial(Lf)}{\partial x} + \tfrac{\partial(Lf)}{\partial u}L^{-1}\mathcal{P} - \tfrac{\partial A}{\partial u}\langle L^{-1}\mathcal{P},\mathcal{P}\rangle - \tfrac{\partial A}{\partial x}\mathcal{P} + \tfrac{\partial L}{\partial u}\langle f - L^{-1}A\mathcal{P}, P\rangle$$
$$+\tfrac{\partial L}{\partial t}L^{-1}\mathcal{P} - \tfrac{\partial L}{\partial u}\langle L^{-1}\mathcal{P}, f - L^{-1}A\mathcal{P}\rangle - \tfrac{\partial L}{\partial x}(f - L^{-1}A\mathcal{P})\}$$
$$+B_1AB_1'\left(\tfrac{\partial\Phi}{\partial u}p_{1,0} + \tfrac{\partial\Phi}{\partial x}\right) + \tfrac{\partial\Phi}{\partial u}p_{0,1} + \tfrac{\partial\Phi}{\partial t}.$$

Differentiation the function $\Phi = B_1\mathcal{P} + \Psi(B_2\mathcal{P}, u, x, t)$ with respect to $\mathcal{P}, u, x, t$ gives:

$$\frac{\partial\Phi}{\partial\mathcal{P}} = Z = B_1 + \Psi_y B_2, \quad \frac{\partial\Phi}{\partial u} = \frac{\partial\Psi}{\partial u}, \quad \frac{\partial\Phi}{\partial x} = \frac{\partial\Psi}{\partial x}, \quad \frac{\partial\Phi}{\partial t} = \frac{\partial\Psi}{\partial t}$$

and substituting the values of these derivatives into $\frac{d\widetilde{\Phi}}{dt}$, one obtains

$$\frac{d\widetilde{\Phi}}{dt} = Z\{\tfrac{\partial(Lf)}{\partial x} + \tfrac{\partial(Lf)}{\partial u}L^{-1}\mathcal{P} - \tfrac{\partial A}{\partial u}\langle L^{-1}\mathcal{P},\mathcal{P}\rangle - \tfrac{\partial A}{\partial x}\mathcal{P} + \tfrac{\partial L}{\partial u}\langle f - L^{-1}A\mathcal{P}, P\rangle +$$
$$+\tfrac{\partial L}{\partial t}L^{-1}\mathcal{P} - \tfrac{\partial L}{\partial u}\langle L^{-1}\mathcal{P}, f - L^{-1}A\mathcal{P}\rangle - \tfrac{\partial L}{\partial x}(f - L^{-1}A\mathcal{P})\}$$
$$+B_1AB_1'\left(\tfrac{\partial\Psi}{\partial u}p_{1,0} + \tfrac{\partial\Psi}{\partial x}\right) + \tfrac{\partial\Psi}{\partial u}p_{0,1} + \tfrac{\partial\Psi}{\partial t}.$$

Since the matrix A is diagonal, one has $B_1AB_1' = ZAB_1'$. Comparing the derivative $\frac{d\widetilde{\Phi}}{dt}$ and the representation of Ω, one has

$$\frac{d\widetilde{\Phi}}{dt} = \Omega.$$

Because of the lemma there are matrices $\Theta_1, \Theta_2 \in C(D)$ such that

$$\frac{d\widetilde{\Phi}}{dt} = -(\Theta_1 S + \Theta_2 \widetilde{\Phi}).$$

As $u(x,t)$ is a solution of system (4.7), one has $S = 0$. Thus, the function $\widetilde{\Phi}(x,t)$ satisfies a homogeneous linear system of partial differential equations

$$\frac{d\widetilde{\Phi}}{dt} + \Theta_2 \widetilde{\Phi} = 0,$$

and the initial conditions $\widetilde{\Phi}(x,0) = 0$. By virtue of the uniqueness of a generalized solution of a linear system [147] the statement of the theorem, namely $\widetilde{\Phi}(x,t) = 0$, is obtained.•

For systems, which are not hyperbolic there is another theorem[7].

Theorem 4.5. *Let the matrix B_2AB_2' in an involutive system (4.7), (4.8) be a diagonal matrix, $B_2AB_1' = 0$, and*

$$L \in C^2(D), \ A \in C^2(D), \ f \in C^2(D), \ \Psi \in C^1(D).$$

Then there exists a unique solution $u(x,t) \in C^2$ of the Cauchy problem of system (4.7), (4.8) with the initial data $u(x,0) = \varphi(x) \in C^2$ satisfying the differential constraints (4.8) at $t = 0$.

3.3 Characteristic curves

Characteristic curves play a significant role in the theory of partial differential equations. Here characteristic curves of the overdetermined system (4.7), (4.8) are studied.

Characteristic curve of the overdetermined system (4.7), (4.8) is defined as follows: a curve $x = x(t)$ along which the derivatives u_x, u_t cannot be found uniquely from the system $(S\Phi)$:

$$S \equiv Lu_t + ALu_x - Lf = 0,$$

$$\Phi = B_1 L u_x + \Psi = 0,$$

and the strip conditions

$$\frac{du}{dt} = \frac{\partial u}{\partial t} + x' \frac{\partial u}{\partial x}.$$

Note that any characteristic curve of the overdetermined system $(S\Phi)$ is a characteristic of the system (S). This gives the equation

$$\det(A - x' E_m) = 0.$$

[7]The proof can be found in [107].

The overdetermined system $(S\Phi)$ with q differential constraints (Φ) has not more than $m-q$ characteristic curves. In fact, the derivative u_t is found uniquely from the system (S). Note that instead the derivatives u_x one can consider the vector Lu_x. The first part B_1Lu_x is uniquely defined by the differential constraints (Φ). For the second part B_2Lu_x there are the equations

$$\begin{aligned} B_1AB_2'B_2Lu_x - (B_1AB_1' - x'E_q)\Psi &= B_1L(f - \tfrac{du}{dt}), \\ (B_2AB_2' - x'E_{m-q})B_2Lu_x - B_2AB_1'\Psi &= B_2L(f - \tfrac{du}{dt}). \end{aligned} \tag{4.15}$$

The rank of the Jacobi matrix of these equations with respect to $y = B_2Lu_x$ is equal to

$$rank\begin{pmatrix} B_1AB_2' - (B_1AB_1' - x'E_q)\Psi_y \\ B_2AB_2' - x'E_{m-q} - B_2AB_1'\Psi_y \end{pmatrix}$$

Because of (4.11),

$$B_1AB_2' - (B_1AB_1' - x'E_q)\Psi_y = -\Psi_y\left(B_2AB_2' - x'E_{m-q} - B_2AB_1'\Psi_y\right). \tag{4.16}$$

This means that the rank of the Jacobi matrix is defined by the rank of the matrix

$$B_2AB_2' - x'E_{m-q} - B_2AB_1'\Psi_y.$$

Therefore, for a characteristic curve

$$rank\left(B_2AB_2' - x'E_{m-q} - B_2AB_1'\Psi_y\right) < m - q,$$

or

$$\det\left(B_2AB_2' - x'E_{m-q} - B_2AB_1'\Psi_y\right) = 0. \tag{4.17}$$

Relations along the characteristic curve follows from equations (4.15), (4.17). For the sake of simplicity these relations are derived for quasilinear differential constraints

$$\Psi = \Psi_y(u, x, t)y + \phi(u, x, t).$$

By virtue of (4.16), equations (4.15) can be rewritten in the form

$$(B_1 + \Psi_yB_2)AB_1'\phi - x'\phi + (B_1 + \Psi_yB_2)L(f - \frac{du}{dt}) = 0, \tag{4.18}$$

$$(B_2AB_2' - x'E_{m-q} - B_2AB_1'\Psi_y)y = B_2AB_1'\phi + B_2L(f - \frac{du}{dt}). \tag{4.19}$$

Because of (4.17) there is linear dependence in the right hand side of (4.19). This linear dependence provides additional relations to (4.18) along the characteristic curve.

Remark 4.6. *If a curve $x = x(t)$ is not a characteristic curve of an overdetermined system $(S\Phi)$, then all derivatives of a solution $u(x, t)$ of this system* *are uniquely defined. Let us obtain these expressions for the case* $B_2AB_1' = 0$. *Because* $x = x(t)$ *is not a characteristic curve of the overdetermined system* $(S\Phi)$, *then there is an inverse matrix* $(B_2AB_2' - x'E_{m-q})^{-1}$. *This means that the values* B_2Lu_x *can be found from the second part of equations (4.15):*

$$B_2Lu_x = (B_2AB_2' - x'E_{m-q})^{-1}B_2L(f - \frac{du}{dt}).$$

The values B_1Lu_x *are found from the differential constraints*

$$B_1Lu_x = -\Psi.$$

Remark 4.7. *The proof of the existence and uniqueness theorem of an overdetermined system* $(S\Phi)$ *is also valid if the initial data are given on any curve, which is not a characteristic curve of the overdetermined system* $(S\Phi)$.

Theorem 4.6. *Let a solution* $u = u(x, t) \in C^1(V)$, $V \subset R^2$ *of a hyperbolic system (4.7) be given. Assume that the characteristic curve* $x = x_j(t)$ *of a system* (S) *is not characteristic of the overdetermined system* $(S\Phi)$, *which satisfies the conditions (4.10), (4.11) and (4.12). If the solution* $u = u(x, t)$ *satisfies the differential constraints* (Φ) *on the curve* $x = x_j(t)$, *then this solution satisfies the same differential constraints in some neighborhood of this curve in domain* V.

The proof of the theorem follows from the property that the curve $x = x_j(t)$ is not a characteristic of the linear system

$$\frac{\partial \widetilde{\Phi}}{\partial t} + B_1AB_1'\frac{\partial \widetilde{\Phi}}{\partial x} + \Theta_2\widetilde{\Phi} = 0.$$

This theorem is a generalization of the theorem in gas dynamics about adjoining the Riemann wave of a motionless state.

There are more detailed theorems about adjoining one solution characterized by differential constraints to another. However, this subject matter lies outside the goals of the present the method of differential constraints.

3.4 Generalized simple waves

In this section one class of solutions, generalizing the class of simple waves is studied.

Let a system of quasilinear differential equations (S) admits $q = m - 1$ quasilinear differential constraints

$$\Phi = B_1Lu_x + \Psi_yB_2Lu_x + \phi = 0, \qquad (4.20)$$

where $\Psi_y = \Psi_y(u, x, t)$ is an $(m - 1) \times m$ matrix, and $\phi = \phi(u, x, t)$. Also assume that

$$B_2AB_1' = 0.$$

A solution satisfying these differential constraints is called a generalized simple wave. A justification for such name follows from the property that a simple wave is described by such differential constraints with $\Psi_y = 0$ and $\phi = 0$.

3.4.1 Compatibility conditions

Conditions (4.11) for the system $(S\Phi)$ to be involutive are

$$\lambda\Psi_y = -B_1 A(B_2' - B_1'\Psi_y), \tag{4.21}$$

and they can be rewritten as

$$A(B_2' - B_1'\Psi_y) = \lambda(B_2' - B_1'\Psi_y), \tag{4.22}$$

where $\lambda = B_2 A B_2'$.

Equations (4.10) become

$$\Omega_1 y^2 + \Omega_2 y + \Omega_3 = 0,$$

where $y = B_2 L u_x$, and

$$\begin{aligned}\Omega_1 = & -Z\tfrac{\partial L}{\partial u}\langle L^{-1}A(B_2' - B_1'\Psi_y), L^{-1}(B_2' - B_1'\Psi_y)\rangle - \tfrac{\partial \Psi_y}{\partial u}L^{-1}A(B_2' - B_1'\Psi_y) - \\ & -Z\tfrac{\partial A}{\partial u}\langle L^{-1}(B_2' - B_1'\Psi_y), (B_2' - B_1'\Psi_y)\rangle + .ZAB_1'\tfrac{\partial \Psi_y}{\partial u}L^{-1}(B_2' - B_1'\Psi_y) + \\ & +Z\tfrac{\partial L}{\partial u}\langle L^{-1}(B_2' - B_1'\Psi_y), L^{-1}A(B_2' - B_1'\Psi_y)\rangle,\end{aligned}$$

$$\begin{aligned}\Omega_2 = & Z\tfrac{\partial L}{\partial u}\langle f, L^{-1}B_2'\rangle - Z\tfrac{\partial L}{\partial u}\langle f, L^{-1}B_1'\Psi_y\rangle + Z\tfrac{\partial L}{\partial u}\langle L^{-1}AB_1'\phi, L^{-1}B_2'\rangle - \\ & -Z\tfrac{\partial L}{\partial u}\langle L^{-1}AB_1'\phi, L^{-1}B_1'\Psi_y\rangle + Z\tfrac{\partial L}{\partial u}\langle L^{-1}AB_2', L^{-1}B_1'\phi\rangle - \\ & -Z\tfrac{\partial L}{\partial u}\langle L^{-1}AB_1'\Psi_y, L^{-1}B_1'\phi\rangle + Z\tfrac{\partial L}{\partial t}L^{-1}B_2' - Z\tfrac{\partial L}{\partial t}L^{-1}B_1'\Psi_y + \\ & +\tfrac{\partial \Psi_y}{\partial u}L^{-1}AB_1'\phi + \tfrac{\partial \Psi_y}{\partial u}f - \tfrac{\partial \phi}{\partial u}L^{-1}AB_2' + \tfrac{\partial \phi}{\partial u}L^{-1}AB_1'\Psi_y + \tfrac{\partial \Psi_y}{\partial t} + \\ & +Z\tfrac{\partial A}{\partial u}\langle L^{-1}B_2', B_1'\phi\rangle - Z\tfrac{\partial A}{\partial u}\langle L^{-1}B_1'\Psi_y, B_1'\phi\rangle + Z\tfrac{\partial A}{\partial u}\langle L^{-1}B_1'\phi, B_2'\rangle - \\ & -Z\tfrac{\partial A}{\partial u}\langle L^{-1}B_1'\phi, B_1'\Psi_y\rangle - Z\tfrac{\partial A}{\partial x}B_2' + Z\tfrac{\partial A}{\partial x}B_1'\Psi_y - ZAB_1'\tfrac{\partial \Psi_y}{\partial u}L^{-1}B_1'\phi + \\ & +ZAB_1'\tfrac{\partial \phi}{\partial u}L^{-1}B_2' - ZAB_1'\tfrac{\partial \phi}{\partial u}L^{-1}B_1'\Psi_y + ZAB_1'\tfrac{\partial \Psi_y}{\partial x} - \\ & -Z\tfrac{\partial L}{\partial u}\langle L^{-1}B_2', L^{-1}AB_1'\phi\rangle + Z\tfrac{\partial L}{\partial u}\langle L^{-1}B_1'\Psi_y, L^{-1}AB_1'\phi\rangle - \\ & -Z\tfrac{\partial L}{\partial u}\langle L^{-1}B_1'\phi, L^{-1}AB_2'\rangle + Z\tfrac{\partial L}{\partial u}\langle L^{-1}B_1'\phi, L^{-1}AB_1'\Psi_y\rangle + \\ & +Z\tfrac{\partial L}{\partial x}L^{-1}AB_2' - Z\tfrac{\partial L}{\partial x}L^{-1}AB_1'\Psi_y + ZL\tfrac{\partial f}{\partial u}L^{-1}B_2' - ZL\tfrac{\partial f}{\partial u}L^{-1}B_1'\Psi_y\end{aligned}$$

$$\begin{aligned}\Omega_3 = & -Z\tfrac{\partial L}{\partial u}\langle L^{-1}AB_1'\phi, L^{-1}B_1'\phi\rangle - Z\tfrac{\partial L}{\partial u}\langle f, L^{-1}B_1'\phi\rangle - \\ & -Z\tfrac{\partial L}{\partial t}L^{-1}B_1'\phi + \tfrac{\partial \phi}{\partial u}f + \tfrac{\partial \phi}{\partial u}L^{-1}AB_1'\phi + \tfrac{\partial \phi}{\partial t} - Z\tfrac{\partial A}{\partial u}\langle L^{-1}B_1'\phi, B_1'\phi\rangle + \\ & +Z\tfrac{\partial A}{\partial x}B_1'\phi - ZAB_1'\tfrac{\partial \phi}{\partial u}L^{-1}B_1'\phi + ZAB_1'\tfrac{\partial \phi}{\partial x} + \\ & +Z\tfrac{\partial L}{\partial u}\langle L^{-1}B_1'\phi, L^{-1}AB_1'\phi\rangle - Z\tfrac{\partial L}{\partial x}L^{-1}AB_1'\phi - ZL\tfrac{\partial f}{\partial u}L^{-1}B_1'\phi + ZL\tfrac{\partial f}{\partial x}.\end{aligned}$$

Because Ω_1, Ω_2 and Ω_3 do not depend on y, the conditions of involutiveness require that

$$\Omega_1 = 0,\ \Omega_2 = 0,\ \Omega_3 = 0. \tag{4.23}$$

Let us simplify the expressions of $\Omega_1, \Omega_2, \Omega_3$. By using (4.22) the expression for Ω_1 can be rewritten as

$$\begin{gathered}\Omega_1 = -\lambda\tfrac{\partial \Psi_y}{\partial u}L^{-1}(B_2' - B_1'\Psi_y) - B_1\tfrac{\partial A}{\partial u}\langle L^{-1}(B_2' - B_1'\Psi_y), B_2'\rangle + \\ +B_1\tfrac{\partial A}{\partial u}\langle L^{-1}(B_2' - B_1'\Psi_y), B_1'\Psi_y\rangle - \Psi_y\left(\tfrac{\partial \lambda}{\partial u}L^{-1}(B_2' - B_1'\Psi_y)\right) + \\ B_1AB_1'\tfrac{\partial \Psi_y}{\partial u}L^{-1}(B_2' - B_1'\Psi_y).\end{gathered}$$

Differentiating (4.21) with respect to u gives

$$\Psi_y\left(\frac{\partial \lambda}{\partial u}\xi\right) + \lambda\frac{\partial \Psi_y}{\partial u}\xi = B_1\frac{\partial A}{\partial u}\langle \xi, B_1'\Psi_y\rangle + B_1AB_1'\frac{\partial \Psi_y}{\partial u}\xi - B_1\frac{\partial A}{\partial u}\langle \xi, B_2'\rangle,$$

where ξ is an arbitrary m–dimensional vector–cclumn. Substituting $\xi = L^{-1}$ $(B_2' - B_1'\Psi_y)$ and comparing with the value of Ω_1, one obtains

$$\Omega_1 \equiv 0.$$

By the same way the underlined terms in Ω_2 can be simplified

$$\begin{gathered}Z\tfrac{\partial A}{\partial u}\langle L^{-1}B_1'\phi, B_2'\rangle - Z\tfrac{\partial A}{\partial u}\langle L^{-1}B_1'\phi, B_1'\Psi_y\rangle - Z\tfrac{\partial A}{\partial x}B_2' + Z\tfrac{\partial A}{\partial x}B_1'\Psi_y - \\ -ZAB_1'\tfrac{\partial \Psi_y}{\partial u}L^{-1}B_1'\phi + ZAB_1'\tfrac{\partial \Psi_y}{\partial x} = -\lambda\tfrac{\partial \Psi_y}{\partial u}L^{-1}B_1'\phi + \lambda\tfrac{\partial \Psi_y}{\partial x}.\end{gathered}$$

Remark 4.8. *In continuum mechanics often*

$$\frac{\partial A}{\partial x} = 0,\ \ \frac{\partial L}{\partial t} = \frac{\partial L}{\partial x} = 0,\ \ \frac{\partial f}{\partial x} = 0,\ \ \Psi_y = 0.$$

In this case, conditions (4.21) become $B_1AB_2' = 0$, *and taking this into account, the expressions for* Ω_2 *and* Ω_3 *become*

$$\begin{gathered}\Omega_2 = B_1\tfrac{\partial L}{\partial u}\langle f, L^{-1}B_2'\rangle + B_1\tfrac{\partial L}{\partial u}\langle L^{-1}AB_1'\phi, L^{-1}B_2'\rangle \\ +\lambda B_1\tfrac{\partial L}{\partial u}\langle L^{-1}B_2', L^{-1}B_1'\phi\rangle + (B_1AB_1' - \lambda E_q)\tfrac{\partial \phi}{\partial u}L^{-1}B_2' \\ +B_1\tfrac{\partial A}{\partial u}\langle L^{-1}B_2', B_1'\phi\rangle \\ -B_1\tfrac{\partial L}{\partial u}\langle L^{-1}B_2', L^{-1}AB_1'\phi\rangle - \lambda B_1\tfrac{\partial L}{\partial u}\langle L^{-1}B_1'\phi, L^{-1}B_2'\rangle \\ +B_1L\tfrac{\partial f}{\partial u}L^{-1}B_2'\end{gathered}$$

$$\begin{gathered}\Omega_3 = -B_1\tfrac{\partial L}{\partial u}\langle L^{-1}AB_1'\phi, L^{-1}B_1'\phi\rangle - B_1\tfrac{\partial L}{\partial u}\langle f, L^{-1}B_1'\phi\rangle \\ +\tfrac{\partial \phi}{\partial u}f + \tfrac{\partial \phi}{\partial u}L^{-1}AB_1'\phi + \tfrac{\partial \phi}{\partial t} - B_1\tfrac{\partial A}{\partial u}\langle L^{-1}B_1'\phi, B_1'\phi\rangle \\ -B_1AB_1'\tfrac{\partial \phi}{\partial u}L^{-1}B_1'\phi + B_1AB_1'\tfrac{\partial \phi}{\partial x} + \\ +B_1\tfrac{\partial L}{\partial u}\langle L^{-1}B_1'\phi, L^{-1}AB_1'\phi\rangle - B_1L\tfrac{\partial f}{\partial u}L^{-1}B_1'\phi.\end{gathered}$$

For example, if $\phi = 0$, then $\Omega_3 \equiv 0$ and the expression of Ω_2 is reduced to

$$\Omega_2 = B_1 \frac{\partial L}{\partial u} \langle f, L^{-1} B_2' \rangle + B_1 L \frac{\partial f}{\partial u} L^{-1} B_2'.$$

In [103] the method of differential constraints is applied to two models from continuum mechanics, written in terms of Riemann invariants: namely traffic flow, and rate-type models.

3.4.2 Integration method

A generalized simple wave satisfies the system of ordinary differential equations along the curve $\frac{dx}{dt} = \lambda$:

$$\begin{aligned} B_1 L \tfrac{du}{dt} &= B_1 (A - \lambda E_m) B_1' \phi + B_1 L f, \\ B_2 L \tfrac{du}{dt} &= B_2 L f, \ \tfrac{dx}{dt} = \lambda, \end{aligned} \tag{4.24}$$

This system of ordinary differential equations is a system of the relations along the characteristic curve $x' = \lambda$ of the overdetermined system $(S\Phi)$.

Equations (4.24) gives an idea of how to use the method of characteristics for constructing a solution of a Cauchy problem for the overdetermined system $(S\Phi)$. Let $u_0(x) \in C^1$ satisfy the differential constraints

$$(B_1 + \Psi_y(u_0(a), a, 0) B_2) L(u_0(a), a, 0) u_0'(a) + \phi(u_0(a), a, 0) = 0.$$

There exists a unique solution $(v(a,t), x(a,t))$ of the Cauchy problem of the system of ordinary differential equations (4.24) with the initial data at $t = 0$:

$$v = u_0(a), \ x = a.$$

The dependence $x = x(a,t)$ can be solved with respect to $a = a(x,t)$ in some neighborhood V of a point $(x_0, 0) \in V$. Let us prove that $u(x,t) = v(a(x,t), t)$ is a solution of the overdetermined system $(S\Phi)$ in V.

Exchanging the variables (x,t) with (a,t), one has

$$\frac{\partial}{\partial t} = \frac{\partial}{\partial t} + \lambda \frac{\partial}{\partial t}, \ \frac{\partial}{\partial a} = x_a \frac{\partial}{\partial x},$$

where $x_a = \frac{\partial x}{\partial a}$. In this case the left hand side of the differential constraints can be rewritten as

$$B_1 L u_x + \Psi_y B_2 L u_x + \phi = \frac{1}{x_a} (B_1 L v_a + \Psi_y B_2 L v_a + x_a \phi).$$

Denoting by $\frac{d}{dt}$ the partial derivative $\frac{\partial}{\partial t}$ in the variables (a,t), and setting $H = (B_1 + \Psi_y B_2) L v_a + x_a \phi$, one obtains

$$\begin{aligned} \tfrac{dH}{dt} &= Z L \tfrac{dv_a}{dt} + Z \tfrac{\partial L}{\partial u} \langle \tfrac{dv}{dt}, v_a \rangle + \lambda Z \tfrac{\partial L}{\partial x} v_a + Z \tfrac{\partial L}{\partial t} v_a + \tfrac{\partial \Psi_y}{\partial u} \tfrac{dv}{dt} B_2 L v_a \\ &+ \lambda \tfrac{\partial \Psi_y}{\partial x} B_2 L v_a + \tfrac{\partial \Psi_y}{\partial t} B_2 L v_a + \tfrac{dx_a}{dt} \phi + x_a \tfrac{\partial \phi}{\partial u} \tfrac{dv}{dt} + \lambda x_a \tfrac{\partial \phi}{\partial x} + x_a \tfrac{\partial \phi}{\partial t}. \end{aligned}$$

Differentiating (4.24) with respect to a gives:

$$\begin{aligned}&\tfrac{\partial}{\partial a}(B_1(A-\lambda E_m)B_1'\phi+B_1Lf)=\tfrac{\partial}{\partial a}(B_1L\tfrac{dv}{dt})\\&=B_1L\tfrac{dv_a}{dt}+B_1\tfrac{\partial L}{\partial u}\langle v_a,\tfrac{dv}{dt}\rangle+x_aB_1\tfrac{\partial L}{\partial x}\tfrac{dv}{dt},\ \tfrac{\partial}{\partial a}(B_2Lf)\\&=\tfrac{\partial}{\partial a}(B_2L\tfrac{dv}{dt})=B_2L\tfrac{dv_a}{dt}+B_2\tfrac{\partial L}{\partial u}\langle v_a,\tfrac{dv}{dt}\rangle+x_aB_2\tfrac{\partial L}{\partial x}\tfrac{dv}{dt},\end{aligned}$$

From the last equalities one can find

$$\begin{aligned}ZL\tfrac{dv_a}{dt}&=ZL\tfrac{\partial f}{\partial u}v_a+x_aZL\tfrac{\partial f}{\partial x}-\tfrac{\partial\lambda}{\partial a}\phi-\lambda\tfrac{\partial\phi}{\partial x}v_a-x_a\lambda\tfrac{\partial\phi}{\partial x}+B_1\tfrac{\partial A}{\partial u}\langle v_a,B_1'\phi\rangle+\\&+x_aB_1\tfrac{\partial A}{\partial x}B_1'\phi+B_1AB_1'\tfrac{\partial\phi}{\partial u}v_a+x_aB_1AB_1'\tfrac{\partial\phi}{\partial x}+Z\tfrac{\partial L}{\partial u}\langle v_a,L^{-1}(\lambda E_m-A)B_1'\phi\rangle\\&+x_aZ\tfrac{\partial L}{\partial x}L^{-1}(\lambda E_m-A)B_1'\phi.\end{aligned}\tag{4.25}$$

If one set $y=B_2Lv_a$, then

$$v_a=L^{-1}B_2'y+L^{-1}B_1'(H-\Psi_yy-x_a\phi)$$

Substituting the values of v_a, $\frac{dv}{dt}$ and $ZL\frac{dv_a}{dt}$ into the expression of $\frac{dH}{dt}$, and using the involutive conditions (4.23), one finds that $H=H(a,t)$ satisfies the linear system of differential equations

$$\frac{dH}{dt}=GH$$

with some matrix $G=G(a,t)$. Because of uniqueness of the solution of the Cauchy problem, and since $H(a,0)=0$, one has

$$H(a,t)=0.$$

This means that the differential constraints are satisfied in V.

Rewriting equations (4.24) in the variables (x,t), one finds that

$$\begin{aligned}B_1L(u_t+\lambda u_x)&=B_1(A-\lambda E_m)B_1'\phi+B_1Lf,\\B_2L(u_t+\lambda u_x)&=B_2Lf.\end{aligned}$$

Substituting $\phi=-(B_1+\Psi_yB_2)Lu_x$ into the last equations, using $\lambda=B_2AB_2'$, $B_2AB_1'=0$, and the involutive conditions (4.21), one has

$$\begin{aligned}B_1(Lu_t+ALu_x)&=B_1Lf,\\B_2(Lu_t+ALu_x)&=B_2Lf.\end{aligned}$$

Thus, the system (S) is also satisfied in V.

3.4.3 Centered rarefaction waves

Here the method of differential constraints is applied to a problem where initial data are given on a characteristic curve. This problem plays a key role in the problem of the decay of an arbitrary discontinuity in continuum mechanics.

Let us consider a system (S), which admits solutions of generalized simple wave type, characterized by (4.20). The problem is to find a solution of the system (S), which takes the values $u(x_0(t), t) = u_\lambda(t) \in C^1$ on the characteristic curve $x = x_0(t)$ of the overdetermined system $(S\Phi)$. Here $x_0' = \lambda \equiv B_2 A B_2'$, $x_0(0) = 0$ and the values $x_0(t)$ and $u_\lambda(t)$ satisfy the relations along this characteristic. The existence of such a solution of system (S), satisfying these conditions can be established in the following way[8]. In a neighborhood of the point $x = 0$ there is a differentiable function $\varphi(x)$, which satisfies the differential constraints (Φ), and $\varphi(0) = u_\lambda(0)$. According to the previous constructions there exists a solution of the Cauchy problem of the overdetermined system $(S\Phi)$. This solution is obtained by integrating the system of ordinary differential equations (4.24) with the initial values $\varphi(x)$. Because of the uniqueness of a solution of the Cauchy problem of system of ordinary differential equations, the characteristic curve passing through the point $(0, 0)$ coincides with the curve $x = x_0(t)$ and $u(x_0(t), t) = u_\lambda(t)$.

Similarly, one can construct a solution of a problem with the initial data on a characteristic curve of the overdetermined system $(S\Phi)$ and with a singularity of the centered rarefaction wave at the point $(0, 0)$.

There exists unique solution of the system (S) in some domain $V \in R^2$ that satisfies the following conditions.

1. On the characteristic curve $\Pi : x = x_0(t)$ the value $u(x_0(t), t) = u_\lambda(t)$ satisfy (4.24).

2. The point $(0, 0) \in \Pi \subset V$ is singular: the solution is multiply defined at this point. The value $u = u_0(a)$ of the solution at this point depends on the parameter a, $(u_0(0) = u_\lambda(0))$ and it defines the curve in the space R^m satisfying the equations

$$\begin{gathered} (B_1 + \Psi_y(u_0(a), 0, 0) B_2) L(u_0(a), 0, 0) u_0'(a) = 0, \\ \frac{\partial \lambda}{\partial u}(u_0(a), 0, 0) u_0'(a) < 0, \ \ (0 \le a \le a_0). \end{gathered} \tag{4.26}$$

The solution of this problem generalizes the well-known centered rarefaction wave in gas dynamics: equations (4.26) define an analogue of the (p, u)-diagram.

4. Generalized simple waves in gas dynamics

This section is devoted to generalized simple waves for one-dimensional unsteady and two-dimensional steady gas dynamics equations. For an isentropic flow these equations can be reduced to equations written in terms of

[8] If there exists a solution of a hyperbolic system (S) satisfying the differential constraints (Φ) on the characteristic curve $x = x_0(t)$, where λ is not an eigenvalue of the matrix $B_1 A B_1'$, then the last theorem of the previous section guarantees that this solution satisfies the differential constraints (Φ) in a neighborhood of the $x = x_0(t)$.

Riemann invariants. A hyperbolic and homogeneous system written in terms of Riemann invariants has simple wave solutions, which are also called Riemann waves. For nonisentropic flows there are no Riemann invariants. In this case generalized simple waves could substitute the Riemann waves.

Note that some applications of the method of differential constraints to gas dynamics equations one can be find in [147, 160, 116], and in the references given therein.

4.1 One-dimensional gas dynamics equations

An unsteady one-dimensional flow of a gas is described by the equations

$$\begin{aligned} &u_t + uu_x + \rho^{-1}p_x = 0, \\ &\rho_t + u\rho_x + \rho u_x = 0, \\ &p_t + up_x + A(\rho, p)u_x = 0. \end{aligned} \tag{4.27}$$

Here ρ is the density, u is the velocity, p is the pressure, η is the entropy, and c is the sound speed ($c^2 = A/\rho$). For a polytropic gas $A = \gamma p,\ \gamma > 1$, and $\eta = g(p\rho^{-\gamma})$. Without loss of generality one can set $\eta = p\rho^{-\gamma}$.

System (4.27) can be rewritten in the matrix form

$$S \equiv L\mathbf{u}_t + \Lambda L\mathbf{u}_x = 0,$$

with

$$\mathbf{u} = \begin{pmatrix} u \\ \rho \\ p \end{pmatrix},\quad L = \begin{pmatrix} 0 & -A & \rho \\ \rho c & 0 & 1 \\ -\rho c & 0 & 1 \end{pmatrix},\quad \Lambda = \begin{pmatrix} u & 0 & 0 \\ 0 & u+c & 0 \\ 0 & 0 & u-c \end{pmatrix}.$$

Since system (4.27) is strictly hyperbolic the differential constraints of first order for it must be quasilinear. The well-known Riemann waves (or simple waves) are obtained by assuming that $u = u(\rho),\quad p = p(\rho)$. It can be shown that the Riemann waves belong to the class of solutions, which is characterized by the following differential constraints

$$p_x - \alpha^2\rho_x = 0,\quad p_x + \rho\alpha u_x = 0,$$

where $\alpha = \pm c$. Here the matrix $B_2 = (0, 1, 0)^t$ for $\alpha = c$, and $B_2 = (0, 0, 1)^t$ for $\alpha = -c$. The first differential constraint leads to the property that the entropy in the Riemann waves is constant. It is more convenient to rewrite the second differential constraint in the form

$$u_x + \rho^{-1}\alpha\rho_x = 0.$$

Let us study the more general class of solutions characterized by the differential constraints

$$p_x - \alpha^2\rho_x = \psi,\quad u_x + \rho^{-1}\alpha\rho_x = \phi, \tag{4.28}$$

where $\psi = \psi(t, x, u, \rho, p)$ and $\phi = \phi(t, x, u, \rho, p)$. The involutive conditions $\Omega_2 = 0,\ \Omega_3 = 0$ for this class of solutions are

$$-\psi_p \gamma p - \psi_\rho \rho + \psi_u \alpha + \psi(\gamma + 1) = 0, \tag{4.29}$$

$$\psi_t + u\psi_x - \psi_u(\psi\rho^{-1} + \alpha\phi) = 0, \tag{4.30}$$

$$-4\phi_p \alpha\gamma p\rho - 4\phi_\rho \alpha\rho^2 + 4\phi_u \gamma p - 3\psi\gamma + \phi\alpha\rho(3-\gamma) = 0, \tag{4.31}$$

$$\begin{aligned} &-\psi_p\psi - \psi_u\phi - \psi_x - \phi_p\psi\alpha\rho + \phi_p\phi\gamma p\rho + \phi_\rho\phi\rho^2 - \\ &\phi_t\rho + \phi_u\psi - \phi_u\phi\alpha\rho - \phi_x\rho(\alpha + u) - \phi^2\rho = 0. \end{aligned} \tag{4.32}$$

If $\psi = 0$ (an isentropic flow), then ϕ cannot equal zero, but only for $\gamma = 3$ and $\gamma = 5/3$ (a mon-atomic gas). However, for an isentropic flow the one-dimensional gas dynamic equations are transformed to the Darboux equation [9]. For $\gamma = 3$ or $\gamma = 5/3$ the general solution of the Darboux equation is expressed through the D'Alambert solution [147]. Further the case $\psi \neq 0$ is studied.

The general solution of equation (4.29) is

$$\psi = \rho^{\gamma+1}\Psi(t, x, \xi, \eta)$$

with some function $\Psi(t, x, \xi, \eta)$, where $\xi = u + \frac{2\alpha}{\gamma-1},\ \eta = p\rho^{-\gamma}$. Substituting Ψ into equation (4.31) one finds the general solution of this equation to be:

$$\phi = \rho^{(3-\gamma)/4}\Phi(t, x, \xi, \eta) - \frac{3\alpha\rho^{\gamma+1}}{p(3\gamma - 1)}\Psi(t, x, \xi, \eta),$$

where the function $\Phi(t, x, \xi, \eta)$ is an arbitrary function of its arguments. After substituting the representations of ψ and ϕ into (4.30), one obtains

$$\Psi_t + (\xi - \frac{2\alpha}{\gamma - 1})\Psi_x + \left(\rho^\gamma \frac{1}{3\gamma - 1}\Psi - \alpha\rho^{(3-\gamma)/4}\Phi\right)\Psi_\xi = 0. \tag{4.33}$$

Splitting this equation with respect to ρ, where it is essentially used that $\gamma > 1$, one has

$$\Psi_t = 0,\ \Psi_x = 0,\ \Psi_\xi = 0.$$

After substituting the representations of ψ and ϕ into (4.32), the equation becomes

$$a_1\rho_1^5 + a_2\rho_1^{4+3\gamma} + a_3\rho_1^{3+6\gamma} + a_4\rho_1^{2+5\gamma} + a_5\rho_1^{2+\gamma} + a_6\rho_1^{3\gamma} = 0, \tag{4.34}$$

where $\rho_1 = \rho^{1/4}$, the coefficients a_i, $(i = 1, 2, ..., 6)$ are expressed through the functions Φ and Ψ, and their derivatives[10]. Hence, they only depend on

[9] See, for example, [147].

[10] Since expressions of the coefficients are cumbersome, they are not presented here. All calculations are done in REDUCE [69].

(t, x, ξ, η). Analysis of the linear functions (powers of ρ_1) gives that for $\gamma > 1$, the degrees $4 + 3\gamma$, $3 + 6\gamma$ and $2 + 5\gamma$ have different values and they differ from the degrees $5,\ 2+\gamma$, and 3γ. Thus, splitting equation (4.34) with respect to ρ gives $a_2 = 0,\ a_3 = 0,\ a_4 = 0$. The last equalities lead to the equations

$$\eta\Psi_\eta = \frac{3\gamma}{3\gamma - 1}\Psi,\ \eta\Phi_\eta = \frac{3(\gamma - 3)}{3\gamma - 1}\Phi,\ \Phi_\xi = 0. \tag{4.35}$$

Because of $\Phi_\xi = 0$, equation (4.34) can also be split with respect to ξ, giving

$$\Phi_x = 0. \tag{4.36}$$

The general solution of (4.35), (4.36) is

$$\Psi = k\eta^\beta,\ \Phi = h(t)\eta^q,$$

where $\beta = \dfrac{3\gamma}{3\gamma - 1}$, $q = \dfrac{3(\gamma - 3)}{3\gamma - 1}$, and k is constant. After that, equation (4.34) is reduced to the equation

$$\rho_1^{3-\gamma}(\gamma + 1)\eta^q h^2 + 4h' = 0.$$

If $\gamma \neq 3$, then further splitting of this equation with respect to ρ gives $h = 0$. If $\gamma = 3$, then $q = 0$ and $h = (t + k_1)^{-1}$, where k_1 is constant. The constant k_1 is not essential, because of shifting with respect to time.

Theorem 4.7. *The general solution of the involutive conditions $\Omega_2 = 0$ and $\Omega_3 = 0$ for generalized simple waves (4.28) (for nonisentropic flows $\psi \neq 0$ and arbitrary polytropic exponent $\gamma > 1$) is*

$$\psi = k\rho^{\beta_1} p^\beta,\ \phi = -\frac{3\gamma}{(3\gamma - 1)\alpha\rho}\psi, \tag{4.37}$$

$$\left(\beta_1 = 1 - \frac{\gamma}{(3\gamma - 1)},\quad \beta = 1 + \frac{1}{(3\gamma - 1)},\quad k \neq 0\right).$$

Remark 4.9. *It can be shown that the differential constraints*

$$p_x = \varphi_1,\ u_{xx} + \rho^{-1}\alpha\rho_{xx} = \varphi_2, \tag{4.38}$$

where the functions φ_1 and φ_2 depend on $t, x, u, \rho, p, u_x, \rho_x$, are admitted by the one-dimensional gas dynamics equations (4.27) only if the functions φ_1 and φ_2 are as follows

$$\begin{aligned}\varphi_1 &= \tfrac{p}{3\rho}\rho_x + \rho\alpha\tfrac{1-3\gamma}{3\gamma}u_x,\\ \varphi_2 &= -\alpha\rho^{-1}\rho_{xx} + \tfrac{5}{3}\alpha\rho^{-2}\rho_x^2 - \tfrac{3\gamma+1}{6}\alpha^{-1}u_x^2.\end{aligned} \tag{4.39}$$

Integration of these differential constraints (once) leads to the differential constraints (4.28) with (4.37). The constant k is a constant of integration.

Similar to simple waves, further analysis of generalized simple waves includes integration along characteristics and the construction of a centered rarefaction wave.

A generalized simple wave satisfies the system of ordinary differential equations (4.24) along the characteristics

$$\frac{dx}{dt} = u - \alpha, \tag{4.40}$$

which have the representations

$$\frac{dp}{dt} = \frac{\alpha\psi}{(3\gamma - 1)}, \ \frac{du}{dt} = \frac{\psi}{\rho(3\gamma - 1)}, \ \frac{d\rho}{dt} = \frac{3\gamma}{(3\gamma - 1)}\frac{\psi}{\alpha}. \tag{4.41}$$

Since

$$\frac{d}{dt}(\alpha + \gamma u) = 0, \ \ \frac{d}{dt}(p\rho^{-1/3}) = 0,$$

one finds that along characteristics $p = c_1\rho^{1/3}$, and $u = -\gamma^{-1}\alpha + c_2$, where the constants c_1 and c_2 depend on the characteristic curve. The dependence of x on ρ along characteristic is determined by the equation

$$\frac{dx}{d\rho} = \frac{(3\gamma - 1)}{3\gamma}\frac{\alpha(u - \alpha)}{\psi},$$

which can be integrated explicitly after substituting the expressions for α, ψ, u, and p through ρ. The dependence ρ on time t is found by integrating the last equation in (4.41), namely

$$\frac{d\rho}{dt} = \frac{3\gamma_1}{(3\gamma - 1)}kc_1{}^{\beta - 1/2}\rho^q, \tag{4.42}$$

where $\gamma_1 = \pm\sqrt{\gamma}$ and $q = (\beta + 1)/3 + \beta_1$. The sign of γ_1 is determined by the sign of α: $\alpha = \gamma_1 p^{1/2}\rho^{-1/2}$.

The constants of integration are defined by the initial values at $t = 0$, namely by:

$$u_o(\xi) = u(0, \xi), \ \rho_o(\xi) = \rho(0, \xi), \ p_o(\xi) = p(0, \xi). \tag{4.43}$$

The functions $u_o(\xi)$, $\rho_o(\xi)$, $p_o(\xi)$ have to satisfy the differential constraints (4.28) using the functions (4.37). Note that in the initial data one can choose one arbitrary function, since the other functions are defined by the system of ordinary differential equations.

According to the existence theorem, there exists unique local solution for $t > 0$ of the overdetermined system (4.27), (4.28) with the initial data (4.43). This solution will exists up to the occurrence of a gradient catastrophe. The occurrence of the intersection of characteristics requires special analysis.

A generalized simple waves can be used to obtain a nonisentropic centered rarefaction wave. These solutions are constructed by integrating (4.40), (4.41) with singular initial data, which satisfy the equations

$$p_a - \alpha^2 \rho_a = 0, \quad \rho u_a + \alpha \rho_a = 0.$$

These equations are equations (4.26) for the overdetermined system (4.27), (4.28), (4.37). In gas dynamics the solution of these equations is called (p, u)-diagram. Note that the (p, u)-diagram for the nonisentropic case is the same as for isentropic centered rarefaction waves.

4.2 Two-dimensional gas dynamic equations

The system of equations describing a steady plane gas flow is:

$$\begin{array}{ll} u\tau_x + v\tau_y - \tau u_x - \tau v_y = 0, & uu_x + vu_y + \tau p_x = 0, \\ uv_x + vv_y + \tau p_y = 0, & up_x + vp_y + A(u_x + v_y) = 0. \end{array} \tag{4.44}$$

Here (u, v) are coordinates of the velocity vector along the x and y axes, p is the pressure, $\tau = 1/\rho$ is the specific volume, c is the sound speed, $A = \gamma p = c^2/\tau$. Note that if the flow of a gas is supersonic $q^2 - c^2 \geq 0$, then system (4.44) is hyperbolic. To write system of equations (4.44) in the form of system (4.7)

$$L\mathbf{u}_x + \Lambda L\mathbf{u}_y = 0$$

the following matrices are used

$$L = \begin{pmatrix} \gamma p & 0 & 0 & \tau \\ 0 & u & v & \tau \\ 0 & -v & u & 0 \\ 0 & 0 & 0 & 1 \end{pmatrix}, \quad \Lambda = \begin{pmatrix} v/u & 0 & 0 & 0 \\ 0 & v/u & 0 & 0 \\ 0 & 0 & uv/h & \tau(q^2 - c^2)/h \\ 0 & 0 & \gamma p/h & uv/h \end{pmatrix},$$

where the bold symbol $\mathbf{u}$ is the vector $\mathbf{u} = (\tau, u, v, p)^t$, $h \equiv u^2 - c^2$, $q^2 = u^2 + v^2$.

Simple waves, which are called Prandtl-Meyer flows, belong to the solutions defined by the differential constraints

$$\Phi \equiv B_1 L\mathbf{u}_y + \Psi(B_2 L\mathbf{u}_y, \mathbf{u}, y, x) = 0. \tag{4.45}$$

with

$$B_2 = \begin{pmatrix} 0 & 0 & 1 & 0 \end{pmatrix} \; or \; B_2 = \begin{pmatrix} 0 & 0 & 0 & 1 \end{pmatrix}. \tag{4.46}$$

In what follows the general case of differential constraints (4.45) with the matrix B_2 defined by (4.46) will be studied.

Analysis of the conditions (4.11) shows that the differential constraints (4.45) exist only for supersonic flows, where the gas dynamic equations are hyperbolic. Hence, the differential constraints are quasilinear. The general solution of the involutive conditions $\Omega_2 = 0,\ \Omega_3 = 0$ defines the following differential constraints admitted by the gas dynamics equations:

$$-vu_y + uv_y + \alpha p_y = 0,\ \gamma p\tau_y + \tau p_y = (\gamma - 1)u\tau^{-\gamma} H(I, \eta),$$
$$uu_y + vv_y + \tau p_y = u[(\gamma - 1)I/(\gamma\eta\tau) - \tau^{-\gamma}]H(I, \eta),$$

where $\alpha^2 = \tau(q^2 - c^2)/(\gamma p)$, $I = q^2/2 + \frac{\gamma p\tau}{\gamma-1}$, $\eta = p\tau^\gamma$, and the function H is an arbitrary function of its arguments. In this case the relations along the characteristic curve (4.24) are

$$\frac{dy}{dx} = \frac{uv - A\alpha}{u^2 - c^2},\ \frac{d\tau}{dx} = \frac{(\gamma - 1)(\alpha u - \tau v)}{(u^2 - c^2)\tau^\gamma} H,\ \frac{dp}{dx} = 0,$$
$$\frac{du}{dx} = \frac{(\gamma - 1)(\alpha u - \tau v)}{2(u^2 - c^2)\tau^{\gamma+1}} uH,\ \frac{dv}{dx} = \frac{(\gamma - 1)(\alpha u - \tau v)}{2(u^2 - c^2)\tau^{\gamma+1}} vH.$$

Along the characteristic curves the following values

$$\frac{v}{u},\ \frac{u^2}{\tau},\ p,\ \frac{uv - A\alpha}{u^2 - c^2}$$

are invariant. Due to the last invariant, the characteristic curves are straight lines.

For centered generalized simple waves the initial data satisfy the conditions

$$\gamma p\tau_a + \tau p_a = 0,\ uu_a + vv_a + \tau p_a = 0,$$
$$-vu_a + uv_a + \alpha p_a = 0.$$

These relations are the same as the relations at a corner in Prandtl–Meyer flow past a convex corner.

4.3 Example of differential constraint of higher order

One example of differential constraints of order greater than one was given in the previous section. Here another example for one-dimensional gas dynamics is given. An isentropic flow of an ideal gas with plane symmetry is described by the equations

$$u_t^1 + \lambda_1 u_x^1 = 0,\ \ u_t^2 + \lambda_2 u_x^2 = 0, \tag{4.47}$$

where $u^1,\ u^2$ are the Riemann invariants, $\lambda_i = (u^1 + u^2)/2 - (-1)^i\psi$, $(i = 1, 2)$, $\psi = \psi(u^1 - u^2)$. The differential constraints

$$\Phi_i = \mu^i u_x^i + \frac{u_{xx}^i}{(u_x^i)^2} + (-1)^i \frac{(u_x^1 - u_x^2)^2}{u_x^1 u_x^2} \frac{2\psi' - 1}{4\psi} = 0,\ (i = 1, 2),$$

are admitted by system (4.47). Here the functions $\mu^i = \mu^i(u^1, u^2)$, $(i = 1, 2)$ satisfy the equations

$$\mu^2 = \mu^1 + \frac{4\psi}{2\psi' - 1}\frac{\partial\mu^1}{\partial u^2},\ \frac{\partial\mu^1}{\partial u^2} - \frac{\partial\mu^2}{\partial u^1} = 0.$$

These differential constraints were obtained [153, 108] as a Lie-Bäcklund generator $X = \Phi_1 \partial_{u^1} + \Phi_2 \partial_{u^2}$ admitted by system (4.47).

For system (4.47) it can also be shown that the method of differential constraints gives a wider class of solutions than the approach using Lie-Bäcklund transformations. For example, the differential constraints

$$u_x^1 = \Phi(u^1 - u^2), \quad u_x^2 = \Phi(u^1 - u^2),$$

where $2\psi\Phi' - \Phi = 0$, cannot be a source of any generator of the contact group of transformations, admitted by system (4.47).

5. Systems of quasilinear differential equations with more than two independent variables

This section is devoted to applications of the method of differential constraints to system of quasilinear differential equations

$$S \equiv \frac{\partial u}{\partial t} + \sum_{i=1}^{n} Q_i \frac{\partial u}{\partial x_i} - f = 0, \tag{4.48}$$

where $t = x_{n+1}$, $Q_i = Q_i(x, t, u)$ are $m \times m$ matrices, $f = f(x, t, u)$ is a vector.

Let us join the q differential constraints of first order

$$\Phi_k(x, t, u, u_{x_1}, \ldots, u_{x_n}) = 0, \quad (k = 1, 2, \ldots, q) \tag{4.49}$$

to system (4.48). The manifold

$$\{(x, t, u, p) |\ S(x, t, u, p) = 0, \Phi(x, t, u, p) = 0\}$$

is denoted by $(S\Phi)$.

5.1 Involutive conditions

Because all derivatives of the function u with respect to t can be defined from the overdetermined system $(S\Phi)$, the basis of the independent variables (x, t) is a weakly quasi-regular basis of the overdetermined system $(S\Phi)$. If the overdetermined system $(S\Phi)$ is involutive, then there is the nonsingular $(n + 1) \times (n + 1)$ matrix

$$\Gamma_{n+1} = \begin{pmatrix} \Gamma_n & 0 \\ 0 & 1 \end{pmatrix},$$

with an $n \times n$ matrix $\Gamma_n = (\gamma_{ij})$, defining the change of the independent variables

$$\widetilde{x}_k = \gamma_{\alpha k} x_\alpha, \quad \frac{\partial}{\partial x_k} = \gamma_{k\alpha} \frac{\partial}{\partial \widetilde{x}_k}, \quad d\widetilde{x}_k = \gamma_{\alpha k} dx_\alpha, \quad (k = 1, 2, , ..., n)$$

such that the basis $(\widetilde{x}, t)$ is a quasi-regular basis. Let the involutive overdetermined system (4.48), (4.49) be written in a quasi-regular basis and the differential constraints (Φ) be numerated so that the following results hold:

$$
\begin{aligned}
q_j &= rank \begin{pmatrix} \Phi_{,j} & \Phi_{,j+1} & \dots & \Phi_{,n} \end{pmatrix} \\
&= rank \begin{pmatrix} \Phi_{1,j} & \Phi_{1,j+1} & \dots & \Phi_{1,n} \\ \dots & \dots & \dots & \dots \\ \Phi_{q_j,j} & \Phi_{q_j,j+1} & \dots & \Phi_{q_j,n} \end{pmatrix}, \\
j &= 1, 2, \dots, n; \ (q_1 \geq q_2 \geq ... \geq q_n \geq q_{n+1} \equiv 0),
\end{aligned}
$$

where

$$
\Phi_{,j} = \begin{pmatrix} \Phi_{1,j} \\ \dots \\ \Phi_{q,i} \end{pmatrix} = \frac{\partial \Phi}{\partial u_{x_j}}, \quad \Phi_{k,j} = \frac{\partial \Phi_k}{\partial u_{x_j}}, \quad (k = 1, 2, ..., n).
$$

Denote by ρ the *max* k for which $q_k = q$ (either $\rho = m$ or $q_{\rho+1} < q_\rho = q$). The space of parametric first order derivatives G_1 of the overdetermined system $(S\Phi)$ is defined by the matrix

$$
\begin{pmatrix} Q_1 & Q_2 & ... & Q_n & E_m \\ \Phi_{,1} & \Phi_{,2} & ... & \Phi_{,n} & 0 \end{pmatrix}.
$$

The dimensions of the excisions $(G_1)^{(j)}$ are

$$
\begin{aligned}
\tau_{j-1} &= (n - j + 2)m - rank \begin{pmatrix} Q_j & Q_{j+1} & ... & Q_n & E_m \\ \Phi_{,j} & \Phi_{,j+1} & ... & \Phi_{,n} & 0 \end{pmatrix} \\
&= (n - j + 2)m - q_j, \\
&(j = 1, 2, ..., n); \ \tau_n = 0, \ \tau_{n+1} \equiv 0,
\end{aligned}
$$

Therefore,

$$
\sum_{j=0}^{n+1} \tau_j = \frac{n(n+1)}{2} - \sum_{j=0}^{n} q_j.
$$

The arbitrariness of the general solution of the system $(S\Phi)$ is defined by the Cartan characters

$$
\sigma_0 = 0, \sigma_{n+1} = 0, \sigma_j = m - q_j + q_{j+1}, \quad (j = 1, 2, \dots, n).
$$

If the initial coordinate system (x, t) is not quasi-regular, it can be transformed to a quasi-regular system with the help of the matrix Γ_{n+1}.

There is the following theorem [107].

Theorem 4.8. *The overdetermined system (4.48), (4.49) is involutive at the point* $(x_0, u_0, p_0) \in (S\Phi)$ *if there exist rectangular* $(q - q_k) \times q_j$ *matrices*

$(k = \rho + 1, \ldots, n;\ j = 1, 2, \ldots, n)$ *such that the equations*

$$\left(\sum_{\alpha=1}^{n} \Lambda_{k\alpha} D_\alpha \Phi \right)_{|(S\Phi)} = 0, \tag{4.50}$$

$$(D_{n+1}\Phi + \sum_{\alpha=1}^{n} \Lambda_\alpha D_\alpha \Phi) - \sum_{\alpha=1}^{n} \Phi_{,\alpha} D_\alpha S)_{|(S\Phi)} = 0.$$

are satisfied. A solution of the involutive overdetermined system (4.48), (4.49) is defined up to σ_i, $(i = 1, \ldots, n-1)$ *arbitrary functions of i arguments.*

Here the matrices Λ_{kj}, Λ_j $(k = \rho + 1, \ldots, n;\ j = 1, 2, \ldots, n)$ are found from the equations

$$\sum_{\mu=1}^{n} \Lambda_{k\mu} \gamma_{\mu j} = \delta_{kj} \widetilde{E}_k - \widetilde{M}_{kj};\ \sum_{\mu=1}^{n} \Lambda_\mu \gamma_{\mu j} = -\widetilde{M}_{n+1j},$$

where the matrices $\widetilde{M}_{kj}$ and $\widetilde{E}_k$ are obtained from the matrices M_{kj} and E_k by joining zero-columns from the right and left sides

$$\widetilde{M}_{kj} = \begin{pmatrix} M_{kj} & 0 \end{pmatrix},\ (k = \rho + 1, \ldots, n;\ j = 1, 2, \ldots, n),$$
$$\widetilde{E}_k = \begin{pmatrix} 0 & E_{q-q_k} \end{pmatrix},\ (k = \rho + 1, \ldots, n, n+1).$$

The proof of the theorem is similar to the case of two independent variables.

5.2 Differential constraints admitted by the gas dynamics equations

Here the method of differential constraints is applied to multi dimensional gas dynamics equations. The first example shows how the well-known irrotational flows of a gas can be considered from the point of view of the method of differential constraints. The second example defines a class of solutions of two-dimensional gas dynamic equations.

5.2.1 Irrotational gas flows

The isentropic flows of a gas are described by the equations

$$\begin{aligned} \frac{\partial \theta}{\partial t} + (\mathbf{v}, \nabla\theta) + a(\theta)\,(\nabla, \mathbf{v}) = 0, \\ \frac{\partial \mathbf{v}}{\partial t} + (\mathbf{v}, \nabla)\mathbf{v} + \nabla\theta = 0. \end{aligned} \tag{4.51}$$

Here $\mathbf{v} = (v_1, v_2, v_3)$ is the velocity, c is the sound speed, and $d\theta = -\tau^{-1}c^2(\tau)d\tau$. For example, for a polytropic gas $a = (\gamma - 1)\theta$.

An irrotational gas flow is determined by the differential constraints

$$\frac{\partial v_1}{\partial x_3} - \frac{\partial v_3}{\partial x_1} = 0,\ \frac{\partial v_2}{\partial x_3} - \frac{\partial v_3}{\partial x_2} = 0,\ \frac{\partial v_1}{\partial x_2} - \frac{\partial v_2}{\partial x_1} = 0. \tag{4.52}$$

One can check that the overdetermined system (4.51), (4.52) is involutive, $q_1 = q_2 = q = 3$, $q_3 = 2$, $\rho = 2$, the basis (x_1, x_2, x_3) is quasi-regular, and matrices M_{4j}, M_{3j} are

$$-M_{41} = \begin{pmatrix} v_1 & 0 & 0 \\ 0 & v_1 & 0 \\ 0 & -v_3 & v_1 \end{pmatrix}, \ -M_{42} = \begin{pmatrix} v_2 & 0 & 0 \\ 0 & v_2 & 0 \\ v_3 & 0 & v_2 \end{pmatrix}, \ -M_{43} = \begin{pmatrix} v_3 & 0 \\ 0 & v_3 \\ 0 & 0 \end{pmatrix},$$

$$M_{31} = \begin{pmatrix} 0 & -1 \end{pmatrix}, \ M_{32} = \begin{pmatrix} 1 & 0 \end{pmatrix}, \ M_{33} = \begin{pmatrix} 0 & 0 \end{pmatrix}.$$

5.2.2 One differential constraint admitted by unsteady two-dimensional gas dynamics equations

An isentropic gas flow is described by equations (4.51), which in the two-dimensional case have the form

$$\begin{aligned} S_1 \equiv u_t + uu_x + vu_y + \theta_x = 0, \\ S_2 \equiv v_t + uv_x + vv_y + \theta_y = 0, \\ \theta_t + u\theta_x + v\theta_y + a(\theta)\,(u_x + v_y) = 0. \end{aligned} \tag{4.53}$$

Here the individual coordinates (u, v) for the velocity, and (x, y) for the space coordinates are used.

Let us find a first order differential constraint

$$\Phi(x, y, t, u, v, \theta, u_x, v_x, \theta_x, u_y, v_y, \theta_y) = 0 \tag{4.54}$$

admitted by equations (4.53).

The second part of equations (4.50) becomes

$$\begin{aligned} b_1\Phi_{u_x} - a\Phi_{\theta_x} = 0,\ b_1\Phi_{v_x} = 0,\ b_1\Phi_{\theta_x} - \Phi_{u_x} = 0, \\ b_2\Phi_{u_y} = 0,\ b_2\Phi_{v_y} - a\Phi_{\theta_y} = 0,\ b_2\Phi_{\theta_y} - \Phi_{v_y} = 0, \\ b_1\Phi_{u_y} + b_2\Phi_{u_x} - a\Phi_{\theta_y} = 0,\ b_1\Phi_{v_y} + b_2\Phi_{v_x} - a\Phi_{\theta_x} = 0, \\ b_1\Phi_{\theta_y} + b_2\Phi_{\theta_x} - \Phi_{v_x} - \Phi_{u_y} = 0, \end{aligned} \tag{4.55}$$

where the matrices M_{31}, M_{32} have the representation

$$M_{31} = u + b_1,\ M_{32} = v + b_2.$$

Since the differential constraint (4.54) is essentially differential, the only solution of equations (4.55) is

$$\Phi_{u_x} = 0,\ \Phi_{v_x} = -\Phi_{u_y},\ \Phi_{\theta_x} = 0,\ \Phi_{v_y} = 0,\ \Phi_{\theta_y} = 0,\ b_1 = 0,\ b_2 = 0.$$

Without loss of generality one can assume that $\Phi_{u_y} = 1$, then the differential constraint (4.54) is reduced to

$$\Phi = u_y - v_x + \phi(x, y, t, u, v, \theta) = 0.$$

The first part of equations (4.50) is

$$(D_t\Phi + uD_x\Phi + vD_y\Phi + D_xS_2 - D_yS_1)_{|(S\Phi)} = 0.$$

After transition on the manifold $(S\Phi)$, and splitting with respect to the parametric derivatives $u_x, v_x, \theta_x, v_y, \theta_y$, one finds that $\varphi = \phi(\theta)$, and the function $\phi(\theta)$ satisfies the equation

$$\phi'(\theta)a(\theta) - \phi(\theta) = 0.$$

For example, for a polytropic gas the general solution of the last equation is

$$\phi(\theta) = c_0\theta^{\frac{1}{\gamma-1}},$$

where c_0 is an arbitrary constant. Hence, the differential constraint admitted by equations (4.50) is

$$\Phi = u_y - v_x + c_0\theta^{\frac{1}{\gamma-1}} = 0.$$

Remark 4.10. *Because of the scale, corresponding to the generator $t\partial_t + x\partial_x + y\partial_y$ admitted by the gas dynamics equations, and the involution $t' = -t$, a nonzero constant c_0 can be transformed to unity.*

Remark 4.11. *Two-dimensional nonisentropic gas dynamic equations admit one differential constraint only in the case that*

$$\Phi = u_y - v_x = 0.$$

Solutions satisfying this differential constraint define the well-known irrotational flows.

6. Differential constraints and one-parameter Lie-Bäcklund group of transformations

Relations between solutions of system of quasilinear equations (4.48) characterized by differential constraints and invariant solutions of an admitted one-parameter Lie-Bäcklund group of transformations are considered in this section. Under some conditions it is shown that invariant solutions of the Lie-Bäcklund group of transformations belong to a class of solutions that is characterized by differential constraints. The arbitrariness of the general solution of these invariant solutions is defined.

The theory of Lie-Bäcklund group of transformations [71] is an extension of the theory of Lie group of transformations. A one-parameter Lie-Bäcklund group of transformations is given by the Lie equations

$$u_\tau = \Phi(x, t, u, p). \tag{4.56}$$

Here $\Phi = (\Phi_1, \Phi_2, \ldots, \Phi_m)$, p denotes the derivatives p_α of order up to $|\alpha| \leq q$, and τ is the group parameter. It is assumed that the vector-valued function Φ is either analytic (in C^{ω}) or infinitely differentiable (in C^{∞}). The canonical infinitesimal generator of this group is

$$X = \Phi \partial_u + \sum_{\alpha} \zeta_\alpha \partial_{p_\alpha}, \tag{4.57}$$

where the coordinates ζ_α of X are

$$\zeta_\alpha = D^\alpha \Phi. \tag{4.58}$$

Systems of differential equations (4.48) admit a one-parameter Lie-Bäcklund group of transformations $G^1(\Phi)$ if, and only if, the invariance conditions

$$X S_{|(\overset{q+1}{S})} = 0 \tag{4.59}$$

hold. Here $(\overset{q+1}{S})$ denotes the manifold defined by the equations

$$S = 0, \quad D^\alpha S = 0, \quad |\alpha| = 1, 2, \ldots, q.$$

The coordinates of X and $D^\alpha S$ can be represented as follows

$$\zeta_j = D_j \Phi = \sum_{|\alpha|=m} \frac{\partial \Phi}{\partial p_\alpha} p_{\alpha,j} + \varphi_j, \quad (j = 1, \ldots, n+1), \tag{4.60}$$

$$D^\alpha S = p_{\alpha,n+1} + \sum_{j=1} Q_j p_{\alpha,j} + \psi_j, \quad (|\alpha| = q). \tag{4.61}$$

Substituting these representations into (4.59), one obtains

$$\begin{gathered}\textstyle\sum_{j=1}^n \sum_{|\alpha|=q} \left(Q_j \frac{\partial \Phi}{\partial p_\alpha} - \frac{\partial \Phi}{\partial p_\alpha} Q_j \right)_{|(\overset{q}{S})} p_{\alpha,j} + \\ \left(\varphi_{n+1} + \sum_{j=1}^n Q_j \varphi_j - \sum_{|\alpha|=q} \frac{\partial \Phi}{\partial p_\alpha} \psi_\alpha + \frac{\partial S}{\partial u} \Phi \right)_{|(\overset{q}{S})} = 0.\end{gathered}$$

Since the φ_j and ψ_α, $(j = 1, \ldots, n+1, \ |\alpha| = q)$, do not depend on the derivatives p_α with $|\alpha| = q+1$, it follows that (4.59) is equivalent to the equations

$$\sum_{j=1}^n \sum_{|\alpha|=q} \left(Q_j \frac{\partial \Phi}{\partial p_\alpha} - \frac{\partial \Phi}{\partial p_\alpha} Q_j \right)_{|(\overset{q}{S})} p_{\alpha,j} = 0, \tag{4.62}$$

$$\left(\varphi_{n+1} + \sum_{j=1}^n Q_j \varphi_j - \sum_{|\alpha|=q} \frac{\partial \Phi}{\partial p_\alpha} \psi_\alpha + \frac{\partial S}{\partial u} \Phi \right)_{|(\overset{q}{S})} = 0. \tag{4.63}$$

A solution $u = \phi(x,t)$ of (4.48) invariant with respect to the one-parameter Lie-Bäcklund group of transformations $G^1(\Phi)$ satisfies the differential constraints

$$X(u - \phi(x,t))_{|(U^q)} = \Phi(x, t, \phi(x,t), p(x,t)) = 0, \tag{4.64}$$

where (U^q) denotes the prolongation of the manifold $U = \{u - \phi(x,t) = 0\}$ up to order q. Here $p(x,t)$ is the set of partial derivatives of the function $\phi(x,t)$ up to order q. Thus, an invariant solution satisfies the overdetermined system of equations (4.48) and

$$\Phi(x, t, u, p) = 0. \tag{4.65}$$

Let us study the problem of compatibility of the overdetermined system (4.48), (4.65). Assume that there is a set of real numbers $\gamma_1, \gamma_2, \ldots, \gamma_n$ such that

$$\det\left(\sum_{j_1,\ldots,j_q=1} \frac{\partial \Phi}{\partial(\partial^q u/\partial x_{j_1} \ldots \partial x_{j_q})} \right)_{(\Phi)} \neq 0, \tag{4.66}$$

where (Φ) is the manifold defined by (4.65).

The vector-valued function Φ is assumed to be either analytic or sufficiently smooth for the existence of a solution of the overdetermined system (4.48), (4.65).

The overdetermined system of differential equations (4.48), (4.65) is involutive if and only if there exist matrices M_j, $j = 1, \ldots, n$, such that

$$\sum_{j=1}^{n} \sum_{|\alpha|=q} \left(M_j \frac{\partial \Phi}{\partial p_\alpha} + \frac{\partial \Phi}{\partial p_\alpha} Q_j \right)_{|(\Phi)} p_{\alpha,j} = 0, \tag{4.67}$$

$$\mathcal{L}(\Phi)_{|(\overset{q}{S},\Phi)} \equiv \left(D_{n+1}\Phi - \sum_{j=1}^{n} M_j D_j \Phi - \sum_{|\alpha|=q} \frac{\partial \Phi}{\partial p_\alpha} D^\alpha S \right)_{|(\overset{q}{S},\Phi)} = 0. \tag{4.68}$$

where $(\overset{q}{S}, \Phi)$ is the intersection of the manifolds $(\overset{q}{S})$ and (Φ). Substituting $D_j\Phi$ and $D^\alpha S$, $(j \leq n+1,\ |\alpha| = q)$, and applying (4.67), conditions (4.68) are reduced to the form

$$\mathcal{L}(\Phi)_{|(\overset{q}{S},\Phi)} = \left(\varphi_{n+1} - \sum_{j=1}^{n} M_j \varphi_j - \sum_{|\alpha|=q} \frac{\partial \Phi}{\partial p_\alpha} \psi_\alpha \right)_{|(\overset{q}{S},\Phi)} = 0. \tag{4.69}$$

Thus, taking $M_j = -Q_j$ one obtains the following theorem.

Theorem 4.9. *Let $G^1(\Phi)$ be a one-parameter Lie-Bäcklund group of transformations admitted by system (4.48) with the infinitesimal generator (4.57)* *satisfying (4.66). Then the overdetermined system (4.48), (4.65) is involutive with the Cartan characters*

$$\sigma_j = \frac{m}{q!} r \left(\frac{(n-j+q)!}{(n-j)} - \frac{(n-j+q-1)!}{(n-j-1)!} \right), \quad (j = 1, \ldots, n-1),$$

$$\sigma_n = 0, \quad \sigma_0 = \frac{m}{(n-1)!} \sum_{k=1}^{q} \frac{(n+k-1)!}{k!)}.$$

Remark 4.12. *When finding particular solutions of (4.48) by the method of differential constraints, it is usually sufficient to require only finite smoothness of the functions Φ. In the theory of Lie-Bäcklund transformations it is required that the functions Φ belong at least to the class C^∞.*

Remark 4.13. *(on Lie groups of point transformations). An infinitesimal generator of a Lie group of point transformations*

$$X = \sum_{j=1}^{n+1} \xi(x,t,u)\partial_{x_j} + \eta(x,t,u)\partial_u,$$

is equivalent to the canonical Lie-Bäcklund operator

$$X = \left(\eta - \xi^{n+1} f + \sum_{j=1}^{n} (\xi^{n+1} Q_j - \xi^j E_m) p_j \right) \frac{\partial}{\partial u} + \ldots \qquad (4.70)$$

The invariance conditions (4.62) are satisfied identically for the operator (4.70). The failure of the condition (4.66) means that for any set of real numbers $\gamma_1, \gamma_2, \ldots, \gamma_n$

$$\det \left(\sum_{j=1}^{n} \gamma_j (\xi^{n+1} Q_j - \xi^j E_m) \right) = 0. \qquad (4.71)$$

This condition is rarely satisfied. For example, for the equations of gas dynamics there are only two such generators among the basis generators [130]:

$$p\partial_p + \rho\partial_\rho, \quad \partial_p,$$

where ρ is the density and p is the pressure.

Remark 4.14. *Comparing the determining equations (4.62), (4.63) and the compatibility conditions (4.67), (4.69), it is clear that not all differential constraints (4.65) can define a one-parameter Lie-Bäcklund group of transformations.*

Remark 4.15. *By noticing that*

$$\left(\varphi_{n+1} + \sum_{j=1}^{n} Q_j\varphi_j - \sum_{|\alpha|=q} \frac{\partial \Phi}{\partial p_\alpha}\psi_\alpha + \frac{\partial S}{\partial u}\Phi\right)_{|(\Phi)} = \left(\varphi_{n+1} + \sum_{j=1}^{n} Q_j\varphi_j - \sum_{|\alpha|=q} \frac{\partial \Phi}{\partial p_\alpha}\psi_\alpha\right)_{|(\Phi)},$$

one can obtain the idea of the method of B-symmetries [82], where the determining equation is

$$\left(D_{n+1}\Phi + \sum_{j=1}^{n} Q_j D_j \Phi - \sum_{|\alpha|=q} \frac{\partial \Phi}{\partial p_\alpha} D^\alpha S\right)_{|(S)} + B\Phi = 0$$

with some matrix B.

7. One class of solutions

In this section one class of solutions of a hyperbolic quasilinear system of equations

$$S \equiv \frac{\partial u}{\partial t} + \sum_{i=1}^{n} Q_i(u)\frac{\partial u}{\partial x_i} - f(t, u) = 0 \tag{4.72}$$

is studied. For the sake of simplicity it is assumed that $n = 2$.

Since $f = f(t, u)$ this system of equations has solutions satisfying the differential constraints

$$\frac{\partial u}{\partial x_1} = 0, \quad \frac{\partial u}{\partial x_2} = 0.$$

These solutions are found by integrating the system of ordinary differential equations

$$\frac{\partial u}{\partial t} = f(t, u).$$

In this section they are called simple solutions.

Assume that $u^-(x_1, x_2, t)$ is a simple solution and the solution $u(x_1, x_2, t)$ continuously joins to the simple solution through the characteristic surface $\Pi = \{h(x_1, x_2, t) = 0\}$. Let us consider the problem of finding necessary conditions for $u(x_1, x_2, t)$ on the characteristic surface.

For any curve $(x_1(\tau), x_2(\tau), t(\tau)) \in \Pi$:

$$\xi_1\frac{dx_1}{d\tau} + \xi_2\frac{dx_2}{d\tau} + \xi_3\frac{dt}{d\tau} = 0,$$

where $(\xi_1, \xi_2, \xi_3) = \left(h_{x_1}, h_{x_2}, h_t\right)$. The conditions are considered at some point $(x_1^o, x_2^o, t^o) \in \Pi$. Without loss of generality it is assumed that $\xi_1(x_1^o, x_2^o, t^o) \neq 0$.

Let $\Pi_1 = \{(x_1^{(1)}(\tau), x_2^{(1)}(\tau), t^{(1)}(\tau))\} \in \Pi$ and $\Pi_2 = \{(x_1^{(2)}(\tau), x_2^{(2)}(\tau), t^{(2)}(\tau))\} \in \Pi$ be two curves belonging to Π, defined by the formulae

$$x_2^{(1)}(\tau) = x_2^o + \tau,\ t^{(1)}(\tau) = t^o,$$
$$x_2^{(2)}(\tau) = x_2^o,\ t^{(2)}(\tau) = t^o + \tau.$$

Because of continuous joining, one has

$$\frac{d_1 u}{d\tau} = \frac{d_1 u^-}{d\tau},\ \frac{d_2 u}{d\tau} = \frac{d_2 u^-}{d\tau},$$

where $\frac{d_i}{d\tau}$ is a derivative along Π_i, $(i = 1, 2)$. Hence,

$$\frac{\partial u}{\partial t} = \frac{\xi_3}{\xi_1}\frac{\partial u}{\partial x_1} + f,\ \frac{\partial u}{\partial x_2} = \frac{\xi_2}{\xi_1}\frac{\partial u}{\partial x_1}.$$

Substituting these relations into (4.72), one obtains

$$Q_\xi \frac{\partial u}{\partial x_1} = 0. \tag{4.73}$$

Here $Q_\xi = \xi_3 E_m + Q_{\widehat{\xi}}$, $Q_{\widehat{\xi}} = \xi_1 Q_1 + \xi_2 Q_2$, and E_m is the $m \times m$ unit matrix. Because of the hyperbolicity of system (4.72), there is a matrix L such that

$$Q_{\widehat{\xi}} = L^{-1} A_{\widehat{\xi}} L$$

with a diagonal matrix $A_{\widehat{\xi}}$. Notice that the matrix L can be chosen only to the dependent on u and the relation $\zeta = \frac{\xi_2}{\xi_1}$. The diagonal entries of the matrix $A_{\widehat{\xi}}$ will be denoted by λ_i, $(i = 1, 2, ..., m)$. Because the surface Π is a characteristic surface, there is an eigenvalue λ_i such that $\frac{\xi_3}{\xi_1} = -\lambda_i$. Without loss of generality one can write $\frac{\xi_3}{\xi_1} = -\lambda_m$.

If λ_m is a root of multiplicity one, then from equation (4.73) one has

$$BL\frac{\partial u}{\partial x_1} = 0, \tag{4.74}$$

where the matrix B has entries

$$B_{ij} = \delta_{ij},\ (i = 1, 2, ..., m-1,\ j = 1, 2, ..., m).$$

Conversely, relations (4.73) can be derived from (4.74), even without an assumption of the multiplicity of the root λ_m.

Thus, a solution $u(x_1, x_2, t)$ continuously joining to the simple solution has to satisfy the differential constraints:

$$BLp_1 = 0,\ p_2 - \zeta p_1 = 0 \tag{4.75}$$

on the characteristic surface $\Pi = \{h(x_1, x_2, t) = 0\}$. Here $p_1 = \frac{\partial u}{\partial x_1}$ and $p_2 = \frac{\partial u}{\partial x_2}$.

Solutions of system (4.72) satisfying the differential constraints (4.75) in some domain V (not only on a surface) are called generalized simple waves. Here one of the equations $BLp_1 = 0$ serves for defining the function ζ.

Let us establish the conditions that guarantee system (4.72) will possess a generalized simple wave solution.

For further consideration we need to introduce matrices B_1, B_2 and B_3 with the entries

$$(B_1)_{ij} = \delta_{ij},\ (i = 1, 2, ..., m-2,\ \ j = 1, 2, ..., m),$$
$$(B_2)_{1j} = \delta_{(m-1),j},\ (B_3)_{1j} = \delta_{mj},\ (j = 1, 2, ..., m).$$

Without loss of generality it is assumed that

$$\alpha \equiv B_2 \frac{\partial L}{\partial \zeta} p_1 \neq 0.$$

Hence, the equation

$$B_2 L p_1 = 0$$

can be used to define the variable $\zeta = \zeta(u, p_1)$. The left-hand sides of the other differential constraints (4.75) are denoted as follows

$$\Psi = B_1 L p_1,\ \ \Phi = p_2 - \zeta p_1 = 0.$$

Because of the theorem, the overdetermined system (4.72), (4.75) is involutive if and only if there exist matrices N_{21}, N_{31}, M_{22}, M_{32} such that they satisfy the system of linear algebraic equations

$$N_{21}\Phi_1 + N_{22}\Psi_1 = 0,\ N_{21}\Phi_2 + N_{22}\Psi_2 + M_{22}\Phi_1 = \Psi_1,$$

$$M_{22}\Phi_2 = \Psi_2,\ N_{31}\Phi_1 + N_{32}\Psi_1 = \begin{pmatrix} \Phi_1 \\ \Psi_1 \end{pmatrix} Q_1,$$

$$N_{31}\Phi_2 + N_{32}\Psi_2 + M_{32}\Phi_1 = \begin{pmatrix} \Phi_1 \\ \Psi_1 \end{pmatrix} Q_2 + \begin{pmatrix} \Phi_2 \\ \Psi_2 \end{pmatrix} Q_1,\ M_{32}\Phi_1 = \begin{pmatrix} \Phi_2 \\ \Psi_2 \end{pmatrix} Q_2,$$

With these matrices the differential equations

$$N_{21} D_{x_1}\Phi + N_{22} D_{x_1}\Psi + M_{22} D_{x_2} - D_{x_2}\Psi = 0, \tag{4.76}$$

$$D_t \begin{pmatrix} \Phi \\ \Psi \end{pmatrix} + N_{31} D_{x_1}\Phi + N_{32} D_{x_1}\Psi + M_{32} D_{x_1}\Phi - \begin{pmatrix} \Phi_1 \\ \Psi_1 \end{pmatrix}$$
$$D_{x_1} S - \begin{pmatrix} \Phi_2 \\ \Psi_2 \end{pmatrix} D_{x_2} S = 0 \tag{4.77}$$

have to be satisfied identically on the manifold defined by equations (4.72), (4.75). Here

$$\Psi_1 = \frac{\partial \Psi}{\partial p_1} = B_1\left(E_m - \alpha^{-1}\frac{\partial L}{\partial \zeta} p_1 \otimes B_2\right)L, \ \Psi_2 = \frac{\partial \Psi}{\partial p_2} = 0,$$
$$\Phi_1 = \frac{\partial \Phi}{\partial p_1} = -\zeta E_m + \alpha^{-1} p_1 \otimes B_2 L, \ \Phi_2 = \frac{\partial \Phi}{\partial p_2} = E_m,$$

and $\otimes$ is a dyad of two vectors.

Using the properties

$$B_1' B_1 + B_2' B_2 + B_3' B_3 = E_m, \ B_1' B_1 = E_{m-2}, \ B_2' B_2 = B_3' B_3 = 1,$$
$$B_i B_j' = 0, \ B_i A_\zeta B_j' = 0, \ (i, j = 1, 2, 3; \ i \neq j),$$

one can find the general solution of the linear system of equations

$$M_{22} = 0, \ M_{32} = \begin{pmatrix} Q_2 \\ 0 \end{pmatrix}, \ N_{21} = \Psi_1, \ N_{31} = \begin{pmatrix} \Phi_1 Q_2 + Q_1 - Q_2 \Phi_1 \\ \Psi_1 Q_2 \end{pmatrix},$$
$$N_{22} = \zeta E_{m-2}, \ N_{32} = \begin{pmatrix} 0 \\ B_1 A_\zeta B_1' \end{pmatrix}.$$

Substituting these matrices into equations (4.76) and (4.76), they are reduced to

$$B_i\left(\frac{\partial L}{\partial u} < f, L^{-1} B_3' > + L\frac{\partial f}{\partial u} L^{-1} B_3'\right) = 0, \ (i = 1, 2), \tag{4.78}$$

where $\frac{\partial L}{\partial u} < \xi, \eta >$ is a vector with the coordinates

$$\left(\frac{\partial L}{\partial u} < \xi, \eta >\right)_i = \sum_{\alpha=1}^{m} \sum_{\beta=1}^{m} \frac{\partial L_{i\alpha}}{\partial u_\beta} \xi_\beta \eta_\alpha .$$

Thus, the conditions (4.78) guarantee that system (4.72) will have a generalized simple wave solution in the case of more than two independent variables.

Chapter 5

INVARIANT AND PARTIALLY INVARIANT SOLUTIONS

This chapter is devoted to giving a concise form of the basic algorithms that form the core of group analysis[1]. The main concept for constructing exact solutions for differential equations using this method is the concept of a Lie group. Since there is a direct relation between a Lie group and a Lie algebra, this allows us to use the power of linear algebra. For finding solutions one exploits an admitted Lie group. Different approaches to the notion of a Lie group to be admitted are discussed in detail in the next chapter. In this chapter the admitted Lie group is a Lie group for which the coefficients of the corresponding generator satisfy the determining equations. The problem of finding an admitted Lie group is the first step in the application of group analysis to constructing exact solutions.

The algebraic structure of the admitted Lie group introduces an algebraic structure into the set of all solutions. This algebraic structure is used to obtain invariant and partially invariant solutions. The main feature of these classes of solutions is that they reduce the number of the independent and dependent variables. In this sense the problem of finding them is simpler than for the general solution.

Partially invariant solutions are more difficult to construct than invariant solutions. It is worth noting that the theory of partially invariant solutions is still developing. For example, just recently the notions of regular and irregular partially invariant solutions were introduced for classification of partially invariant solutions. It was also shown that for constructing partially invariant solutions there is no necessity for the Lie group to be admitted. An application of a

[1]The author studied group analysis in lectures given by L.V.Ovsiannikov. A detailed analysis of many problems in group analysis can be found in his classical book [130]. A history of the development of the method is to be found in [73, 22]. The modern state of group analysis is reviewed in the CRC Handbook of Lie Group Analysis of differential equations [72].

new method of using partially invariant solutions for finding exact solutions is discussed.

Most differential equations include arbitrary elements: constants and functions of the independent and dependent variables. Hence, the admitted Lie group depends on these elements. A transformation that preserves the equations, while only changing their arbitrary elements is called an equivalence transformation. A Lie group of equivalence transformations allows one to choose a simple representation of the arbitrary elements of a physical problem.

A solution invariant with respect to a one-parameter Lie group satisfies the differential constraints which are equivalent to the property of invariance. The search for a one-parameter Lie group admitted by the original system of equations and the conditions of the invariance led to the notions of nonclassical, weak and conditional symmetries.

Through involving derivatives in the transformation, the notion of a Lie group of point transformations is generalized. In the general case the Bäcklund theorem asserts that all tangent transformations of finite order are prolongations of contact or point transformations.

The limitations dictated by the Bäcklund theorem can be overcome by considering the admitted transformations or extending the space of the derivatives involved in the transformations up to infinity. The second alternative leads to the Lie-Bäcklund transformations.

1. The main definitions

A local Lie group of transformations plays a key role in group analysis. Relating a local Lie group of transformations with a system of equations one arrives at an admitted Lie group, which is also called a symmetry group. The main definitions and properties of admitted Lie groups are studied in this section.

1.1 Local Lie group of transformations

Let V be an open set in $Z = R^N(z)$, Δ be a symmetric interval in R^1. Assume that the point transformations

$$z_i' = g^i(z; a), \quad (i = 1, 2, \ldots, N) \tag{5.1}$$

are invertible. Here $z \in V \subset Z$ and the parameter $a \in \Delta$. It is also convenient to use the notation $g_a(z) = g(z, a)$.

For differential equations the variable z is separated into two parts $z = (x, u) \in V \subset Z = R^n(x) \times R^m(u)$, $N = n+m$. Here $x = (x_1, x_2, \ldots, x_n)$ are the independent variables and $u = (u^1, u^2, \ldots, u^m)$ the dependent variables. In this case the invertible transformations are represented as follows

$$x_i' = f^i(x, u; a), \quad u'^j = \varphi^j(x, u; a), \tag{5.2}$$
$$(i = 1, 2, \ldots, n; \quad j = 1, 2, \ldots, m).$$

Definition 5.1. *A set of transformations (5.1) is a local one–parameter Lie group G^1 if it has the following properties:*

1^0. $g(z, 0) = z$ for all $z \in V$.

2^0. $g(g(z, a), b) = g(z, a + b)$ for all $a, b, a + b \in \Delta, z \in V$.

3^0. If $a \in \Delta$ and $g(z, a) = z$ for all $z \in V$, then $a = 0$.

The set

$$g_z(\Delta) = \{z' \in V \mid z' = g(z, a),\ a \in \Delta\}$$

is called an orbit of the point $z \in V$.

A Lie group[2] G^1 is called a continuous group of the class C^k if the function $g(z, a)$ belongs to the class $C^k(V)$. In applied group analysis all functions are considered to be sufficiently many times continuously differentiable.

To the group G^1 one can relate the infinitesimal generator

$$X = \zeta^i(z)\partial_{z_i},$$

where

$$\zeta(z) = (\zeta^1(z), \zeta^2(z), ..., \zeta^N(z)) = \frac{dg}{da}(z, 0).$$

The infinitesimal generator X is invariant under any invertible change of the variables z:

$$\widehat{z} = q(z).$$

In fact, in the new variables $\widehat{z}$ the group $G^1 : \widehat{z}' = \widehat{g}(\widehat{z}, a)$ is given by the formula

$$\widehat{g}(\widehat{z}, a) = (q \circ g \circ q^{-1})(\widehat{z}).$$

Thus

$$\widehat{\zeta}^\beta(\widehat{z}) = \frac{d\widehat{g}^\beta}{da}_{|a=0} = \frac{\partial q^\beta}{\partial z_i}(z)\frac{\partial g^i}{\partial a}(\widehat{z}, 0) = \zeta^i(\widehat{z})\frac{\partial q^\beta}{\partial z_i}(z) = X(q^\beta(z)).$$

where $z = q^{-1}(\widehat{z})$. Since

$$\zeta^i(z)\partial_{z_i} = \zeta^i(z)\frac{\partial q^\beta(z)}{\partial z_i}\partial_{\widehat{z}_\beta} = X(q^\beta(z))\partial_{\widehat{z}_\beta} = \widehat{\zeta}^\beta(\widehat{z})\partial_{\widehat{z}_\beta},$$

one has

$$X = \zeta^i(z)\partial_{z_i} = \widehat{\zeta}^\beta(\widehat{z})\partial_{\widehat{z}_\beta} = \widehat{X}.$$

For example, let the function $\varphi^N(z)$ be any solution of the equation $X\varphi = 1$. Because there are $N - 1$ functionally independent solutions $\varphi^j(z)$, $(j = 1, 2, ..., N - 1)$ of the equation $X\varphi = 0$, the functions $\varphi^j(z)$, $(j = 1, 2, ..., N)$

---

[2]For brevity a local Lie group of transformations will also be called a Lie group or simply a group.

form an invertible change of variables. Taking this change of variables, one finds that for any generator X there is a coordinate system such that

$$X = \partial_{\widetilde{z}_N}.$$

A local Lie group of continuous transformations (5.1) is completely defined by the solution of the Cauchy problem:

$$\frac{dz_i'}{da} = \zeta^i(z') \tag{5.3}$$

$$z' = z. \tag{5.4}$$

Here the initial data (5.4) are taken at the point $a = 0$. Equations (5.3) are called Lie equations. Before proving this statement let us prove the following lemma.

Lemma 5.1. *Assume that $\zeta \in C^1(\Omega)$ and $g(z,a)$ is a solution of the Cauchy problem (5.3), (5.4). For any point $z_o \in \Omega$ there exist a ball $B_r(z_o) \subset \Omega$, symmetric interval $\Delta \subset R$ and a number $M > 0$ such that*

$$||g(z,a) - z|| \geq |a|||\zeta(z)|| - Ma^2, \ \forall (z,a) \in B_r(z_o) \times \Delta. \tag{5.5}$$

The transformation $g_a : B_r(z_o) \to g_a(B_r(z_o))$ is one-to-one.

Proof.

Since the mappings ζ and $\frac{\partial \zeta}{\partial z}$ are continuous, there is a radius r_1 such that they are defined and bounded in the ball $B_{r_1}(z_o)$. Because $g(z,a)$ is continuous, and $g(z,0) = z$, there exists a ball $B_{r_2}(z_o) \subset \Omega$, and a symmetric interval $\Delta_1 \subset R$ such that

$$||g(z,a) - z|| < r_1, \ \forall (z,a) \in B_{r_2}(z_o) \times \Delta_1.$$

Let us apply the Taylor formula to the function $g(z,a)$

$$g(z,a) = z + a\zeta(z) + a^2\psi(z,a),$$

where the function $\psi(z,a) = \frac{1}{2}\frac{\partial^2 g}{\partial a^2}(z,\widetilde{a})$ at some point $\widetilde{a} \in (0,a)$. The derivative $\frac{\partial^2 g}{\partial a^2}(z,a)$ is found by differentiating (5.3) with respect to the parameter a:

$$\frac{\partial^2 g}{\partial a^2}(z,a) = \frac{\partial \zeta}{\partial z_i}(g(z,a))\zeta^i(g(z,a)).$$

Therefore, there is a constant M such that $||\psi(z,a)|| < M$ and $\forall (z,a) \in B_{r_2}(z_o) \times \Delta_1$

$$||g(z,a) - z|| \geq |a|||\zeta(z)|| - Ma^2.$$

Let us consider the mapping $\Psi(z,a) = (g(z,a),a)$, $(z,a) \in B_{r_2}(z_o) \times \Delta_1$. Since $\zeta \in C^1(\Omega)$, and $\frac{\partial g^j}{\partial z_i}(z_o,0) = \delta_{ij}$, then $\Psi(z,a) \in C^1(B_{r_2}(z_o) \times \Delta_1)$ and the Jacobian

$$\frac{\partial \Psi}{\partial(z,a)}(z_o,0) = \det\left(\frac{\partial g}{\partial z}(z_o,0)\right) = 1 \neq 0.$$

By virtue of the inverse function theorem there exists a neighborhood $U \subset B_{r_2}(z_o) \times \Delta_1$ of the point $(z_o,0) \in U$ such that Ψ is invertible in U. Taking a radius r and an open interval Δ with the condition $B_r(z_o) \times \Delta \subset U$, one obtains the proof of the last statement in the lemma. •

Theorem 5.1. *(Lie). Let $\zeta \in C^1(\Omega)$ with $\zeta(z_0) \neq 0$ for some $z_0 \in \Omega$. The solution of the Cauchy problem (5.3), (5.4) generates a local Lie group of transformations with the infinitesimal operator $X = \zeta^i(z)\partial_{z_i}$.*

Proof.

At first, one has to choose the set V and the interval Δ for checking the properties of a Lie group. Let the ball $B_r(z_o)$ and the interval Δ be defined as in the lemma. Since the function ζ is continuous, without loss of generality the radius r can be chosen to satisfy the condition (the radius r is decreased, if necessary)

$$||\zeta(z)|| > \frac{1}{2}||\zeta(z_o)||, \ \forall z \in B_r(z_o).$$

The interval Δ is also decreased (if necessary) to satisfy the condition

$$\frac{1}{2M}||\zeta(z_o)|| > |a|, \ \forall a \in \Delta.$$

The transformations $g(z,a)$ are considered in the set $B_r(z_o) \times \Delta$. Note that in the set $B_r(z_o) \times \Delta$ the transformations $g(z,a)$ satisfy the inequality

$$||g(z,a) - z|| \geq |a|\left(||\zeta(z)|| - M|a|\right).$$

After obtaining the set $V = B_r(z_o)$ and the interval Δ, one can check the properties of a Lie group.

Because $g(z,0) = z$, property 1^0 is trivial.

To check property 2^0 one notes that $g(z,a+b)$ and $g(g(z,a),b)$ satisfy the same Cauchy problem:

$$\frac{dy}{db} = \zeta(y), \ y(0) = g(z,a).$$

Thus, because of the uniqueness of the Cauchy problem $g(z,a+b) = g(g(z,a),b)$.

Let $a \in \Delta$, and $g(z,a) = z$ for all $z \in B_r(z_o)$. If $a \neq 0$, one obtains the contradiction

$$M|a| \geq ||\zeta(z)|| > \frac{1}{2}||\zeta(z_o)|| > M|a|.$$

•

The Lie theorem establishes a one-to-one correspondence between the Lie group of transformations G^1 and the infinitesimal generator $X = \zeta(z)\partial_z$.

The space $Z = R^n(x) \times R^m(u)$ is prolonged by introducing the variables $p = (p^k_\alpha)$. Here $\alpha = (\alpha_1, \alpha_2, \dots, \alpha_n)$ is a multiindex. For the multiindex the notations $|\alpha| \equiv \alpha_1 + \alpha_2 + \dots + \alpha_n$ and $\alpha, i \equiv (\alpha_1, \alpha_2, \dots, \alpha_{i-1}, \alpha_i + 1, \alpha_{i+1}, \dots, \alpha_n)$ are used. The variable p^k_α plays the role of the derivative $\frac{\partial^{|\alpha|}u^k}{\partial x^\alpha}$. The space J^l of the variables

$$x = (x_i),\ u = (u^k),\ p = (p^k_\alpha),$$
$$(i = 1, 2, \dots, n;\ \ k = 1, 2, \dots, m;\ \ |\alpha| \le l)$$

is called an l-th prolongation of the space Z. This space can be provided with a manifold structure. For convenience we agree that $J^0 \equiv Z$.

Let the infinitesimal generator

$$X = \xi^i(x, u)\partial_{x_i} + \eta^j(x, u)\partial_{u^j}$$

be an infinitesimal generator of a Lie group G^1 of transformations (5.2).

Definition 5.2. *The generator*

$$\underset{l}{X} = X + \sum_{j,\alpha} \eta^j_\alpha \partial_{p^j_\alpha}$$

with the coefficients

$$\eta^j_{\alpha,k} = D_k\eta^j_\alpha - \sum_i p^j_{\alpha,i} D_k\xi^i \tag{5.6}$$

is called an l–th prolongation of the generator X.

Here the operators

$$D_k = \frac{\partial}{\partial x_k} + \sum_{j,\alpha} p^j_{\alpha,k} \frac{\partial}{\partial p^j_\alpha}$$

are operators of the total derivatives with respect to x_k, $(k = 1, 2, \dots, n)$.

The generator $\underset{l}{X}$ induces the local Lie group of transformations in the space J^l

$$x' = f(x, u; a),\ \ u' = \varphi(x, u; a),\ \ p' = \psi(x, u, p; a) \tag{5.7}$$

with the prolonged generator $\underset{l}{X}$. This group is called an l–th prolongation of the group G^1 (5.2) and it is denoted by $\underset{l}{G}^1$ or simply by $\underset{l}{G}$.

Formulae (5.6) for the coefficients of the prolonged operator are defined by the following. Let a function $u = u_o(x)$ be given. Substituting it into the first part of transformations (5.2) and applying the inverse function theorem, one finds that

$$x = \psi(x', a).$$

The transformed function $u_a(x')$ is given as follows

$$u_a(x') = \phi(\psi(x', a), u_o(\psi(x', a)); a). \tag{5.8}$$

Hence, the derivatives $p_a(x')$ of the transformed function $u_a(x')$ are defined through the derivatives of the functions $u_o(x)$, $\phi(x, u; a)$ and $\psi(x', a)$. Because of the inverse function theorem the derivatives of the function $\psi(x', a)$ are also defined through the derivatives of the functions $u_o(x)$ and $f(x, u; a)$. Hence, the derivatives $p_a(x')$ are defined through the derivatives of the functions $u_o(x)$, $\phi(x, u; a)$ and $f(x, u; a)$. Differentiating $p_a(x')$ with respect to the parameter a and setting $a = 0$ the coefficients (5.6) are obtained.

To avoid taking derivatives of the function $\psi(x', a)$ one can use another equivalent representation of (5.8)

$$u_a\,(f(x, u_o(x); a)) = \phi\,(f(x, u_o(x), a), u_o(x); a)\,.$$

For example, differentiating the last expressions with respect to x_i, one has

$$\psi^j_\alpha\,(f^\alpha_{x_i} + f^\alpha_{u^\beta} p^\beta_i) = \varphi^j_{x_i} + \varphi^j_{u^\beta} p^\beta_i, \ \ (j = 1, 2, ..., m; \ i = 1, 2, ..., n).$$

Differentiating them with respect to the parameter a and using property 1^0 of a Lie group, one gets

$$\eta^j_i + p^j_\alpha\,(\xi^\alpha_{x_i} + \xi^\alpha_{u^\beta} p^\beta_i) = \eta^j_{x_i} + \eta^j_{u^\beta} p^\beta_i, \ \ (j = 1, 2, ..., m; \ i = 1, 2, ..., n).$$

Rewriting this in the form (5.6) one obtains

$$\eta^j_i = D_i\eta^j - p^j_\alpha D_i\xi^\alpha, \ \ (j = 1, 2, ..., m; \ i = 1, 2, ..., n).$$

In what follows, the operation of prolongation is denoted by the sign $\widetilde{\ }$.

Notice that, according to the constructions, the orbit of the points $(x, u_o(x), p_o(x))$ with respect to the prolonged Lie group $\underset{1}{G}$ consists of the points $(x', u_a(x'), p_a(x'))$. Here $p_o(x)$ are derivatives of the function $u_o(x)$.

Lemma 5.2. *Assume that* $F = F(x, u)$. *For any generator*

$$X = \xi^i(x, u)\partial_{x_i} + \eta^j(x, u)\partial_{u^j}$$

the relation

$$D_k(XF) = \widetilde{X}(D_kF) + (D_k\xi^\beta)D_\beta F. \tag{5.9}$$

is satisfied.

Proof.
The proof of the lemma follows immediately after noting that

$$D_k F_{x_i} = \partial_{x_i} (D_k F), \ D_k F_{u^j} = \partial_{u^j} (D_k F),$$
$$F_{x_\beta} D_k \xi^\beta = (D_\beta F)(D_k \xi^\beta) - F_{u^j} p^j_\beta D_k \xi^\beta. \quad \bullet$$

Theorem 5.2. *The operations of prolongation and an invertible change of variables*

$$y = Y(x,u), \ v = V(x,u),$$

commute.
Proof.
Let the inverse mapping be denoted by

$$x = H(y,v), \ u = U(y,v).$$

This means

$$y \equiv Y(H(y,v), U(y,v)), \ v \equiv V(H(y,v), U(y,v)),$$
$$x \equiv H(Y(x,u), V(x,u)), \ u \equiv U(Y(x,u), V(x,u)).$$

The following notations are used for the derivatives

$$p^j_\alpha = \frac{\partial u^j}{\partial x_\alpha}, \ q^j_\alpha = \frac{\partial v^j}{\partial y_\alpha}.$$

The derivative q^j_α can be expressed as $q^j_\alpha = Q^j_\alpha(x,u,p)$. In further proofs there is no necessity to have exact expressions of the functions $Q^j_\alpha(x,u,p)$. Since the infinitesimal generator is invariant with respect to an invertible change,

$$X = \xi^i(x,u)\partial_{x_i} + \eta^j(x,u)\partial_{u^j} =$$
$$X' = \widehat{\xi}^i(y,v)\partial_{y_i} + \widehat{\eta}^j(y,v)\partial_{v^j} = X(Y_i)\partial_{y_i} + X(V^j)\partial_{v^j}.$$

For the prolongations of the generators X and X' one has

$$\widetilde{X} = X + \eta^j_\alpha \partial_{p^j_\alpha}, \ \widetilde{(X')} = X' + \widehat{\eta}^j_\alpha \partial_{q^j_\alpha}.$$

Notice that

$$(\widetilde{X})' = \widetilde{X}(Y_i)\partial_{y_i} + \widetilde{X}(V^j)\partial_{v^j} + \widetilde{X}(Q^j_\alpha)\partial_{q^j_\alpha}.$$

Thus, comparing $(\widetilde{X})'$ and $\widetilde{(X')}$, the theorem is proved if one can obtain the following result

$$\widehat{\eta}^j_\alpha = \widetilde{X}(Q^j_\alpha).$$

To prove the last equality the following constructions are used. Let the function $u = \varphi(x)$ be represented by $v = \psi(y)$ in the variables (y, v):

$$V(x, \varphi(x)) = \psi(Y(x, \varphi(x))), \ U(y, \psi(y)) = \varphi(H(y, \psi(y))).$$

Differentiating these identities, one obtains

$$D_i V^k = q^k_\alpha D_i Y^\alpha, \ \ D'_j U^k = p^k_\alpha D'_j H^\alpha.$$

Here $D_i = D_{x_i} = \partial_{x_i} + p^k_i \partial_{u^k}$ is the total derivative with respect to x_i in the variables (x, u), and $D'_j = D_{y_j} = \partial_{y_j} + q^k_j \partial_{v^k}$ is the total derivative with respect to y_j in the variables (y, v). These total derivatives are related by the formula

$$D'_j = D'_j(H^\alpha) D_\alpha.$$

In fact, let $\widetilde{F}(y, v) = F(H(y, v), U(y, v))$, then

$$\begin{aligned} D'_j \widetilde{F} = F_{x_\alpha} D'_j(H^\alpha) + F_{u^k} D'_j(U^k) = F_{x_\alpha} D'_j(H^\alpha) + F_{u^k} p^k_\alpha D'_j H^\alpha \\ = (D'_j H^\alpha)(D_\alpha F). \end{aligned}$$

For example,

$$D'_j y_t = \delta_{jt} = D'_j(H^\alpha) D_\alpha(Y^t). \tag{5.10}$$

By the definition

$$\begin{gathered} \widehat{\eta}^k_i = D'_i(X(V^k)) - q^k_t D'_i(X(Y^t)) = \\ D'_i(H^\alpha) D_\alpha(X(V^k)) - q^k_t D'_i(H^\alpha) D_\alpha(X(Y^t)) = \\ D'_i(H^\alpha) \left(D_\alpha(X(V^k)) - q^k_t D_\alpha(X(Y^t)) \right). \end{gathered}$$

Applying the previous lemma to $D_\alpha(X(V^k))$ and $D_\alpha(X(Y^t))$, one obtains

$$\widehat{\eta}^k_i = D'_i(H^\alpha) \left[\widetilde{X}(D_\alpha V^k) - q^k_t \widetilde{X}(D_\alpha Y^t) + \underline{(D_\alpha \xi^\beta) D_\beta V^k - q^k_t (D_\alpha \xi^\beta) D_\beta Y^t} \right]$$

Because of $D_i V^k = q^k_\alpha D_i Y^\alpha$, the underlined terms are equal to zero. Due to

$$\widetilde{X}(D_\alpha V^k) = \widetilde{X}(Q^k_\beta) D_\alpha Y^\beta + q^k_t \widetilde{X}(D_\alpha Y^t),$$

and (5.10), one obtains

$$\widehat{\eta}^k_i = D'_i(H^\alpha) \left(D_\alpha Y^\beta \right) \widetilde{X}(Q^k_\beta) = \widetilde{X}(Q^k_i).$$

•

1.2 Invariant manifolds

Most equations studied in applications compose a regularly assigned manifold. This subsection is devoted to the definitions concerning invariant regularly assigned manifolds.

Assume that $\psi : V \to R^s$ is a mapping of the class $C^1(V)$. This mapping has a rank on the set V if the rank of the Jacobi matrix $\dfrac{\partial \psi}{\partial z}$ has the same rank at all points $z \in V$. The equation $\psi(z) = 0$ is called regular if the mapping ψ

has the rank s. If the mapping $\psi(z)$ is regular and $\psi(z_0) = 0$ for some $z_0 \in V$, then the equation $\psi(z) = 0$ generates in some neighborhood U of the point z_0 the manifold

$$\Psi = \{z \in U \mid \psi(z) = 0\},$$

which is called a regularly assigned manifold. If the manifold $\psi(z) = 0$ is regularly assigned, then in a neighborhood of the point z_0 there exists a change of the variables $\widehat{z} = q(z)$ such that this manifold is assigned by the equations

$$\widehat{z}_i = 0, \ (i = 1, 2, ..., s).$$

This statement can be obtained by choosing a change of the variables z where the first s functions are

$$q^i = \psi^i(z), \ (i = 1, 2, ..., s).$$

Definition 5.3. *A function $\Phi(z)$, $z \in V$ is called an invariant of a Lie group G^1 if*

$$\Phi(g(z, a)) = \Phi(z), \ \forall z \in V, \ \forall a \in \Delta.$$

Theorem 5.3. *A function $\Phi(z)$ is an invariant of a group $G^1(X)$ with a generator X if and only if*

$$X\Phi(z) = 0.$$

Using the Lie equations the proof of this theorem is easily obtained.

Definition 5.4. *A manifold Ψ is invariant with respect to a Lie group G^1 if after its transformation, any point $z \in \Psi$ belongs to the same manifold Ψ.*

Theorem 5.4. *A regularly assigned manifold Ψ is an invariant manifold with respect to a Lie group $G^1(X)$ (5.1) if and only if*

$$X\psi^k(z)_{|\Psi} = 0, \ (k = 1, 2, \ldots, s). \tag{5.11}$$

Here $|_\Psi = 0$ means that the equations $X\psi^k(z)$ are considered on the manifold Ψ.

Proof.

Assume that a manifold Ψ is invariant with respect to a Lie group $G^1(X)$. If $z \in \Psi$, then $\psi^k(g(z, a)) = 0$, $(k = 1, 2, ..., s)$. Differentiating with respect to the parameter a, one obtains (5.11).

Let us prove a converse statement of the theorem. Assume that Ψ is a regularly assigned manifold and equations (5.11) are satisfied. By changing $\widehat{z} = q(z)$ with

$$q^k = \psi^k(z), \ (k = 1, 2, ..., s),$$

the Lie group $G^1(X)$ is represented as $\widehat{z}' = \widehat{g}(\widehat{z}, a)$, where

$$\widehat{g}(\widehat{z}, a) = q(g(q^{-1}(\widehat{z}), a)).$$

Since the coordinates of the infinitesimal generator X under this change of the variables z are $\widehat{\zeta}^\beta(\widehat{z}) = X(q^\beta(z))$, equations (5.11) give

$$\widehat{\zeta}^k(0, 0, ..., 0, \widehat{z}_{s+1}, ..., \widehat{z}_N) = X\psi^k(z)_{|\Psi} = 0, \quad (k = 1, 2, \ldots, s). \qquad (5.12)$$

The proof of the theorem is obtained if one can prove that

$$\widehat{g}^k(0, 0, ..., 0, \widehat{z}_{s+1}, ..., \widehat{z}_N, a) = 0, \quad (k = 1, 2, \ldots, s). \qquad (5.13)$$

Because of the Lie equations the orbit of the point $\widehat{z} = (0, 0, ..., 0, \widehat{z}_{s+1}, ..., \widehat{z}_N) \in \Psi$ satisfies the Cauchy problem

$$\begin{gathered}\frac{d\widehat{g}^i}{da}(0, 0, ..., 0, \widehat{z}_{s+1}, ..., \widehat{z}_N, a) = \widehat{\zeta}^i\left(\widehat{g}(0, 0, ..., 0, \widehat{z}_{s+1}, ..., \widehat{z}_N, a\right), \\ \widehat{g}^k(0, 0, ..., 0, \widehat{z}_{s+1}, ..., \widehat{z}_N, 0) = 0, \ \ \widehat{g}^j(0, 0, ..., 0, \widehat{z}_{s+1}, ..., \widehat{z}_N, 0) = \widehat{z}_j \\ (i = 1, 2, ..., N; \ k = 1, 2, ..., s; \ \ j = s+1, s+2, ..., N)\end{gathered} \qquad (5.14)$$

Notice that if $\widehat{g}^k(0, 0, ..., 0, \widehat{z}_{s+1}, ..., \widehat{z}_N, a) = 0, \ (k = 1, 2, ..., s)$, then because of (5.12) the first s equations of (5.14) are satisfied. For the remaining functions $\widehat{g}^j(0, 0, ..., 0, \widehat{z}_{s+1}, ..., \widehat{z}_N, a), \ (j = s+1, s+2, ..., N)$, one has the well-posed Cauchy problem

$$\begin{gathered}\frac{d\widehat{g}^j}{da}(0, 0, ..., 0, \widehat{z}_{s+1}, ..., \widehat{z}_N, a) = \widehat{\zeta}^j\left(\widehat{g}(0, 0, ..., 0, \widehat{z}_{s+1}, ..., \widehat{z}_N, a\right), \\ \widehat{g}^j(0, 0, ..., 0, \widehat{z}_{s+1}, ..., \widehat{z}_N, 0) = \widehat{z}_j, \\ (j = s+1, s+2, ..., N).\end{gathered}$$

Due to the uniqueness of the Cauchy problem (5.14) one obtains (5.13).•

1.3 Admitted Lie group

A system of l-th order differential equations is considered as a manifold in J^l. Let this manifold be assigned by the equations

$$F^k(x, u, p) = 0, \quad (k = 1, 2, \ldots, s).$$

The system and the manifold

$$(S) = \{(x, u, p) \in J^l \mid F^k(x, u, p) = 0, \ \ (k = 1, 2, \ldots, s)\} \qquad (5.15)$$

are denoted by (S). It is assumed that (S) is a regularly assigned manifold. For differential equations it is also assumed

$$rank\left(\frac{\partial(F)}{\partial(u, p)}\right) = s.$$

The last assumption excludes dependence between the independent variables. By virtue of the inverse function theorem this assumption allows system (S) to be solved with respect to s terms, which are the dependent functions and their derivatives. These terms are called main variables, and the other variables are called parametric variables. For the sake of simplicity we study systems of equations where the main variables only consist of the derivatives. This excludes nondifferential equations: all equations of the system (S) are assumed to be essentially differential.

Let $\underset{l}{G}$ be l-th prolongation of a Lie group $G^1(X)$. An invariant $\Phi : U \to R$ of the group $\underset{l}{G}$ with $\Phi_p \neq 0$ is called a differential invariant. Here U is some open set in J^l.

Definition 5.5. *A local Lie group of transformations $G^1(X)$ (5.2) is admitted by the system (S) (or the system (S) admits the Lie group $G^1(X)$) if the manifold (S) is an invariant manifold with respect to the prolonged group $\underset{l}{G}$. The generator X of the admitted Lie group G^1 is also called admitted by the system (S).*

The main feature of the admitted Lie group G^1 is: any solution $u_o(x)$ of the system (S) is transformed to the solution $u_a(x')$ of the same system[3]. Here the functions $u_o(x)$ and $u_a(x')$ are related by the formulae (5.8). In fact, since $u_o(x)$ is a solution of (S), then $(x, u_o(x), p_o(x)) \in (S)$, where $p_o(x)$ are derivatives of the function $u_o(x)$. The manifold (S) is invariant with respect to the prolonged Lie group $\underset{l}{G}$. This means that the orbit of the point $(x, u_o(x), p_o(x))$ belongs to the manifold (S). According to the construction of the prolonged group, the orbit of this point consists of the points $(x', u_a(x'), p_a(x'))$. Thus, $u_a(x')$ is also a solution of the system (S).

Since any element $u_a(x')$ is a solution, the admitted group can produce new solutions from a particular solution $u_o(x)$. There are systems for which an admitted group can produce the general solution from one particular[4]. Here is a simple example.

Let us consider the Lie group

$$x' = x, \quad u' = ue^a.$$

The prolonged Lie group is

$$x' = x, \quad u' = ue^a, \quad p' = pe^a,$$

where p and p' denote the derivatives $\frac{du}{dx}$ and $\frac{du'}{dx'}$, respectively. It is easy to check that this group is admitted by the ordinary differential

[3]The ability of an admitted Lie group to transform any solution into a solution is very important, and it is also used for defining the admitted Lie group. A discussion on this topic can be found in [72] and the references therein. The equivalence of the two definitions will be also discussed in Chapter 6.

[4]These systems are called automorphic [130].

equation

$$\frac{du}{dx} = 2ux.$$

The function $u_o(x) = e^{x^2}$ is a solution of this equation. The transformed function is $u_a(x) = e^{a+x^2}$. Noting that the change of the variables

$$x' = x, \ u' = -u$$

leaves the equation $\frac{du}{dx} = 2ux$ unchanged[5], from the particular solution e^{x^2} the general solution $u_a(x) = Ce^{x^2}$ is obtained. Here $C = \pm e^a$.

1.4 Algorithm of finding an admitted Lie group

By virtue of the theorem for a regularly assigned manifold to be invariant, the algorithm for finding a local one–parameter Lie group (5.2), admitted by a system of differential equations (S), consists of the following five steps.

At the first step, the form of the admitted generator

$$X = \xi^i(x,u)\partial_{x_i} + \eta^j(x,u)\partial_{u^j},$$

with the unknown coefficients $\xi^i(x,u)$, $\eta^j(x,u)$ is given. For the second step one constructs the prolonged infinitesimal operator $\underset{l}{X}$. The coefficients of the operator $\underset{l}{X}$ are defined by the prolongation formulae (5.6). At the third step the prolonged operator $\underset{l}{X}$ is applied to each equation of the system (S). The next step is a transition onto the manifold (S): the main variables are substituted through the parametric variables. As the result the system of differential equations

$$DS: \qquad \underset{l}{X}F^k(x,u,p)_{|(S)} = 0, \quad (k = 1, 2, \ldots, s).$$

is obtained. These equations are linear homogeneous differential equations for the unknown coefficients $\xi^i(x,u)$, $\eta^j(x,u)$. They are called the determining equations. Since the coefficients of the generator X do not depend on the derivatives p^k_α, the determining equations can be split with respect to the parametric derivatives. The split system of equations is an overdetermined system. It is also denoted by DS. The general solution of the determining equations DS generates a principal Lie algebra LS of the system (S). The set of transformations, which is finitely generated by one-parameter Lie groups corresponding to the generators $X \in LS$ is called a principal Lie group admitted by the system (S). This group is denoted by GS.

Since, in the given approach for finding an admitted Lie group, the main role belongs to manifolds, this approach is called a geometrical approach. In the

[5]Such transformations are called discrete symmetries.

geometrical approach there is no necessity to have the existence of a solution of the system (S). In other approaches one needs the existence of a solution of the system (S). Relations between different approaches are discussed in [96, 130, 127, 73, 150]. Notice that all these approaches give the same result for involutive systems.

1.5 Example of finding an admitted Lie group

Let us study the following nonlinear equation

$$(S): \qquad u_{tt} = k(u)u_{xx}, \quad (k(u) > 0, \;\; k'(u) \neq 0). \tag{5.16}$$

In this case $n = 2,\ m = 1,\ l = 2$, the form of an infinitesimal generator is

$$X = \xi^t(t, x, u)\partial_t + \xi^x(t, x, u)\partial_x + \eta(t, x, u)\partial_u.$$

The prolonged generator is

$$\underset{2}{X} = X + \eta^{u_t}\partial_{u_t} + \eta^{u_x}\partial_{u_x} + \eta^{u_{xx}}\partial_{u_{xx}} + \eta^{u_{tx}}\partial_{u_{tx}} + \eta^{u_{tt}}\partial_{u_{tt}}$$

with the coefficients

$$\eta^{u_x} = D_x\eta - u_x D_x\xi^x - u_t D_x\xi^t = \eta_x + u_x\eta_u - u_x(\xi^x_x + u_x\xi^x_u) - u_t(\xi^t_x + u_x\xi^t_u),$$

$$\eta^{u_t} = D_t\eta - u_x D_t\xi^x - u_t D_t\xi^t = \eta_t + u_t\eta_u - u_x(\xi^x_t + u_t\xi^x_u) - u_t(\xi^t_t + u_t\xi^t_u),$$

$$\eta^{u_{xx}} = D_x\eta^{u_x} - u_{xx}D_x\xi^x - u_{xt}D_x\xi^t =$$
$$= \eta_{xx} + 2u_x\eta_{xu} + u_{xx}\eta_u + (u_x)^2\eta_{uu} - 2u_{xx}\xi^x_x - u_x\xi^x_{xx} -$$
$$-2(u_x)^2\xi^x_{xu} - 3u_x u_{xx}\xi^x_u - (u_x)^3\xi^x_{uu} - 2u_{tx}\xi^t_x - u_t\xi^t_{xx} -$$
$$-2u_x u_t\xi^t_{xu} - (u_t u_{xx} + 2u_x u_{tx})\xi^t_u - (u_x)^2 u_t\xi^t_{uu}.$$

$$\eta^{u_{tt}} = D_t\eta^{u_t} - u_{xt}D_t\xi^x - u_{tt}D_t\xi^t =$$
$$= \eta_{tt} + 2u_t\eta_{tu} + u_{tt}\eta_u + (u_t)^2\eta_{uu} - 2u_{tt}\xi^t_t - u_t\xi^t_{tt} -$$
$$-2(u_t)^2\xi^t_{tu} - 3u_t u_{tt}\xi^t_u - (u_t)^3\xi^t_{uu} - 2u_{xt}\xi^x_t - u_x\xi^x_{tt} -$$
$$-2u_x u_t\xi^x_{tu} - (u_x u_{tt} + 2u_t u_{xt})\xi^x_u - (u_t)^2 u_x\xi^x_{uu}.$$

Here the operators

$$D_t = \partial_t + u_t\partial_u + u_{tt}\partial_{u_t} + u_{tx}\partial_{u_x} + \ldots,$$

$$D_x = \partial_x + u_x\partial_u + u_{tx}\partial_{u_t} + u_{xx}\partial_{u_x} + \ldots,$$

are the operators of the total derivatives with respect to t and x, respectively. The determining equations DS are

$$[\eta^{u_{tt}} - k\eta^{u_{xx}} - u_{xx}k'\eta]_{|(S)} = 0.$$

Substituting the coefficients $\eta^{u_{tt}}$, $\eta^{u_{xx}}$ and the derivative: $u_{tt} = ku_{xx}$, one obtains

$$-ku_x^3\xi^x_{uu} - ku_x^2u_t\xi^t_{uu} + ku_x^2(\eta_{uu} - 2\xi^x_{xu}) + u_xu_t^2\xi^x_{uu} + 2u_xu_t(\xi^x_{tu} - k\xi^t_{xu}) -$$

$$-2ku_xu_{xx}\xi^x_u - 2u_xu_{xt}k\xi^t_u + u_x(2k\eta_{xu} + \xi^x_{tt} - k\xi^x_{xx}) + u_t^3\xi^t_{uu} + u_t^2(-\eta_{uu} + 2\xi^t_{tu}) +$$

$$+2ku_tu_{xx}\xi^t_u - u_t(2\eta_{tu} - \xi^t_{tt} + k\xi^t_{xx}) + u_{xx}(k'\eta - 2k\xi^x_x + 2k\xi^t_t) + 2u_{xt}(\xi^x_t - k\xi^t_x)$$

$$-\eta_{tt} + k\eta_{xx} = 0$$

The last equation can be split with respect to the parametric derivatives u_x, u_t, u_{xx}, u_{xt}. After a simple reduction, one has

$$\begin{aligned} \eta_{uu} = 0, \ \ \xi^x_u = 0, \ \ \xi^t_u = 0, \ \ 2k\eta_{xu} + \xi^x_{tt} - k\xi^x_{xx} = 0, \\ 2\eta_{tu} - \xi^t_{tt} + k\xi^t_{xx} = 0, k'\eta - 2k(\xi^x_x - \xi^t_t) = 0, \\ k\eta_{xx} - \eta_{tt} = 0. \ \ \xi^x_t - k\xi^t_x = 0, \end{aligned} \quad (5.17)$$

From these equations (after differentiating some with respect to u) the relations

$$\xi^x_t = 0, \quad \xi^t_x = 0$$

are obtained. Since $k' \neq 0$, the function η is found from (5.17)

$$\eta = 2\frac{k}{k'}(\xi^x_x - \xi^t_t).$$

By virtue of the equation $\eta_{uu} = 0$ one has

$$(\frac{k}{k'})''(\xi^x_x - \xi^t_t) = 0. \quad (5.18)$$

This equation is called a classifying relation: the admitted group depends on the function $k = k(u)$. Here, for the sake of simplicity[6], a particular problem is considered: namely, to find an admitted Lie group of continuous transformations, which is admitted by equation (5.16) for all functions $k = k(u)$. This Lie group is called a kernel of the principal groups GS. Since $k = k(u)$ is arbitrary, the determining equations DS also can be split with respect to k, k', k''. Thus, one gets

$$\eta = 0, \ \ \xi^x_x - \xi^t_t = 0, \ \ \xi^t_{tt} = 0, \ \ \xi^x_{xx} = 0,$$

A solution of the last equations is

$$\eta = 0, \ \ \xi^x = c_1x + c_2, \ \ \xi^t = c_1t + c_3.$$

[6]The determining equations (5.17) will be studied further in section 8.

The generator corresponding to these coefficients is

$$X = c_1X_1 + c_2X_2 + c_3X_3$$

with

$$X_1 = \partial_t, \quad X_2 = \partial_x, \quad X_3 = t\partial_t + x\partial_x. \tag{5.19}$$

The generators X_i $(i = 1, 2, 3)$ compose a basis of the kernel of the admitted Lie algebras. The Lie groups of transformations corresponding to these basis generators are:

$$\begin{aligned} X_1: &\quad t' = t + a, \; x' = x, \; u' = u. \\ X_2: &\quad t' = t, \; x' = x + a, \; u' = u, \\ X_3: &\quad t' = te^a, \; x' = xe^a, \; u' = u, \end{aligned}$$

1.6 Lie algebra of generators

Let us consider two generators

$$X_i = \zeta_i^\alpha(z)\partial_{z_\alpha}, \quad (i = 1, 2).$$

Definition 5.6. *The generator*

$$X_3 = \zeta_3^\alpha(z)\partial_{z_\alpha}$$

with the coefficients

$$\zeta_3^\alpha = X_1(\zeta_2^\alpha) - X_2(\zeta_1^\alpha)$$

is called a commutator[7] *of the generators* X_1 *and* X_2.

The commutator is denoted by

$$X_3 = [X_1, X_2].$$

For a function $F(z) \in C^2$ one has

$$[X_1, X_2]F = X_1(X_2F) - X_2(X_1F). \tag{5.20}$$

The operation of commutation is bilinear

$$\begin{aligned} [\alpha X_1 + \beta X_2, X_3] &= \alpha[X_1, X_3] + \beta[X_2, X_3], \\ [X_1, \alpha X_2 + \beta X_3] &= \alpha[X_1, X_2] + \beta[X_1, X_3], \end{aligned}$$

antisymmetric

$$[X_1, X_2] = -[X_2, X_1],$$

and satisfies the Jacobi identity

$$[[X_1, X_2], X_3] + [[X_2, X_3], X_1] + [[X_3, X_1], X_2] = 0.$$

[7]Compare with the Poisson bracket.

Definition 5.7. *A vector space with an operation of commutation, which satisfies the properties of bilinearity, antisymmetry, the Jacobi identity, and acts into this space, is called a Lie algebra.*

In particular, for a space of generators.

Definition 5.8. *A vector space L of generators is a Lie algebra if the commutator $[X_1, X_2]$ of any two generators $X_1 \in L$ and $X_2 \in L$ belongs to L.*

Lemma 5.3. *A commutator is invariant with respect to any change of variables.*

Proof.

Let $\widehat{z} = q(z)$ be a change of variables. Since generators are invariant with respect to this operation, $X = X' = X(q_i)\partial_{\widehat{z}_i}$, $Y = Y' = Y(q_i)\partial_{\widehat{z}_i}$. Hence,

$$\begin{aligned}
[X', Y'] &= \left(X'(Y(q_i)) - Y'(X(q_i))\right)\partial_{\widehat{z}_i} = (X(Y(q_i)) - Y(X(q_i)))\,\partial_{\widehat{z}_i} \\
&= [X, Y](q_i)\partial_{\widehat{z}_i} = [X, Y]'.
\end{aligned}$$

•

Theorem 5.5. *If a regularly assigned manifold Ψ is invariant with respect to generators X and Y, then it is invariant with respect to their commutator $[X, Y]$.*

Proof.

Since the operation of commutation is invariant with respect to a change of variables, without loss of generality one can use the assignment of the manifold Ψ given by the equations

$$z_1 = 0,\ z_2 = 0,\ \dots,\ z_s = 0. \tag{5.21}$$

Necessary and sufficient conditions for this manifold to be invariant with respect to the generators $X = \zeta_1^j(z)\partial_{z_j}$, $Y = \zeta_2^j(z)\partial_{z_j}$ are

$$\zeta_i^j(0, 0, \dots, z_{s+1}, \dots, z_N) = 0,\ \ \zeta_i^j(0, 0, \dots, z_{s+1}, \dots, z_N) = 0,$$
$$(j = 1, 2, \dots, s;\ i = 1, 2). \tag{5.22}$$

The coefficients of the commutator are

$$[X, Y]^j = X\zeta_2^j - Y\zeta_1^j = \zeta_1^\alpha \frac{\partial \zeta_2^j}{\partial z_\alpha} - \zeta_2^\alpha \frac{\partial \zeta_1^j}{\partial z_\alpha} =$$
$$\sum_{\alpha=1}^{s}\left(\zeta_1^\alpha \frac{\partial \zeta_2^j}{\partial z_\alpha} - \zeta_2^\alpha \frac{\partial \zeta_1^j}{\partial z_\alpha}\right) + \sum_{\alpha=s+1}^{N}\left(\zeta_1^\alpha \frac{\partial \zeta_2^j}{\partial z_\alpha} - \zeta_2^\alpha \frac{\partial \zeta_1^j}{\partial z_\alpha}\right),\ (j = 1, 2, \dots, N).$$

On the manifold Ψ the first sum vanishes because of (5.21), the second sum vanishes since the operations of taking derivatives with respect to z_j, $(j = 1, 2, \dots, s)$ and transition onto the manifold Ψ commute. Thus,

$$[X, Y]^j(0, 0, \dots, z_{s+1}, \dots, z_N) = 0,\ (j = 1, 2, \dots, s).$$

•

Theorem 5.6. *The operation of prolongation commutes with the operation of taking a commutator.*

Proof.

There exists a change of variables such that the operator X has the representation $X = \partial_{x_1}$. Since the operations of commutation and prolongation are invariant with respect to a change of variables, it is enough to prove the theorem in these variables. In these coordinates

$$[X, Y] = \frac{\partial \xi^i}{\partial x_1}\partial_{x_i} + \frac{\partial \eta^k}{\partial x_1}\partial_{u^k},$$

the prolonged operator $\widetilde{X} = X$, and

$$\widetilde{Y} = Y + \eta_i^k(x, u, p)\partial_{p_i^k} = \xi^\beta(x, u)\partial_{x_\beta} + \eta^\alpha(x, u)\partial_u^\alpha + \eta_i^k(x, u, p)\partial_{p_i^k},$$

where $\eta_i^k = D_i\eta^k - p_\beta^k D_i\xi^\beta$. Hence,

$$[\widetilde{X}, \widetilde{Y}] = [X, \widetilde{Y}] = [X, Y] + [X, \eta_i^k\partial_{p_i^k}] = [X, Y] + \frac{\partial \eta_i^k}{\partial x_1}\partial_{p_i^k}.$$

The prolongation of the commutator is

$$[\widetilde{X, Y}] = [X, Y] + \widehat{\eta}_i^k\partial_{p_i^k},$$

where

$$\widehat{\eta}_i^k = D_i\left(\frac{\partial \eta^k}{\partial x_1}\right) - p_\beta^k D_i\left(\frac{\partial \xi^\beta}{\partial x_1}\right) = \frac{\partial}{\partial x_1}\left(D_i\eta^k - p_\beta^k D_i\xi^\beta\right) = \frac{\partial \eta_i^k}{\partial x_1}.$$

Thus,

$$[\widetilde{X, Y}] = [\widetilde{X}, \widetilde{Y}].$$

•

In particular, if the generators X and Y are admitted by a system of differential equations (S), then

$$\widetilde{X}F^k(x, u, p)_{|(S)} = 0, \ \widetilde{Y}F^k(x, u, p)_{|(S)} = 0, \quad (k = 1, 2, \ldots, s).$$

Hence, the manifold (S) is invariant with respect to $[\widetilde{X}, \widetilde{Y}]$:

$$[\widetilde{X}, \widetilde{Y}]F^k(x, u, p)_{|(S)} = 0, \quad (k = 1, 2, \ldots, s).$$

Since $[\widetilde{X, Y}] = [\widetilde{X}, \widetilde{Y}]$, we have the following theorem.

Theorem 5.7. *If a system (S) admits generators X and Y, then it admits their commutator $[X, Y]$.*

This theorem means that the vector space LS of all admitted generators is a Lie algebra (admitted by the system (S)). This algebra is called a principal algebra. To construct exact solutions one uses subalgebras of the admitted algebra.

Let L be a Lie algebra of generators.

Definition 5.9. *A vector subspace $L' \subset L$ of a Lie algebra L is called a subalgebra if it is a Lie algebra, i.e., for arbitrary vectors X_1 and X_2 from L' their commutator $[X_1, X_2]$ belongs to L'.*

Therefore any subalgebra is a Lie algebra. Among all subalgebras there are subalgebras which play a special role.

Definition 5.10. *Let $I \subset L$ be a subspace of a Lie algebra L with the property:*

$$[X, Y] \in I, \quad \forall X \in I, \ \forall Y \in L.$$

The subspace I is called an ideal.

Definition 5.11. *Two Lie algebras L' and L'' are similar if there exists a change of variables that transforms one into another.*

Hence, if two Lie algebras L' and L'' are similar, then the generators $X = \zeta^\beta(z)\partial_{z_\beta} \in L'$ and $\widehat{X} = \widehat{\zeta}^\beta(\widehat{z})\partial_{\widehat{z}_\beta} \in L''$ of these algebras are related by the formulas

$$\widehat{\zeta}^\beta(\widehat{z}) = X(q^\beta(z))_{|z=q^{-1}(\widehat{z})}.$$

A set of all subalgebras can be classified with respect to automorphisms (isomorphic homomorphisms from L into L). These constructions are studied in the next subsection.

1.7 Classification of subalgebras

Since any solution of a system of differential equations is mapped by a transformation from admitted Lie group into a solution of the same system, the problem of separating solutions into classes of essentially different solutions appears. For this problem one needs to study what happens to a solution along an orbit of a one-parameter Lie group.

Assume that $Y = \zeta^\alpha(z)\partial_{z_\alpha}$. Solving the Cauchy problem

$$\frac{dg}{dt}(z, t) = \zeta(g(z, t)), \ \ g(z, 0) = z \tag{5.23}$$

in a neighborhood of some point z_0, where $\zeta(z_0) \neq 0$, one finds a Lie group of transformations

$$\widehat{z} = g(z, t).$$

The properties of the function $g(z, t)$ are as follows. First of all there is the relation

$$g^\alpha(g(\widehat{z}, -t), t) = \widehat{z}_\alpha, \ \ (\alpha = 1, 2, ..., N). \tag{5.24}$$

Differentiating the last relations with respect to t, one obtains

$$\frac{\partial g^{\alpha}}{\partial t}(g(\widehat{z},-t),t)-\frac{\partial g^{\alpha}}{\partial z_{\gamma}}(g(\widehat{z},-t),t)\frac{\partial g^{\gamma}}{\partial t}(\widehat{z},-t)=0,$$

or, using the Lie equations,

$$\zeta^{\alpha}(\widehat{z})-\frac{\partial g^{\alpha}}{\partial z_{s}}(g(\widehat{z},-t),t)\frac{\partial g^{s}}{\partial t}(\widehat{z},-t)=0. \tag{5.25}$$

Differentiating equations (5.24) with respect to $\widehat{z}_{\beta}$, one has

$$\frac{\partial g^{\alpha}}{\partial z_{\gamma}}(g(\widehat{z},-t),t)\frac{\partial g^{\gamma}}{\partial z_{\beta}}(\widehat{z},-t)=\delta_{\alpha\beta}. \tag{5.26}$$

This means that two matrices $A=\left(\frac{\partial g^{\alpha}}{\partial z_{\gamma}}(g(\widehat{z},-t),t)\right)$, $B=\left(\frac{\partial g^{\gamma}}{\partial z_{\beta}}(\widehat{z},-t)\right)$ are invertible and that

$$AB=I=BA. \tag{5.27}$$

Note also that differentiating the Lie equations with respect to z_{α}, and substituting $z=g(\widehat{z},-t)$, one finds

$$\frac{\partial^{2}g^{\beta}}{\partial z_{\alpha}\partial t}(g(\widehat{z},-t),t)=\frac{\partial \zeta^{\beta}}{\partial z_{\gamma}}(\widehat{z})\frac{\partial g^{\gamma}}{\partial z_{\alpha}}(g(\widehat{z},-t),t). \tag{5.28}$$

Varying the parameter t, the generator $X=h^{\alpha}(z)\partial_{z_{\alpha}}$ is changed as follows

$$\widehat{X}=h^{\alpha}(g(\widehat{z},-t))\frac{\partial g^{\beta}}{\partial z_{\alpha}}(g(\widehat{z},-t),t)\,\partial_{\widehat{z}_{\beta}}=\phi^{\beta}(\widehat{z},t)\partial_{\widehat{z}_{\beta}}.$$

This means that along the trajectory defined by the generator Y, the coefficients of the operator X are changed according to the last formulae.

Differentiating the coefficients of $\widehat{X}$ with respect to the parameter t (considered in terms of the variables $t,\widehat{z}$), one obtains

$$\begin{gathered}\frac{\partial}{\partial t}\phi^{\beta}(\widehat{z},t)=\frac{\partial}{\partial t}\left(h^{\alpha}(g(\widehat{z},-t))\frac{\partial g^{\beta}}{\partial z_{\alpha}}(g(\widehat{z},-t),t)\right)=\\ -\frac{\partial h^{\alpha}}{\partial z_{\gamma}}(g(\widehat{z},-t))\frac{\partial g^{\gamma}}{\partial t}(\widehat{z},-t)\frac{\partial g^{\beta}}{\partial z_{\alpha}}(g(\widehat{z},-t),t)+\\ h^{\alpha}(g(\widehat{z},-t))[\frac{\partial^{2}g^{\beta}}{\partial z_{\alpha}\partial t}(g(\widehat{z},-t),t)-\frac{\partial g^{\gamma}}{\partial t}(\widehat{z},-t)\frac{\partial^{2}g^{\beta}}{\partial z_{\alpha}\partial z_{\gamma}}(g(\widehat{z},-t),t)]\end{gathered}$$

Using (5.28) gives

$$\begin{gathered}\frac{\partial}{\partial t}\phi^{\beta}(\widehat{z},t)=-\frac{\partial g^{\gamma}}{\partial t}(\widehat{z},-t)[\frac{\partial h^{\alpha}}{\partial z_{\gamma}}(g(\widehat{z},-t))\frac{\partial g^{\beta}}{\partial z_{\alpha}}(g(\widehat{z},-t),t)+\\ h^{\alpha}(g(\widehat{z},-t))\frac{\partial^{2}g^{\beta}}{\partial z_{\alpha}\partial z_{\gamma}}(g(\widehat{z},-t),t)]+h^{\alpha}(g(\widehat{z},-t))\frac{\partial \zeta^{\beta}}{\partial z_{\gamma}}(\widehat{z})\frac{\partial g^{\gamma}}{\partial z_{\alpha}}(g(\widehat{z},-t),t)\end{gathered}$$

The coefficients $\lambda^{\beta}(\widehat{z}, t)$ of the commutator

$$[\widehat{X}, \zeta^{\mu}(\widehat{z})\partial_{\widehat{z}_{\mu}}] = \lambda^{\beta}(\widehat{z}, t)\partial_{\widehat{z}_{\beta}},$$

are

$$\begin{aligned}\lambda^{\beta}(\widehat{z}, t) = h^{\alpha}(g(\widehat{z}, -t))\tfrac{\partial g^{\gamma}}{\partial z_{\alpha}}(g(\widehat{z}, -t), t)\tfrac{\partial \zeta^{\beta}}{\partial z_{\gamma}}(\widehat{z}) \\ -\zeta^{\mu}(\widehat{z})\tfrac{\partial}{\partial \widehat{z}_{\mu}}\left(h^{\alpha}(g(\widehat{z}, -t))\tfrac{\partial g^{\beta}}{\partial z_{\alpha}}(g(\widehat{z}, -t), t)\right) = \\ h^{\alpha}(g(\widehat{z}, -t))\tfrac{\partial g^{\gamma}}{\partial z_{\alpha}}(g(\widehat{z}, -t), t)\tfrac{\partial \zeta^{\beta}}{\partial z_{\gamma}}(\widehat{z}) \\ -\zeta^{\mu}(\widehat{z})\left(\tfrac{\partial h^{\alpha}}{\partial z_{\gamma}}(g(\widehat{z}, -t))\tfrac{\partial g^{\gamma}}{\partial z_{\mu}}(\widehat{z}, -t)\tfrac{\partial g^{\beta}}{\partial z_{\alpha}}(g(\widehat{z}, -t), t)\right) \\ -\zeta^{\mu}(\widehat{z})\left(h^{\alpha}(g(\widehat{z}, -t))\tfrac{\partial^{2} g^{\beta}}{\partial z_{\alpha}\partial z_{\gamma}}(g(\widehat{z}, -t), t)\tfrac{\partial g^{\gamma}}{\partial z_{\mu}}(\widehat{z}, -t)\right).\end{aligned}$$

Substituting $\zeta^{\mu}(\widehat{z})$ from (5.25), and taking into account (5.27), one obtains

$$\frac{\partial}{\partial t}\phi^{\beta}(\widehat{z}, t) = \lambda^{\beta}(\widehat{z}, t).$$

Thus,

$$\frac{d}{dt}\widehat{X} = [\widehat{X}, Y], \ \widehat{X}_{|t=0} = X. \tag{5.29}$$

Let L_r be an r-th dimensional Lie algebra of generators with a basis $\{X_1, X_2, ..., X_r\}$. Any generator X is decomposed through the basis

$$X = x^{\alpha}X_{\alpha}.$$

Since L_r is closed under the commutation, the commutator of any two generators X_i and X_j in the basis is a linear combination of the basis generators

$$[X_i, X_j] = C^{\gamma}_{ij}X_{\gamma}.$$

The constants C^{α}_{ij} are called structure constants.

If $Y = y^{\alpha}X_{\alpha} \in L_r$, then

$$[X, Y] = x^{\alpha}y^{\beta}[X_{\alpha}, X_{\beta}] = x^{\alpha}y^{\beta}C^{\gamma}_{\alpha\beta}X_{\gamma}.$$

Therefore, for the coordinates $x = (x^1, x^2, ..., x^r)$ and $y = (y^1, y^2, ..., y^r)$ of the generators X and Y one can define an operation of commutation $[x, y]$:

$$[x, y]^{\gamma} = x^{\alpha}y^{\beta}C^{\gamma}_{\alpha\beta}, \ (\gamma = 1, 2, ..., r).$$

With this operation the vector space $R^r(x)$ becomes a Lie algebra. This Lie algebra is also denoted by L_r. The Jacobi identity for this Lie algebra

$$[[x, y], z] + [[y, z], x] + [[z, x], y] = 0, \ \forall x, y, z \in R^r$$

is induced by the Lie algebra L_r of generators.

For a finite dimensional Lie algebra L_r equations (5.29) become

$$\frac{d}{dt}\widehat{x}^\gamma = [\widehat{x}, y]^\gamma = \widehat{x}^\alpha y^\beta C^\gamma_{\alpha\beta}, \ \widehat{x}^\gamma_{|t=0} = x^\gamma, \quad (\gamma = 1, 2, ..., r), \tag{5.30}$$

where $\widehat{X} = \widehat{x}^\gamma X_\gamma$. Hence, the problem of studying the coefficients $\phi^\beta(z, t)$ of the generator $\widehat{X}$ along the trajectory is reduced to the problem of studying coordinates of the vector $\widehat{x} = (\widehat{x}_1, \widehat{x}_2, ..., \widehat{x}_r)$.

By virtue of linearity of the problem, (5.30) its solution is an automorphism of R^r

$$\widehat{x} = A_y(t)x.$$

The set of all automorphisms $A_y(t)$ is called a set of inner automorphisms of the Lie algebra L_r. This set is denoted by $Int(L_r)$. The name automorphism is justified by the following property. Let $\widehat{x} = A_y(t)x$ and $\widehat{u} = A_y(t)u$, then

$$\frac{d}{dt}[\widehat{x}, \widehat{u}] = [[\widehat{x}, y], \widehat{u}] + [\widehat{x}, [\widehat{u}, y]] = [[\widehat{x}, \widehat{u}], y],$$
$$[\widehat{x}, \widehat{u}]_{|t=0} = [x, u].$$

Here the Jacobi identity was used. Hence, because of uniqueness of the solution of the Cauchy problem

$$A_y(t)[x, u] = [\widehat{x}, \widehat{u}] = [A_y(t)x, A_y(t)u].$$

Any subalgebra $L_s \subset L_r$ is transformed by $A_y(t)$ into a similar subalgebra. Similar subalgebras of the same dimension composes a class.

Definition 5.12. *The set of all classes (one representative from each class) is called an optimal system of subalgebras.*

Thus, the optimal system of subalgebras of a Lie algebra L with inner automorphisms $A = Int(L)$ is a set of subalgebras $\Theta_A(L)$ such that:

a) there are no two elements of this set which can be transformed into each other by inner automorphisms of the Lie algebra L;

b) any subalgebra of the Lie algebra L can be transformed into one of subalgebras of the set $\Theta_A(L)$

Any subalgebra of a Lie algebra L is completely defined by its basis generators. These basis generators are linear combinations of basis operators of the Lie algebra L. Hence, the subalgebra is completely defined by coefficients of these linear combinations. As was shown, actions of automorphisms are transferred to these coefficients. Except for automorphisms, one has also to take into account a uniform scaling of all generators: any subalgebra is transformed into a similar subalgebra under this operation. Other possibilities appear in choosing a basis of subalgebra. The problem of constructing an optimal system of subalgebras reduces to obtaining the maximum possible number of zero coordinates of the subalgebras bases.

In group analysis it is shown[8] that the problem of finding all automorphisms is reduced to the problem for finding automorphisms A_k for the basis vectors $y = e^k$, $(k = 1, 2, ..., s)$:

$$\frac{d}{dt}\widehat{x}^{\gamma} = \widehat{x}^{\alpha} C^{\gamma}_{\alpha k}, \ \widehat{x}^{\gamma}_{|t=0} = x^{\gamma}, \ (\gamma = 1, 2, ..., r). \tag{5.31}$$

Here $\{e^k\}_{k=1}^r$ is the canonical basis in R^r. The automorphism A_k corresponds to the Lie group of transformations with the generator

$$x^{\alpha} C^{\gamma}_{\alpha k} \partial_{x^{\gamma}}. \tag{5.32}$$

1.8 Classification of subalgebras of algebra (5.19)

The table of commutators of the generators (5.19) is

	X_1	X_2	X_3
X_1	0	0	X_1
X_2	0	0	X_2
X_3	$-X_1$	$-X_2$	0

According to (5.32) the automorphisms are defined by the generators

$$\begin{array}{ll} A_1: & -x^3\partial_{x^1}, \\ A_2: & -x^3\partial_{x^2}, \\ A_3: & x^1\partial_{x^1} + x^2\partial_{x^2} \end{array}$$

or, after integrating the Lie equations, they become

$$\begin{array}{ll} A_1: & \widehat{x}^1 = x^1 - a_1x^3, \ \widehat{x}^2 = x^2, \ \widehat{x}^3 = x^3, \\ A_2: & \widehat{x}^1 = x^1, \ \widehat{x}^2 = x^2 - a_2x^3, \ \widehat{x}^3 = x^3, \\ A_3: & \widehat{x}^1 = e^{a_3}x^1, \ \widehat{x}^2 = e^{a_3}x^2, \ \widehat{x}^3 = x^3, \end{array}$$

where a_1, a_2, a_3 are parameters of automorphisms A_1, A_2, A_3, respectively.

The two dimensional subalgebras can be of two types: subalgebras including the generator X_3, and those without X_3. There is only subalgebra without the generator X_3

$$\{X_1, X_2\}.$$

For the second type of subalgebras, after scaling and taking into account linear combinations, a basis of them can be chosen as follows

$$\{X_3 + q_1X_1 + q_2X_2, \ p_1X_1 + p_2X_2\}, \ (p_1^2 + p_2^2 \neq 0).$$

[8]See in [130].

By the automorphisms A_1 and A_2 the coefficients q_1 and q_2 are transformed to zero. By these automorphisms the coefficients p_1 and p_2 are unchanged. Hence, all classes of these subalgebras are

$$\{X_3, X_1 + p_2 X_2\}, \ \{X_3, X_2\}$$

with an arbitrary p_2. Thus, the list of all dissimilar two dimensional subalgebras consists of the subalgebras

$$\{X_3, X_1 + p_2 X_2\}, \ \{X_3, X_2\}, \ \{X_1, X_2\}.$$

Similarly one can classify all one dimensional subalgebras. They are

$$\{X_3\}, \{X_1 + p_2 X_2\}, \ \{X_2\}.$$

The optimal system of subalgebras of algebra (5.19) consists of the subalgebras

$$\{X_1, X_2, X_3\}, \ \{X_3, X_1 + p_2 X_2\}, \ \{X_3, X_2\}, \ \{X_1, X_2\},$$
$$\{X_3\}, \{X_1 + p_2 X_2\}, \ \{X_2\}.$$

1.9 On classification of high dimensional Lie algebras

Calculations of an optimal system of subalgebras is easy enough for low-dimensional Lie algebras. For high-dimensional Lie algebras one can use a two-step algorithm [131]. This algorithm reduces the problem of constructing an optimal system of subalgebras with high dimensions to a problem with low dimensions.

Let L be a Lie algebra L with the basis $\{X_1, X_2, \ldots, X_r\}$. Assume that the Lie algebra L is decomposed in $L = I \oplus F$, where I is a proper ideal of the algebra L and F is a subalgebra. The set of inner automorphisms $A = Int(L)$ of the Lie algebra L is also decomposed $A = A_I A_F$. This means the following [131].

Let $x \in L$ be decomposed as $x = x_I + x_F$, where $x_I \in I$, and $x_F \in F$. Any automorphism $B \in A$ can be written as $B = B_I B_F$, where $B_I \in A_I$, $B_F \in A_F$. The automorphisms B_I and B_F have the properties:

$$B_I x_F = x_F, \ \forall x_F \in F, \ \forall B_I \in A_I,$$
$$B_F x_I \in I, \ B_F x_F \in F, \ \forall x_I \in I, \ \forall x_F \in F, \ \forall B_F \in A_F.$$

At the first step, an optimal system of subalgebras $\Theta_{A_F}(F) = F_1, F_2, ..., F_p, F_{p+1}$ of the algebra F is formed. Here $F_{p+1} = \{0\}$ and the optimal system of the algebra F is constructed with respect to the automorphisms A_F. For each subalgebra F_j, $(j = 1, 2, ..., p+1)$ one has to find its stabilizator $St(F_j) \subset A$:

$$St(F_j) = \{B \in A \mid B(F_j) = F_j\}.$$

Note that $St(F_{p+1}) = A$.

The second step consists of forming optimal systems $\Theta_{St(F_j)}(I \oplus F_j)$. The optimal system of subalgebras $\Theta_A(L)$ of the algebra L is a collection of $\Theta_{St(F_j)}(I \oplus F_j)$, $(j = 1, 2, ..., p + 1)$.

If the subalgebra F also can be decomposed, then the two-step algorithm can be used when constructing $\Theta_{A_F}(F)$.

Definition 5.13. *Let F be a subalgebra of a Lie algebra L. A maximum subalgebra of the Lie algebra L among all subalgebras such that F is an ideal of these subalgebras is called a normalizator NF of the subalgebra F.*

It is useful to require that an optimal system of subalgebras be normalized [131], that is that along any subalgebra from the optimal system its normalizator is also included in the optimal system.

2. Group classification

Most differential equations include arbitrary elements: constants and functions of the independent and dependent variables. These elements specify a process. For example, in equation (5.16) it is the function $k = k(u)$, in the gas dynamics equations and the theory of plasticity arbitrary elements are state equations. The group classification problem consists of finding all principal Lie groups GS admitted by a system of partial differential equations (S). Part of these groups is admitted for all arbitrary elements. This part is called a kernel of admitted Lie groups. Another part depends on the specification of arbitrary elements. This part contains non–equivalent extensions of the kernel.

The first problem of group classification is the problem of constructing transformations which change arbitrary elements, while preserving the differential structure of the equations themselves. These transformations are called equivalent transformations. The group classification is regarded with respect to such transformations.

2.1 Equivalence transformations

A nondegenerate change of the dependent x, independent variables u, and arbitrary elements ϕ, which transfers a system of differential equations of the given class

$$F^k(x, u, p, \phi) = 0, \ (k = 1, 2, \ldots, s) \tag{5.33}$$

to the system of equations of the same class is called an equivalence transformation. The class is defined by the functions $F^k(x, u, p, \phi)$. Here $(x, u) \in V \subset R^{n+m}$, and $\phi : V \to R^t$.

The problem of finding equivalent transformations consists of the construction a transformation of the space $R^{n+m+t}(x, u, \phi)$ that preserves the equations, while only changing their representative $\phi = \phi(x, u)$. For this purpose a one–parameter Lie group of transformations of the space R^{n+m+t} with the group parameter a is used:

$$x' = f^x(x, u, \phi; a), \ u' = f^u(x, u, \phi; a), \ \phi' = f^\phi(x, u, \phi; a). \qquad (5.34)$$

The generator of this group has the form:

$$X^e = \xi^x \partial_x + \zeta^u \partial_u + \zeta^\phi \partial_\phi \qquad (5.35)$$

where the coordinates are:

$$\xi^i = \xi^i(x, u, \phi), \ \zeta^{u^j} = \zeta^{u^j}(x, u, \phi), \ \zeta^{\phi^k} = \zeta^{\phi^k}(x, u, \phi),$$
$$(i = 1, \dots, n; \ j = 1, \dots, m; \ k = 1, \dots, t).$$

Any solution $u_0(x)$ of system (5.33) with the functions $\phi(x, u)$ is transformed by (5.34) into the solution $u = u_a(x')$ of system (5.33) with the same functions F^k, but with another (transformed) function $\phi_a(x, u)$. The function $\phi_a(x, u)$ is defined as follows. Solving the relations

$$x' = f^x(x, u, \phi(x, u); a), \ \ u' = f^u(x, u, \phi(x, u); a)$$

for (x, u), one obtains

$$x = g^x(x', u'; a), \quad u = g^u(x', u'; a). \qquad (5.36)$$

The transformed function is

$$\phi_a(x', u') = f^\phi(x, u, \phi(x, u); a),$$

where, instead of (x, u), one has to substitute their expressions (5.36). Because of the definition of the function $\phi_a(x', u')$, there is the identity with respect to x and u:

$$\left(\phi_a \circ \left(f^x, f^u\right)\right)(x, u, \phi(x, u); a) = f^\phi(x, u, \phi(x, u); a).$$

The transformed solution $T_a(u) = u_a(x)$ is obtained by solving the relations

$$x' = f^x(x, u_o(x), \phi(x, u_o(x)); a)$$

for x and substituting this solution $x = \psi^x(x'; a)$ into

$$u_a(x') = f^u(x, u_o(x), \phi(x, u_o(x)); a).$$

As for the function ϕ_a, there is the identity with respect to x

$$\left(u_a \circ f^x\right)(x, u_o(x), \phi(x, u_o(x)); a) = f^u(x, u_o(x), \phi(x, u_o(x)); a) \quad (5.37)$$

Formulae for transformations of the partial derivatives $p'_a = f^p(x, u, p, \phi, \dots, a)$ are obtained by differentiating (5.37) with respect to x'.

Lemma 5.4. *The transformations $T_a(u)$, constructed above, form a group.*

A proof of the lemma follows from the main property of a Lie group of transformations and the sequence of the equalities

$$x' = f^x(x, u_o(x), \phi(x, u_o(x)); a), \quad u_a(x') = f^u(x, u_o(x), \phi(x, u_o(x)); a)$$

$$x'' = f^x(x', u_a(x'), \phi_a(x', u_a(x')); b),$$
$$u_b(x'') = f^u(x', u_a(x'), \phi_a(x', u_a(x')); b)$$

$$\begin{gathered}(u_b \circ f^x)(x', u_a(x'), \phi_a(x', u_a(x')); b) = \\ f^u(x', u_a(x'), \phi_a(x', u_a(x')); b) = \\ f^u\left(f^x(x, u_o(x), \phi(x, u_o(x)); a), f^u(x, u_o(x), \phi(x, u_o(x)); a), \right. \\ \left. f^\phi(x, u_o(x), \phi(x, u_o(x)); a); b\right) \\ = f^u(x, u_o(x), \phi(x, u_o(x)); a + b) \\ = (u_{a+b} \circ f^x)(x, u_o(x), \phi(x, u_o(x)); a + b).\end{gathered}$$

Because the transformed function $u_a(x')$ is a solution of system (5.33) with the transformed arbitrary element $\phi_a(x', u')$, the equations

$$F^k(x', u_a(x'), p'_a(x'), \phi_a(x', u_a(x'))) = 0, \quad (k = 1, 2, \ldots, s)$$

are satisfied for an arbitrary x'. Because of a one–to–one correspondence between x and x' one has

$$F^k(f^x(z(x), a), f^u(z(x), a), f^p(z_p(x), a), f^\phi(z(x)))) = 0, (k = 1, 2, \ldots, s)$$

where $z(x) = (x, u_o(x), \phi(x, u_o(x)))$, $z_p(x) = (x, u_o(x), \phi(x, u_o(x)), p_o(x), \ldots)$. Differentiating these equations with respect to the group parameter a, one obtains the determining equations

$$\widetilde{X}^e F^k(x, u, p, \phi)_{|(S)} = 0, \quad (k = 1, 2, \ldots, s). \tag{5.38}$$

The prolonged operator for the equivalence Lie group

$$\widetilde{X}^e = X^e + \zeta^{u_x}\partial_{u_x} + \zeta^{\phi_x}\partial_{\phi_x} + \zeta^{\phi_u}\partial_{\phi_u} + \ldots$$

has the following coordinates. The coordinates related to the dependent functions are

$$\zeta^{u_\lambda} = D^e_\lambda \zeta^u - u_x D^e_\lambda \xi^x, \quad D^e_\lambda = \partial_\lambda + u_\lambda \partial_u + (\phi_u u_\lambda + \phi_\lambda)\partial_\phi,$$

where λ takes the values x_i, $(i = 1, 2, ..., n)$. The coordinates of the prolonged operator $\widetilde{X}^e$, related with the arbitrary elements, are defined by the formulae

$$\zeta^{\phi_\lambda} = \tilde{D}^e_\lambda \zeta^\phi - \phi_x \tilde{D}^e_\lambda \xi^x - \phi_u \tilde{D}^e_\lambda \zeta^u, \quad \tilde{D}^e_\lambda = \partial_\lambda + \phi_\lambda \partial_\phi,$$

where $\lambda = u^j$, $(j = 1, 2, ..., m)$, and $\lambda = x_i$, $(i = 1, 2, ..., n)$. The sign $|(S)$ means that the equations $\widetilde{X}^e F^k(x, u, p, \phi)$ are considered on any solution $u_o(x)$ of equations (5.33).

The set of transformations, which is finitely generated by one-parameter Lie groups corresponding to the generators X^e, is called an equivalence group. This group is denoted by GS^e.

The determining equations (5.38) were obtained by using existence of solution of (5.33). After constructing (5.38) one can use a geometrical approach in which the equivalence group is defined by equations (5.38) without the requirement of the existence a solution of (5.33). In this case the sign $|(S)$ means that the equations $\widetilde{X}^e F^k(x, u, p, \phi)$ are considered on the manifold defined by equations (5.33). The difference between these two approaches consists in defining the sign $|(S)$. Note that the same difference between the geometrical approach and the others lies in the definitions for obtaining an admitted Lie group.

By virtue of the constructing equivalence group GS^e, the following theorem is true.

Theorem 5.8. *The kernel of the principal Lie groups is included in the equivalence group GS^e.*

The kernel and the equivalence group GS^e are considered in the same approach.

Remark 5.1. *In some cases there are additional requirements for the arbitrary elements. For example, it is supposed that the arbitrary elements ϕ^μ do not depend on the independent variables $\dfrac{\partial \phi^\mu}{\partial x_k} = 0$. These conditions have to be appended to the original system of differential equations (5.33). They lead to additional determining equations.*

2.1.1 Examples and remarks about an equivalence group

In the classical approach [130] for equivalence groups it was assumed

$$\frac{\partial \xi^i}{\partial \phi^k} = 0, \quad \frac{\partial \zeta^{u^i}}{\partial \phi^k} = 0, \quad (i = 1, \dots, n;\ j = 1, \dots, m;\ k = 1, \dots, t). \tag{5.39}$$

Comparing this with the general case presented above, these assumptions restrict the definition of an equivalence group. Some examples concerning this are given in this section.

Let us consider the system of equations[9]

$$\frac{\partial u}{\partial t} + \frac{\partial}{\partial x} p(u, v) = 0, \quad \frac{\partial v}{\partial t} = g(u, v). \tag{5.40}$$

[9]This system is studied in [78].

These equations have the additional equivalence group with the generator[10]

$$p\partial_u + x\partial_t.$$

Another example of expansion is the system of two quasilinear differential equations

$$u_y\phi_u + \phi_u(\phi^2 - 2u)u_x + v_x e^{-1/\phi_u} = 0, \quad v_y - v_x v e^{-\phi} = 0,$$

with the arbitrary element $\phi = \phi(u)$. If the coefficients of the generator of the equivalence group $X^e = \xi^x\partial_x + \xi^y\partial_y + \zeta^u\partial_u + \zeta^v\partial_v + \zeta^\phi\partial_\phi$ satisfy the conditions

$$\xi^x_\phi = \xi^y_\phi = \zeta^u_\phi = \zeta^v_\phi = 0,$$

then the equivalence group is three–parameter with the operators

$$x\partial_x + y\partial_y, \quad \partial_x, \quad \partial_y.$$

Using the approach just developed, one more generator is obtained as:

$$\phi\partial_u + v\partial_v + \partial_\phi.$$

The third example is related to the system of equations

$$\begin{aligned} -vU_u + (2V - h)U_v + V = 0,\\ -(2U + v)V_u - hV_v + h'U = 0. \end{aligned} \tag{5.41}$$

This system is obtained from the system of equations

$$u_t - u_x = v, \quad v_t + v_x = h(u) \tag{5.42}$$

which is equivalent to the semilinear wave equation

$$u_{tt} - u_{xx} = h(u).$$

System (5.41) is a system of compatibility conditions for the relations

$$u_x = U(u, v), \quad v_x = V(u, v),$$

and (5.42). The group of equivalence transformations of system (5.41) corresponds to the Lie algebra with the generators

$$X_1^e = -u\partial_u + V\partial_V + h\partial_h, \quad X_2^e = \partial_u,$$
$$X_3^e = 2u\partial_u + v\partial_v + U\partial_U,$$
$$X_4^e = -v\partial_v + (U + v)\partial_U + (h - 2V)\partial_V.$$

The transformations corresponding to the generators X_2^e and X_3^e do not change the variable u and the function $h(u)$. They compose the kernel of admitted groups. The generator X_4^e would be lost if one assumes (5.39). It should be also noted that X_4^e does not belong to the kernel, but it is admitted by system (5.41) for any function $h(u)$.

[10]This generator is additional with respect to the classical approach [130], where the conditions (5.39) are assumed.

2.1.2 Group classification of equation (5.16)

Equation (5.16)

$$u_{tt} - ku_{xx} = 0$$

is now studied further here. The arbitrary element in this equation is $k = k(u)$. Since the function $k(u)$ only depends on the dependent variable u, then searching for the equivalence group one has to append the equations

$$k_t = 0, \quad k_x = 0.$$

The generator of a one–parameter equivalence group is

$$X^e = \xi^t \partial_t + \xi^x \partial_x + \zeta^u \partial_u + \zeta^k \partial_k$$

with the coefficients

$$\xi^t = \xi^t(x,t,u,k), \quad \xi^x = \xi^x(x,t,u,k), \quad \zeta^u = \zeta^u(x,t,u,k),$$
$$\zeta^k = \zeta^k(x,t,u,k).$$

The prolonged operator is

$$\widetilde{X}^e = X^e + \zeta^{u_{tt}} \partial_{u_{tt}} + \zeta^{u_{xx}} \partial_{u_{xx}} + \zeta^{k_u} \partial_{k_u} + \zeta^{k_x} \partial_{k_x} + \zeta^{k_t} \partial_{k_t}.$$

The coefficients of the prolonged operator are

$$\zeta^{u_x} = D_x^e \zeta^u - u_x D_x^e \xi^x - u_t D_x^e \xi^t, \ \zeta^{u_t} = D_t^e \zeta^u - u_x D_t^e \xi^x - u_t D_t^e \xi^t,$$

$$\zeta^{u_{xx}} = D_x^e \zeta^{u_x} - u_{xx} D_x^e \xi^x - u_{xt} D_x^e \xi^t, \ \zeta^{u_{tt}} = D_t^e \zeta^{u_t} - u_{xt} D_t^e \xi^x - u_{tt} D_t^e \xi^t,$$

$$\zeta^{k_x} = \tilde{D}_x^e \zeta_u^k - k_u \tilde{D}_x^e \zeta^u, \ \zeta^{k_t} = \tilde{D}_t^e \zeta^k - k_u \tilde{D}_t^e \zeta^u,$$
$$\zeta^{k_u} = \tilde{D}_u^e \zeta^k - k_u \tilde{D}_u^e \zeta^u.$$

Here the operators D_t^e, D_x^e, $\tilde{D}_x^e$, $\tilde{D}_t^e$, $\tilde{D}_u^e$ are

$$D_t^e = \partial_t + u_t(\partial_u + k_u \partial_k + k_{uu} \partial_{k_u}) + u_{tt} \partial_{u_t} + u_{tx} \partial_{u_x} + \ldots,$$

$$D_x^e = \partial_x + u_x(\partial_u + k_u \partial_k + k_{uu} \partial_{k_u}) + u_{xt} \partial_{u_t} + u_{xx} \partial_{u_x} + \ldots,$$
$$\tilde{D}_x^e = \partial_x, \ \tilde{D}_t^e = \partial_t, \ \tilde{D}_u^e = \partial_u + k_u \partial_k.$$

The determining equations DS^e are

$$[\zeta^{u_{tt}} - k\zeta^{u_{xx}} - \zeta^k u_{xx}]_{|(S)} = 0, \quad \zeta^{k_x} = 0, \ \zeta^{k_t} = 0.$$

After substitutions $\zeta^{u_{tt}}, \zeta^{u_x}, \zeta^{u_{xx}}, \zeta^{k_u}$ and transition onto the manifold $u_t = ku_{xx}$ one gets the equations, which can be split with respect to the variables $u_x, u_t, u_{xt}, u_{xx}, k_u, k_{uu}$. As the result, the determining equations are

$$\xi_x^t = 0, \ \xi_u^t = 0, \ \xi_k^t = 0, \ \xi_{tt}^t = 0, \ \xi_t^x = 0, \ \xi_u^x = 0, \ \xi_k^x = 0, \ \xi_{xx}^x = 0,$$

$$\zeta^u_x = 0, \ \ \zeta^u_t = 0, \ \ \zeta^u_k = 0, \ \ \zeta^u_{uu} = 0, \ \ \zeta^k = 2k(\xi^x_x - \xi^t_t).$$

The general solution of the last equations is

$$\xi^x = (c_1 + c_4)x + c_2, \ \xi^t = c_1 t + c_3, \ \zeta^u = c_5 u + c_6, \ \zeta^k = 2c_4 k,$$

or

$$X^e = c_1 X_1 + c_2 X_2 + c_3 X_3 + c_4 X_4 + c_5 X_5 + c_6 X_6,$$

where

$$X^e_1 = x\partial_x + t\partial_t, \ X^e_2 = \partial_x, \ X^e_3 = \partial_t, \ X^e_4 = x\partial_x + 2k\partial_k, \ \ X^e_5 = u\partial_u, \ X^e_6 = \partial_u.$$

The equivalence group GS^e corresponds to the Lie algebra LS^e with the basis X^e_i, $(i = 1, 2, ..., 6)$.

As it was obtained, the relation which defines extensions of the kernel of principal Lie groups $GS(k)$ is

$$(\frac{k}{k'})'' = 0.$$

Integration of this equation gives

$$k' = \frac{k}{\alpha u + \beta}, \tag{5.43}$$

with $\alpha^2 + \beta^2 \neq 0$.

Further study of determining equations (5.17) is simplified by using the equivalence group GS^e.

If $\alpha = 0$, then the general solution of this equation is $k = \gamma e^{qu}$ with constant γ and $q = 1/\beta$. With the help of the equivalence transformation, which corresponds to the generator X^e_5 and the involution, which corresponds to the change of the sign of the function u, the function $k(u)$ can be transformed to the function $k = \gamma e^u$. Applying the equivalence transformation with the generator X^e_6 enables the function $k = \gamma e^u$ to be transformed into the function $k = e^u$.

If $\alpha \neq 0$, then the general solution of equation (5.43) is $k = \gamma_1(u+\gamma_2)^q$ with the constants γ_1, $\gamma_2 = \beta/\alpha$ and $q = 1/\alpha$. By the equivalence transformation, corresponding to the generator X^e_5, this function can be transformed into the function $k = (u + \gamma_2)^q$. The equivalence transformation with the generator X^e_6 transforms the function $k = (u + \gamma_2)^q$ to the function $k = u^q$.

Thus, one can confine integrating the determining equations (5.17) to two cases of the function $k(u)$: either $k = e^u$ or $k = u^q$.

In the first case, where $k = e^u$, an additional set of transformations (compared with the kernel (5.19)) is the set of transformations with the generator $X_4 = x\partial_x + 2\partial_u$.

In the second case, where $k = u^q$, after substituting the expression for η one gets

$$\xi^x_{xxx} = 0, \ \ \xi^t_{ttt} = 0$$

and the classifying relations

$$(q-4)\xi^x_{xx}=0,\quad (q+4)\xi^t_{tt}=0.$$

The process of integrating the determining equations in this case is separated into the following three cases: $q=4$, $q=-4$, and an arbitrary q.

For an arbitrary q there is an additional (to the kernel (5.19)) set of transformations with the generator $X_5=qx\partial_x+2u\partial_u$. If $q=4$, then there is the transformations with the generator $X_6=x^2\partial_x+xu\partial_u$. For $q=-4$ there is the transformations with the generator $X_7=t^2\partial_t+tu\partial_u$.

The result of the extensions of the kernel is:

e^u	X_4
u^q	X_5
u^4	$X_5,\ X_6$
u^{-4}	$X_5,\ X_7$

3. Multi-parameter Lie group of transformations

Let $O\subset R^r$ be a ball with the center at the zero point of the space R^r. Assume that the pair (O,φ) with the mapping $\varphi: O\times O\to R^r$ satisfies the conditions:

1^0. $\varphi(a,0)=\varphi(0,a)=a$ for all $a\in O$.

2^0. $\varphi(\varphi(a,b),c)=\varphi(a,\varphi(b,c))$ for all $a,b,c\in O$ for which $\varphi(a,b),\varphi(b,c)\in O$.

The pair (O,φ) is called a local Lie group with the multiplication law φ. For the sake of simplicity we assume that $\varphi\in C^\infty(O,O)$. Invertible transformations

$$z_i'=g^i(z;a),\quad (i=1,2,\dots,N) \tag{5.44}$$

with $z\in V\subset Z=R^N(z)$ and the vector-parameter $a\in O$ are studied in this section. The set V is an open set in Z.

Definition 5.14. *A set of transformations (5.44) is a local r–parameter Lie group G^r if:*

1^0. $g(z,0)=z$ for all $z\in V$.

2^0. $g(g(z,a),b)=g(z,\varphi(a,b))$ for all $a,b,\varphi(a,b)\in O$, $z\in V$.

3^0. If for $a\in O$ one has $g(z,a)=z$ for all $z\in V$, then $a=0$.

With the group G^r one relates the infinitesimal generators

$$X_k=\zeta^i_k(z)\partial_{z_i},\ (k=1,2,...,r),$$

where

$$\zeta^i(z)=\frac{\partial g^i}{\partial a_k}(z,0).$$

In applications of multi-parameter groups for constructing invariant and partially invariant solutions it is enough to use their following properties. Any

r-parameter group is a composition of r one-parameter subgroups belonging to it[11]. If a set of generators X_k, $(k = 1, 2, ..., r)$ composes a Lie algebra H_r, then there exists a local Lie group G^r of transformations with the generators X_k, $(k = 1, 2, ..., r)$, and conversely.

Similar to one-parameter Lie groups, there are the following definitions and theorems.

Definition 5.15. *A function $\Phi(z)$ is called an invariant of a group G^r if*

$$\Phi(z') = \Phi(z).$$

Theorem 5.9. *A function $\Phi(z)$ is an invariant of a group $G^r(X)$ with generators X_j, $(j = 1, 2, \ldots, r)$ if and only if*

$$X_j \Phi(z) = 0, \ \ (j = 1, 2, \ldots, r).$$

Definition 5.16. *A manifold Ψ is invariant with respect to a group G^r if any point $z \in \Psi$ is transformed into the same manifold $g(z; a) \in \Psi$.*

A regularly assigned manifold Ψ given by the equations

$$\psi^j(z_1, z_2, ..., z_N) = 0, \ \ (j = 1, 2, ..., s) \tag{5.45}$$

is invariant with respect to an r-parameter group G^r if and only if

$$X_k \psi^j_{|\Psi} = 0, \ \ (k = 1, 2, ..., r; \ j = 1, 2, ..., s).$$

Let H_r be a Lie algebra. A manifold $\Psi \subset R^N$ is called a nonsingular manifold (with respect to the Lie algebra H_r) if

$$r_*(\zeta^i_k(z)_{|\Psi}) = r_*, \ \ \forall z \in \Psi.$$

Here the function $r_*(\zeta^i_k(z))$ defines the rank of the matrix composed of the coefficients $\zeta^i_k(z)$:

$$r_*(\zeta^i_k(z)) = rank(\zeta^i_k(z)).$$

For applications the following theorem plays a very important role.

Theorem 5.10. *If $r_* < N$, then there exist $N - r_*$ functionally independent invariants $J^j(z)$, $(j = 1, 2, ..., N - r_*)$. Any nonsingular regularly assigned by (5.45) invariant manifold Ψ is given by the formula*

$$\Phi^i\left(J^1(z), J^2(z), ..., J^{N-r_*}(z)\right) = 0, \ \ (i = 1, 2, ..., s).$$

The vector function $J = \left(J^1(z), J^2(z), ..., J^{N-r_*}(z)\right)$ is called an universal invariant.

[11]The theory of multi-parameter groups can be found in [130].

4. Invariant solutions

4.1 The main definitions

Let GS be a principal Lie group of transformations, admitted by a system of differential equations (S) and $H \subset GS$ be a subgroup.

Definition 5.17. *A solution $u = U(x)$ of the system (S) is called an H-invariant solution if the manifold $u = U(x)$ is an invariant manifold with respect to any transformation of the group H.*

If an r–parameter group H_r is generated by the operators

$$X_k = \xi_k^i(x,u)\partial_{x_i} + \zeta_k^j(x,u)\partial_{u^j}, \quad (k = 1, 2, \ldots, r),$$

by virtue of the theorem about an invariant manifold the invariant solution $u = U(x)$ has to satisfy the additional differential equations (differential constraints)

$$\zeta_k^j(x,U) - \sum_i \xi_k^i(x,U)\frac{\partial U^j}{\partial x_i} = 0, \quad (k = 1, 2, \ldots, r; \; j = 1, 2, \ldots, m). \tag{5.46}$$

Hence, any invariant solution is a solution of the overdetermined system of equations (S) and (5.46). In the general case, any overdetermined system of partial differential equations has to be studied for compatibility. For H_r-invariant solutions this problem is simpler than in the general case.

Note also that because of (5.46), for an invariant solution one has

$$r_*(\xi_s^i(x, U(x)), \zeta_s^j(x, U(x))) = r_*(\xi_s^i(x, U(x))).$$

Definition 5.18. *Let H_r be an r–parameter Lie group. An invariant solution $u = U(x)$ is called nonsingular if the manifold defined by the formulae $u = U(x)$ is a nonsingular manifold with respect to the group H_r.*

Thus an invariant solution $u = U(x)$ is nonsingular if, and only if,

$$r_*(\xi_s^i(x, U(x))) = r_*.$$

According to the statements of the previous section a nonsingular invariant solution can be represented by the formulae

$$\Phi^i\left(J^1(x,u), J^2(x,u), \ldots, J^{m+n-r_*}(x,u)\right) = 0, \quad (i = 1, 2, \ldots, m).$$

For a nonsingular invariant solution this allows the reduction of the initial system of differential equations (S) to a system with smaller number of the independent variables. There is the following theorem.

Theorem 5.11. *Let a subgroup $H_r \subset GS$ satisfy the condition*

$$r_*(\xi_s^i) = r_*(\xi_s^i, \zeta_s^j) = r_* \leq n. \tag{5.47}$$

Then there are the mappings

$$t_l = t_l(x, u), \quad w^k = w^k(x, u), \quad (l = 1, 2, \ldots, n - r_*; \; k = 1, 2, \ldots, m) \tag{5.48}$$

and a system of differential equations (S/H_r) *for the functions* $W(t) \in R^m$ *with the independent variables* $t = (t_1, t_2, \ldots, t_{n-r_*})$ *satisfying the following properties: any nonsingular invariant solution* $u = U(x)$ *can be defined from the functional equations*

$$w(x, U(x)) = W(t(x, U(x))),$$

where $W(t)$ *is a solution of the system* (S/H_r). *Inversely, if a solution* $W = W(t)$ *of the system* (S/H_r) *produces a manifold in the space* Z

$$w(x, u) = W(t(x, u)), \tag{5.49}$$

which can be written in the form $u = U(x)$, *then* $u = U(x)$ *is an invariant solution of the system* (S).

The system of differential equations (S/H_r) is called a factor system, the number $\sigma = n - r_*$ is the number of the independent variables in the factor system (S/H_r) and it is called a rank of the invariant solution. According to the theorem, the construction of nonsingular invariant solutions is built algorithmically by means of the following steps. At first all $q = m + n - r_* = m + \sigma$ independent invariants $J^k = J^k(x, u), \quad (k = 1, \ldots, m + \sigma)$ are found. The rank of the Jacobi matrix $\dfrac{\partial(J^1, \ldots, J^{m+n-r_*})}{\partial(u^1, \ldots, u^m)}$ has to be equal to m. Without loss of generality one can choose the first m invariants $J^1, \ldots, J^m$ such that the rank of the Jacobi matrix $\dfrac{\partial(J^1, \ldots, J^m)}{\partial(u_1, \ldots, u_m)}$ is equal to m. At the next step one supposes that the first m invariants $J^k, \; (k = 1, \ldots, m)$ depend[12] on the remaining invariants $J^k, \; (k = m + 1, \ldots, m + \sigma)$, i.e.,

$$J^k = W^k(J^{m+1}, \ldots, J^{m+\sigma}), \; (k = 1, \ldots, m). \tag{5.50}$$

Equations (5.50) should be such that they can be solved with respect to all dependent variables $u^i, \; (i = 1, \ldots, m)$. After substituting the representation of the functions $u^i, \; (i = 1, \ldots, m)$ into the initial system of partial differential equations one obtains the system of equations for the unknown functions $W^k(t_1, t_2, \ldots, t_\sigma)$, $(k = 1, \ldots, m)$. This system is called a factor system (S/H_r).

In most applications the invariants $J^{m+1}, J^{m-1}, \ldots, J^{m+\sigma}$ can be chosen independently of the dependent variables

$$J^{m+1} = J^{m+1}(x), \; J^{m+2} = J^{m+2}(x), \ldots, J^{m+\sigma} = J^{m+\sigma}(x).$$

[12] The case $r_* = n$ is also included in this scheme.

In this case

$$u^k = \bar{\psi}^k(x, J^1, J^2, \ldots, J^m), \quad (k = 1, 2, \ldots, m). \tag{5.51}$$

Thus, the representation of the invariant solution is

$$u^k = \psi^k(x, J^{m+1}, J^{m+1}, \ldots, J^{m+\sigma}), \quad (k = 1, 2, \ldots, m), \tag{5.52}$$

where

$$\psi^k(x, J^{m+1}, J^{m+1}, \ldots, J^{m+\sigma}) =$$
$$\bar{\psi}^k\left(x, W^1(J^{m+1}, \ldots, J^{m+\sigma}), \ldots, W^m(J^{m+1}, \ldots, J^{m+\sigma})\right),$$
$$(k = 1, 2, \ldots, m).$$

4.2 Invariant solutions of equation (5.16)

The optimal system of subalgebras of the kernel (5.19) is

$$\{X_1, X_2, X_3\},\ \{X_3, X_1 + p_2X_2\},\ \{X_3, X_2\},\ \{X_1, X_2\},$$
$$\{X_3\}, \{X_1 + p_2X_2\},\ \{X_2\}.$$

For the subgroups corresponding to the subalgebras

$$\{X_1, X_2, X_3\},\ \{X_3, X_1 + p_2X_2\},\ \{X_3, X_2\},\ \{X_1, X_2\}$$

there is only one invariant: u. Hence, the invariant solution for these subgroups is $u = const$.

If the subgroup corresponds to the generator $X_1 + p_2X_2$, the invariants are: $J^1 = u,\ J^2 = x - p_2t$. The representation of the invariant solution is

$$u = W(x - p_2t).$$

Substituting this representation into equation (5.16), one obtains the reduced equation

$$(k(W) - p_2^2)W'' = 0.$$

If the subgroup corresponds to the generator X_2, the invariants are: $J^1 = u$, $J^2 = t$. The representation of the invariant solution is

$$u = W(t).$$

Substituting this representation into equation (5.16), one obtains the reduced equation

$$W'' = 0.$$

If the subgroup corresponds to the generator X_3, the invariants are: $J^1 = u$, $J^2 = x/t$. The representation of the invariant solution is

$$u = W(y),\ y = x/t.$$

Substituting this representation into equation (5.16), one obtains the reduced equation

$$(y^2 - k(W))W'' + 2yW' = 0.$$

5. Group classification of two–dimensional steady gas dynamics equations

The partial differential equations describing the motion of a two–dimensional steady inviscid gas are

$$\begin{aligned}&\rho(uu_x + vu_y) + p_x = 0, \quad \rho(uv_x + vv_y) + p_y = 0,\\ &u\rho_x + v\rho_y + \rho(u_x + v_y + \nu\tfrac{u}{y}) = 0,\\ &up_x + vp_y + A(p,\rho)(u_x + v_y + \nu\tfrac{u}{y}) = 0.\end{aligned} \tag{5.53}$$

Here ρ is the density, p is the pressure, $\varepsilon = \varepsilon(p,\rho)$ is the internal energy, $\nu = 0$ for plane flows and $\nu = 1$ for axially symmetric flows. The coefficient $A(p,\rho)$ is defined by the internal energy $A = (\rho\varepsilon_\rho)^{-1}(p - \rho^2\varepsilon_\rho)$. The group classification is studied with respect to the arbitrary element $A(p,\rho)$. For simplicity the case $A(p,\rho) \neq 0$ is considered.

5.1 Equivalence transformations

The first stage of a group classification requires finding an equivalence group of transformations. Since the arbitrary element A only depends on p and ρ, the system of equations (5.53) has to be supplemented with auxiliary equations

$$A_x = 0, \; A_y = 0, \; A_u = 0, \; A_v = 0.$$

The infinitesimal generator of the equivalence group has the representation

$$X^e = \zeta^x\partial_x + \zeta^y\partial_y + \zeta^u\partial_u + \zeta^v\partial_v + \zeta^\rho\partial_\rho + \zeta^p\partial_p + \zeta^A\partial_A.$$

The coefficients of the generator X^e are assumed to be dependent on x, y, u, v, ρ, p, A. Let us denote $u^1 = u, \; u^2 = v, \; u^3 = \rho, \; u^4 = p$ and $z^1 = x$, $z^2 = y, z^3 = u, \; z^4 = v, \; z^5 = \rho, \; z^6 = p$. The coefficients of the prolonged generator

$$\tilde{X}^e = X^e + \sum_i (\zeta^{u^i_x}\partial_{u^i_x} + \zeta^{u^i_y}\partial_{u^i_y}) + \sum_j \zeta^{A_{z^j}}\partial_{A_{z^j}} + \ldots$$

are

$$\begin{aligned}\zeta^{u^i_x} &= D^e_x\zeta^{u^i} - u^i_x D^e_x\xi^x - u^i_y D^e_x\xi^y,\\ \zeta^{u^i_y} &= D^e_y\zeta^{u^i} - u^i_x D^e_y\xi^x - u^i_y D^e_y\xi^y,\\ \zeta^{A_{z^j}} &= \tilde{D}^e_{z^j}\zeta^A - \sum_\alpha A_{z^\alpha}\tilde{D}^e_{z^j}\zeta^{z^\alpha}.\end{aligned}$$

The operators D^e_x, D^e_y are operators of the total derivatives with respect to x and y, respectively in the space of the independent variables x and y:

$$D^e_x = \partial_x + u_x\partial_u + v_x\partial_v + \rho_x\partial_\rho + p_x\partial_p + (A_x + A_\rho\rho_x + A_p p_x)\partial_A$$

$$D_y^e = \partial_y + u_y\partial_u + v_y\partial_v + \rho_y\partial_\rho + p_y\partial_p + (A_y + A_\rho\rho_y + A_p p_y)\partial_A$$

The operators $\tilde{D}^e_{z^j}$ are operators of the total derivatives with respect to $z^1, \ldots, z^6$ in the space of the independent variables x, y, u, v, ρ and p:

$$\tilde{D}^e_{z^j} = \partial_{z^j} + (\delta_{3j}A_\rho + \delta_{4j}A_p)\partial_A,$$

where δ_{ij} is the Kronecker delta symbol.

After some calculations[13], one obtains the equivalence group of transformations of system (5.53). This group corresponds to the Lie algebra with the generators ($\nu = 0, 1$):

$$\partial_x, \;\; x\partial_x + y\partial_y, \;\; x\partial_x + y\partial_y + u\partial_u + v\partial_v - 2\rho\partial_\rho,$$

$$\rho\partial_\rho + p\partial_p + A\partial_A, \;\; \partial_p.$$

In the case of plane symmetry ($\nu = 0$), there are two more generators

$$\partial_y, \;\; -y\partial_x + x\partial_y - v\partial_u + u\partial_v.$$

5.2 Admitted group

Let an infinitesimal generator of a one-parameter Lie group admitted by system of equations (5.53) is

$$X = \zeta^x\partial_x + \zeta^y\partial_y + \zeta^u\partial_u + \zeta^v\partial_v + \zeta^\rho\partial_\rho + \zeta^p\partial_p,$$

where the coefficients depend on the variables x, y, u, v, ρ and p.

Splitting the determining equations DS with respect to the parametric derivatives constructed for system (5.53), and solving some of them, one obtains

$$\zeta^x = c_1x + c_2y + c_3, \;\; \zeta^y = c_1y - c_2x + c_4,$$
$$\zeta^u = -au + c_2v, \;\; \zeta^v = -av - c_2u,$$
$$\zeta^\rho = (c_5 + 2a)\rho, \;\; \zeta^p = c_5p + c_6.$$

where $a = a(q, \rho, p)$, $q^2 = u^2 + v^2$ and $c_1, c_2, \ldots, c_6$ are arbitrary constants. The remained determining equations are:

$$c_5(\rho A_\rho + pA_p - A) + c_6A_p + 2a\rho A_\rho = 0, \tag{5.54}$$

$$A\frac{\partial a}{\partial q} - q\rho(\rho\frac{\partial a}{\partial \rho} + A\frac{\partial a}{\partial p}) = 0, \tag{5.55}$$

$$\nu c_2 = 0, \;\; \nu c_4 = 0. \tag{5.56}$$

[13] All necessary calculations were carried on a computer using the symbolic manipulation program REDUCE [69].

Equations (5.54), (5.55) and (5.56) comprise classifying relations. If equations (5.54) are satisfied for any function $A(p, \rho)$, then $c_5 = c_6 = a = 0$. Hence, the kernel of the principal Lie algebras is generated by the operators: for axisymmetric flows ($\nu = 1$)

$$X_1 = \partial_x, \quad X_4 = x\partial_x + y\partial_y,$$

and for plane flows ($\nu = 0$)

$$X_1 = \partial_x, \; X_2 = \partial_y, \; X_3 = -y\partial_x + x\partial_y - v\partial_u + u\partial_v,$$
$$X_4 = x\partial_x + y\partial_y.$$

The kernel is extended by specifying the function $A(p, \rho)$.

If $A_\rho \neq 0$, the function $a(q, \rho, p)$ is found from equation (5.54):

$$a = -c_5(h_1)_\rho - c_6(h_2)_p,$$

where

$$h_1 = (2\rho A_\rho)^{-1}(\rho A_\rho + pA_p - A), \quad h_2 = (2\rho A_\rho)^{-1}A_p.$$

Substituting the value of a into equation (5.55), one has

$$c_5V^1 + c_6V^2 = 0 \tag{5.57}$$

with

$$V^1 = \rho h_{1\rho} + Ah_{1p}, \quad V^2 = \rho h_{2\rho} + Ah_{2p}.$$

Since $A_\rho \neq 0$, the variables (ρ, p) can be exchanged with (A, p). Introducing the function $\phi = \rho A_\rho + AA_p$, the functions V^1 and V^2 are rewritten in the form

$$V^1 = (2A_\rho)^{-1}(A\phi_A + p\phi_p - \phi), \; V^2 = (2A_\rho)^{-1}\phi_p,$$

where $\phi = \phi(A, p)$.

While varying p and ρ there are three non-equivalent cases involving changing the vector $\mathbf{V} = (V^1, V^2)$: (a) the vector $\mathbf{V}$ belongs to a two–dimensional vector space (essentially two–dimensional), (b) the vector $\mathbf{V}$ belongs to a one-dimensional space, (c) the vector $\mathbf{V}$ is constant.

In case (a) equation (5.57) leads to $c_5 = 0$, $c_6 = 0$, and hence, $a = 0$. In this case there is no extension of the kernel.

In case (b) there exists a constant vector $(k_1, k_2) \neq 0$ orthogonal to all vectors $\mathbf{V}$: $k_1V^1 + k_2V^2 = 0$.

If $k_1 \neq 0$, then because of the equivalence transformation corresponding to the generator ∂_p the equation $k_1V^1 + k_2V^2 = 0$ can be transformed into $V^1 = 0$. In fact, let $p' = p + k$, where $k_1^{-1}k_2$. Then

$$V^1 + qV^2 = (2A_\rho)^{-1}(A\phi_A + (p + k)\phi_p - \phi) =$$
$$(2A_\rho)^{-1}(A\phi_A + p'\phi_{p'} - \phi).$$

The general solution of the equation $V^1 = 0$ is $\phi = p\Phi(A/p)$, where the function Φ is arbitrary. Thus, the function $A = A(p, \rho)$ satisfies the equation

$$\rho A_\rho + AA_p = p\Phi(A/p). \tag{5.58}$$

Since the vector $\mathbf{V}$ belongs to a one-dimensional space, there is a point (p, ρ) such that $V^2 \neq 0$ (since $V^2 \neq const$). This means that $c_6 = 0$ and $\lambda\Phi' - \Phi \neq 0$, where $\lambda = A/p$. The constant c_5, which is arbitrary, gives the additional generator $h_1 H_1 + Y_1$, where

$$Y_1 = \rho\partial_\rho + p\partial_p, \quad H_1 = u\partial_u + v\partial_v - 2\rho\partial_\rho.$$

If $k_1 = 0$, then because of $(k_1, k_2) \neq 0$, one has $V^2 = 0$. Since $V^1 \neq const$, there is a point (p, ρ) such that $V^1 \neq 0$. In this case equation (5.57) leads to $c_5 = 0$. Integrating the equation $V^2 = 0$, one has

$$\rho A_\rho + AA_p = \Phi(A) \tag{5.59}$$

with an arbitrary function $\Phi(A)$, and

$$V^1 = (2A_\rho)^{-1}(A\Phi' - \Phi).$$

Because $V^1 \neq 0$, there is a restriction for the function $\Phi(A)$:

$$A\Phi' - \Phi \neq 0$$

or $\Phi \neq qA$ with constant q. The extension of the kernel is given by the generator $h_2 H_1 + Y_2$, which corresponds to the constant c_6. Here $Y_2 = \partial_p$.

In case (c) $V^1 = k_1$ and $V^2 = k_2$, where k_1 and k_2 are constant. If $k_2 \neq 0$, then by using the equivalence transformation corresponding to the generator ∂_p one finds that $V^1 = 0$. This case can be included in (5.58) by omitting the restriction $V^2 \neq const$. If $k_2 = 0$, then this case can be included in (5.59) by omitting the restriction $V^1 \neq const$. If $k_1 = k_2 = 0$, then the function $A(p, \rho)$ is a solution of the equation

$$\rho A_\rho + AA_p = \gamma A$$

with constant γ. In this case there exist two additional generators $h_1 H_1 + Y_1$ and $h_2 H_1 + Y_2$.

Assume that $A_\rho = 0$. The kernel is extended by the generator aH_1, where $a = a(q, \rho, p)$ is a solution of equation (5.55). Further extensions are obtained by studying equation (5.54), which becomes

$$c_5(pA_p - A) + c_6 A_p = 0. \tag{5.60}$$

Analysis of this equation is similar to the analysis of the case $A_\rho \neq 0$ with the vector $\mathbf{V} = (pA_p - A, A_p)$.

The results of the group classification are presented in Table 5.1. In this table the function $a = a(q, \rho, p)$ is an arbitrary function satisfying equation (5.55).

Table 5.1. Group classification

N	$A(p,\rho)$	Extensions
1	$f(p)$	aH_1
2	γp	aH_1, Y_1
3	1	aH_1, Y_2
4	$\rho A_\rho + AA_p = \gamma A, \ (A_\rho \neq 0)$	$h_1H_1 + Y_1, h_2H_1 + Y_2$
5	$\rho A_\rho + AA_p = p\Phi(A/p),$ $(\Phi(\xi) \neq k\xi, \ A_\rho \neq 0)$	$h_1H_1 + Y_1$
6	$\rho A_\rho + AA_p = \Phi(A),$ $(\Phi(\xi) \neq k\xi, \ A_\rho \neq 0)$	$h_2H_1 + Y_2$

Table 5.2. Table of commutators

	X_1	X_2	X_3	X_4
X_1	0	0	X_2	X_1
X_2	0	0	$-X_1$	X_2
X_3	$-X_2$	X_1	0	0
X_4	$-X_1$	$-X_2$	0	0

Table 5.3. Inner automorphisms

	x_1	x_2
A_1	$x_1 + a_1x_4$	$x_2 + a_1x_3$
A_2	$x_1 - a_2x_3$	$x_2 + a_2x_4$
A_3	$x_1\cos(a_3) + x_2\sin(a_3)$	$-x_1\sin(a_3) + x_2\cos(a_3)$
A_4	x_1a_4	x_2a_4

5.3 Optimal system of subalgebras

For forming an optimal system of subalgebras of the Lie algebra $L_4 = \{X_1, X_2, X_3, X_4\}$ the two-step algorithm [131] is applied.

The table of commutators is presented in Table 5.2. From the table of commutators one obtains the following decomposition: $L_4 = I \oplus F$, where $I = \{X_1, X_2\}$ is an ideal and $F = \{X_3, X_4\}$ is a subalgebra. The effect of the group of inner automorphisms upon the coordinates of the operator $X = x_iX_i$ is presented in Table 5.3. There is also one involution E_1

$$E_1(x_1, x_2, x_3, x_4) = (-x_1, x_2, -x_3, x_4),$$

which corresponds to the changes in orientation of the coordinate system

$$x' = -x, \ y' = y, \ u' = -u, \ v' = v.$$

Since the subalgebra F is abelian its classification is simple. The optimal system of the algebra F consists of the subalgebras

$$F_1 = \{X_3, X_4\}, \ F_2 = \{X_3 + \alpha X_4\}, \ F_3 = \{X_4\}, \ F_4 = \{0\}.$$

The stabilizator of each subalgebra F_i, $(i = 1, 2, 3, 4)$ is A. The coefficients α and β are arbitrary constants.

For the second step one has to glue the subalgebras F_i, $(i = 1, 2, 3, 4)$ with the ideal I. The process of gluing includes choosing a basis of a subalgebra, nulling as many as possible coordinates of the basis operators by using automorphisms, and then checking the subalgebra conditions. To give a description of the whole process of gluing is not necessary, since it is monotonous. The demonstration will only be given for three dimensional subalgebras.

Let us start from F_1. Three dimensional subalgebra including F_1 has the basis

$$\{\alpha X_1 + \beta X_2,\ X_3,\ X_4\},\ \alpha^2 + \beta^2 \neq 0.$$

The subalgebra conditions

$$[\alpha X_1 + \beta X_2, X_3] = \alpha X_2 - \beta X_1,$$
$$[\alpha X_1 + \beta X_2, X_4] = \alpha X_1 - \beta X_2$$

giving

$$\alpha = c_1\beta,\ \beta = -c_1\alpha,$$
$$(1 - c_2)\alpha = 0,\ (1 + c_2)\beta = 0$$

with some constants c_1 and c_2. These equations contradict to the condition $\alpha^2 + \beta^2 \neq 0$. Hence, there are no such types of subalgebra.

The three dimensional subalgebras including F_2 have the basis

$$\{X_1, X_2, X_3 + \alpha X_4\}. \tag{5.61}$$

The subalgebra conditions are satisfied. Similarly for the subalgebra

$$\{X_1, X_2, X_4\}. \tag{5.62}$$

Thus, there are only two three-dimensional subalgebras (5.61) and (5.62) in the optimal system $\Theta_A(L_4)$.

The normalized optimal system of subalgebras of the Lie algebra L_4 is presented in Table 5.4. Here r denotes the subalgebra dimension. The last column

Table 5.4. Optimal system of subalgebras

N	subalgebra	Nor
	r=4	
1	X_1, X_2, X_3, X_4	= 4.1
	r=3	
1	$X_1, X_2, X_3 + \alpha X_4$	4.1
2	X_1, X_2, X_4	4.1
	r=2	
1	X_1, X_2	4.1

N	subalgebra	Nor
	r=2	
2	X_1, X_4	= 2.2
3	X_3, X_4	= 2.3
	r=1	
1	X_1	3.2
2	$X_3 + \alpha X_4$	2.3
3	X_4	2.3

represents the identificator of the normalizator of the subalgebra in a given line. The first number in the normalizator identificator denotes its dimension, the second number is its number among the subalgebras of the given dimension. Self–normalized subalgebras are marked by the equality sign.

5.4 Invariant solutions

For subalgebras having the operator X_3 or the operator X_4 it is useful to rewrite the gas dynamics equations ($\nu = 0$) and these generators in the cylindrical coordinate system

$$x = r\cos(\theta),\ \ y = r\sin(\theta),$$
$$u = U\cos(\theta) + V\sin(\theta),\ \ v = -U\sin(\theta) + V\cos(\theta).$$

In this coordinates the gas dynamics equations are

$$\begin{gathered} UU_r + r^{-1}VU_\theta + \rho^{-1}p_r = r^{-1}V^2, \\ UV_r + r^{-1}VV_\theta + (r\rho)^{-1}p_\theta = -r^{-1}UV, \\ U\rho_r + r^{-1}V\rho_\theta + \rho(U_r + r^{-1}U + r^{-1}V_\theta) = 0, \\ Up_r + r^{-1}Vp_\theta + A(\rho, p)(U_r + r^{-1}U + r^{-1}V_\theta) = 0. \end{gathered} \tag{5.63}$$

The operators are

$$X_3 = \partial_\theta,\ \ X_4 = r\partial_r.$$

A representation of a solution invariant with respect to X_1 is

$$u = u(y),\quad v = v(y),\quad \rho = \rho(y),\quad p = p(y).$$

Substituting the representation into the gas dynamics equations, one has the reduced system

$$\begin{gathered} \rho v u' = 0,\ \ \rho v v' + p' = 0, \\ (\rho v)' = 0,\ \ v p' + A(p, \rho) v' = 0. \end{gathered}$$

Analysis of a solution of these equations is trivial.

A representation of a solution invariant with respect to $X_3 + \alpha X_4 = \partial_\theta + \alpha r\partial_r$ in the cylindrical coordinate system is

$$U = U(\lambda),\quad V = V(\lambda),\quad \rho = \rho(\lambda),\quad p = p(\lambda),\ \lambda = re^{-\alpha\theta}$$

Substituting the representation into the gas dynamics equations, one obtains

$$\begin{gathered} \lambda(U - \alpha V)U' + \rho^{-1}\lambda p' = V^2,\ \ (U - \alpha V)\left(\alpha\lambda U' + (\lambda V)'\right) = 0, \\ (\lambda(U - \alpha V)\rho)' + \alpha\rho V = 0,\ \ (U - \alpha V)(\rho p' - A(p, \rho)\rho') = 0. \end{gathered}$$

If $U - \alpha V = 0$, then $U = 0$, $\alpha V = 0$, and $p' = \rho V^2/\lambda$. If $(U - \alpha V) \neq 0$, the flow is isentropic and one obtains a system of first order ordinary differential equations solved with respect to all first derivatives.

An invariant solution with respect to X_4 in the cylindrical coordinate system has the representation

$$U = U(\theta), \quad V = V(\theta), \quad \rho = \rho(\theta), \quad p = p(\theta).$$

The factor system for this invariant solution is

$$\begin{array}{ll} V(U' - V) = 0, & \rho(UV + VV') + p' = 0, \\ (V\rho)' + \rho U = 0, & V(\rho p' - A(p,\rho)\rho') = 0. \end{array}$$

Analysis of this reduced system is similar to the previous case.

6. Partially invariant solutions

The notion of a partially invariant solution generalizes the notion of an invariant solution. This generalization extends an area of application of group analysis to the construction of exact solutions of partial differential equations. The algorithm for finding partially invariant solutions comprises of several steps, which are similar to the steps when constructing invariant solutions.

Let LS be a principal Lie algebra of generators, admitted by a system of differential equations (S) and $L_r \subset LS$ be a subalgebra. Invariant and partially invariant solutions with respect to the Lie algebra L_r are called H-solutions. Here H is a Lie group corresponding to the Lie algebra L_r.

Assume that $X_1, ..., X_r$ is a basis of the Lie algebra L_r. The universal invariant J consists of $s = m + n - r_*$ functionally independent invariants

$$J = \left(J^1(x,u), J^2(x,u), ..., J^{m+n-r_*}(x,u)\right),$$

where r_* is the total rank of the matrix composed by the coefficients of the generators X_i, $(i = 1, 2, ..., r)$. If the rank of the Jacobi matrix $\dfrac{\partial(J^1, ..., J^{m+n-r_*})}{\partial(u^1, ..., u^m)}$ is equal to q, then one can choose the first $q \le m$ invariants $J^1, ..., J^q$ such that the rank of the Jacobi matrix $\dfrac{\partial(J^1, ..., J^q)}{\partial(u_1, ..., u_m)}$ is equal to q. An H-solution is characterized by two integer numbers: a rank $\sigma = \delta + n - r_* \ge 0$ and a defect $\delta \ge 0$. These solutions are also called $H(\sigma, \delta)$-solutions. The rank σ and the defect δ must satisfy the inequalities

$$\rho \le \sigma < n, \ \max\{r_* - n, m - q, 0\} \le \delta \le \min\{r_* - 1, m - 1\},$$

where ρ is a maximum number of invariants only dependent on the independent variables. Notice that for invariant solutions the defect $\delta = 0$ and $q = m$.

To construct a representation of an $H(\sigma, \delta)$-solution one needs to choose $l = m - \delta$ invariants, and then to separate the universal invariant into two parts:

$$\overline{J} = (J^1, ..., J^l), \ \overline{\overline{J}} = (J^{l+1}, J^{l+2}, ..., J^{m+n-r_*}).$$

The number l satisfies the inequality $1 \leq l \leq q \leq m$. A representation of the $H(\sigma, \delta)$-solution is obtained by assuming that the first l coordinates $\overline{J}$ of the universal invariant are functions of the invariants $\overline{\overline{J}}$:

$$\overline{J} = W(\overline{\overline{J}}). \tag{5.64}$$

Equations (5.64) form an invariant part of the representation of a solution. The next assumption about a partially invariant solution is: equations (5.64) can be solved with respect to l dependent functions, for example:

$$u^i = \phi^i(u^{l+1}, u^{l+2}, ..., u^m, x), \ (i = 1, ..., l). \tag{5.65}$$

It is very important to notice that the functions W^i, $(i = 1, ..., l)$ are involved in the expressions for the functions ϕ^i, $(i = 1, ..., l)$. The functions $u^{l+1}, u^{l+2}, ..., u^m$ are called superfluous. The rank and the defect of the $H(\sigma, \delta)$-solution are $\delta = m - l$ and $\sigma = m + n - r_* - l = \delta + n - r_*$.

Notice that for $\delta = 0$ the above algorithm is the algorithm for finding a representation of an invariant solution. If $\delta \neq 0$, equations (5.65) do not define all dependent functions. Since a partially invariant solution satisfies the restrictions (5.64), the algorithm cut out some particular solutions from the set of all solutions. These particular solutions are called partially invariant solutions.

After constructing the representation of invariant or partially invariant solution (5.65) it has to be substituted into the original system of equations. The system of equations found for the functions W and the superfluous functions u^k, $(k = l + 1, 2, ..., m)$ is called a reduced system. This system is overdetermined and requires a compatibility analysis. The compatibility analysis for invariant solutions is easier than for partially invariant solutions. Another case that is easier than the general case is the case of partially invariant solutions where $\overline{\overline{J}}$ only depends on the independent variables

$$J^{l+1} = J^{l+1}(x), J^{l+2} = J^{l+2}(x), ..., J^{m+n-r_*} = J^{m+n-r_*}(x).$$

In this case the partially invariant solution is called regular otherwise it is irregular [134]. The number $\sigma - \rho$ is called a measure of irregularity.

The process of studying compatibility consists of reducing the overdetermined system of partial differential equations to an involutive system. During this process different subclasses of $H(\sigma, \delta)$ of partially invariant solutions can be obtained. Some of these subclasses can be $H_1(\sigma_1, \delta_1)$-solutions with the subalgebra $H_1 \subset H$. In this case $\sigma_1 \geq \sigma$, $\delta_1 \leq \delta$ [130]. The study of compatibility of partially invariant solutions with the same rank $\sigma_1 = \sigma$ but with the smaller defect $\delta_1 < \delta$ is simpler than studying the compatibility for an $H(\sigma, \delta)$-solution. In many applications there is a reduction of $H(\sigma, \delta)$-solution to $H'(\sigma, 0)$. In this case the $H(\sigma, \delta)$-solution is called reducible to an invariant solution. The problem of reduction to an invariant solution is important since invariant solutions are studied first. There are a few general theorems [130] of reduction of partially invariant solutions to invariant. One of these theorems is presented in Chapter 3.

It should be also noted that the notion of compatibility, as in method of differential constraints, plays one of the key roles in constructing partially invariant solutions.

7. Extension of the algorithm of finding partially invariant solutions for a non admitted Lie group

The notion of compatibility plays one of the key roles in constructing partially invariant solutions. Since the notion of compatibility does not depend on the representation of a solution, one can extend the algorithm of finding partially invariant solutions to the case of a non admitted Lie group. In this section an example of such a partially invariant solution of the Navier-Stokes equations is presented.

Unsteady motion of incompressible viscous fluid is governed by the Navier-Stokes equations

$$\mathbf{u}_t + \mathbf{u} \cdot \nabla \mathbf{u} = -\nabla p + \Delta \mathbf{u}, \quad \nabla \cdot \mathbf{u} = 0, \tag{5.66}$$

where $\mathbf{u} = (u, v, w)$ is the velocity field, p is the fluid pressure, ∇ is the gradient operator in the three-dimensional space (x, y, z), and Δ is the Laplacian. Let us consider a solution with respect to the subalgebra corresponding to the generators[14]

$$\begin{aligned} &X_1 = \partial_x, \; X_4 = t\partial_x + \partial_u, \; X_6 = t\partial_z + \partial_w, \\ &\alpha X_5 + X_{11} = \alpha(t\partial_y + \partial_v) + t\partial_t + x\partial_x + y\partial_y + z\partial_z. \end{aligned} \tag{5.67}$$

Notice that the last generator $\alpha X_5 + X_{11}$ is not admitted by the Navier-Stokes equations. Nevertheless one can seek for a partially invariant solution with respect to the Lie group corresponding to this algebra.

Invariants of this Lie group are

$$v - y/t, \quad w - z/t, \quad p, \quad s = y/t - \alpha \ln t.$$

The representation of a regular partially invariant solution is

$$v = V(s) + y/t, \; w = W(s) + z/t, \; p = P(s),$$

and the function $u(t, x, y, z)$ depends on all independent variables. Substituting this representation into the Navier-Stokes equations, one obtains

$$u_t + uu_x + (V + yt^{-1})u_y + (W + zt^{-1})u_z = (u_{xx} + u_{yy} + u_{zz}), \tag{5.68}$$

[14] This subalgebra is taken from the optimal system of subalgebras admitted by the gas dynamic equations [133]. All regular partially invariant submodels of gas dynamic equations with respect to four dimensional subalgebras of the admitted Lie algebra were studied in [135].

$$
\begin{gathered}
t(P' + (V - \alpha)V' + V) - V'' = 0,\\
t((V - \alpha)W' + W) - W'' = 0.
\end{gathered}
\tag{5.69}
$$

$$
u_x + t^{-1}(V' + 2) = 0. \tag{5.70}
$$

Equation (5.70) is integrated to

$$
u = -xt^{-1}(V' + 2) + U,
$$

where $U = U(t, s, \tilde{z})$ is an arbitrary function arising from integration, and $\tilde{z} = zt^{-1}$. Since V and W only depend on s, equations (5.69) can be split with respect to t:

$$
V'' = 0, \quad W'' = 0, \quad VW' - \alpha W' + W = 0, \tag{5.71}
$$

$$
P' - \alpha V' + VV' + V = 0. \tag{5.72}
$$

The general solution of (5.71) is

$$
V = q_1 s + q_2, \quad W = q_3(s + \alpha - q_2)
$$

where q_1, q_2 and q_3 are arbitrary constants, satisfying the condition

$$
q_3(q_1 + 1) = 0. \tag{5.73}
$$

Equation (5.72) serves for finding the pressure[15]

$$
p = -\frac{q_1(q_1 + 1)}{2}s^2 + (\alpha q_1 - q_2(q_1 + 1))s.
$$

The remaining unsolved equation is equation (5.68), which after splitting with respect to x leads to the pair of equations:

$$
\begin{gathered}
(q_1 + 2)(q_1 + 3) = 0,\\
t^2 U_t + t\left((q_1 s + q_2 + \alpha)U_s + q_3(s + \alpha - q_2)U_{\tilde{z}} - (q_1 + 2)U\right) = U_{ss} + U_{\tilde{z}\tilde{z}}.
\end{gathered}
$$

By the first equation, either $q_1 = -2$ or $q_1 = -3$. Hence, because of (5.73), one finds that $q_3 = 0$.

Thus, there are two types of solutions of the Navier-Stokes equations which are partially invariant with respect to the Lie group corresponding to the generators (5.67). The first type of solutions ($q_1 = -2$) is defined by the formulae

$$
u = U(t, s, \tilde{z}), \; v = -y/t + 2\alpha \ln t + q_2, \; w = z/t,
$$

[15]Notice that the pressure is defined by the Navier-Stokes equations up to an arbitrary function of time.

where the function $U(t, s, \tilde{z})$ satisfies the equation

$$t^2U_t + t(-2s + q_2 + \alpha)U_s = U_{ss} + U_{\tilde{z}\tilde{z}}.$$

The second type of solutions ($q_1 = -3$) is defined by the formulae

$$u = xt^{-1} + U,\ v = -2yt^{-1} + 3\alpha \ln t + q_2,\ w = z/t,$$

where the function $U(t, s, \tilde{z})$ satisfies the equation

$$t^2U_t + t\left((-3s + q_2 + \alpha)U_s + U\right) = U_{ss} + U_{\tilde{z}\tilde{z}}.$$

8. Some classes of partially invariant solutions

In this section some classes of partially invariant solutions are presented. Since the analysis of compatibility is very cumbersome and depends on the study of a system, the main attention in this section is given to the construction of a representation of these solutions.

8.1 The Navier–Stokes equations

The unsteady motion of an incompressible viscous fluid is governed by the Navier-Stokes equations

$$\mathbf{u}_t + \mathbf{u} \cdot \nabla \mathbf{u} = -\nabla p + \Delta \mathbf{u}, \quad \nabla \cdot \mathbf{u} = 0, \tag{5.74}$$

where $\mathbf{u} = (u_1, u_2, u_3) = (u, v, w)$ is the velocity field, p is the fluid pressure, ∇ is the gradient operator in the three-dimensional space $\mathbf{x} = (x_1, x_2, x_3) = (x, y, z)$ and Δ is the Laplacian. A group classification of the Navier-Stokes equations in the three-dimensional case has been done in [20][16]. The Lie group admitted by the Navier-Stokes equations is infinite[17]. Its Lie algebra can be presented in the form of the direct sum $L^\infty \oplus L^5$, where the infinite-dimensional ideal L^∞ is generated by the operators

$$X_{\phi_i} = \phi_i(t)\partial_{x_i} + \phi_i'(t)\partial_{u_i} - \phi_i''(t)x_i\partial_p,\ X_\psi = \psi(t)\partial_p$$

with arbitrary functions $\phi_i(t) : R \to R$, $(i = 1, 2, 3)$, and $\psi(t) : R \to R$. The subalgebra L^5 has the following basis:

$$\begin{aligned} &Y = 2t\partial_t + x_i\partial_{x_i} - u_i\partial_{u_i} - 2p\partial_p,\ Z_0 = \partial_t, \\ &Z_{ik} = x_i\partial_{x_k} - x_k\partial_{x_i} + u_i\partial_{u_k} - u_k\partial u_i,\ (i < k \le 3). \end{aligned}$$

The Galilean algebra L^{10} is contained in $L^\infty \oplus L^5$. Several articles [141, 21, 98, 14, 58, 79, 139][18] are devoted to invariant solutions of the Navier-Stokes

[16] A classification of the two-dimensional Navier-Stokes equations was studied in [140].

[17] There is still no complete classification of the subalgebras of the Lie algebra $L^\infty \oplus L^5$. Classification of infinite–dimensional subalgebras of this algebra was studied in [84].

[18] Reviews devoted to invariant solutions of the Navier-Stokes equations can be found in [141, 47, 48, 101, 22] and references therein. Applications of group analysis to different models in hydrodynamics are discussed in [4].

equations. While partially invariant solutions of the Navier–Stokes equations have been less well studied[19], there has been substantial progress in studying such classes of solutions of the inviscid gas dynamic equations [130, 134, 135, 160, 111, 114, 132, 29, 65].

There are other approaches for constructing exact solutions of the Navier–Stokes equations. Let us mention two of them: namely, nonclassical symmetry reductions and direct method [101, 100], and the method where a profile of velocity is assumed to be linear [159, 172].

8.1.1 One class of solutions

Let us consider the four-dimensional Lie algebra L_4, which is generated by the operators

$$X_1 = \phi_1 \partial_x + \phi_1' \partial_u - x\phi_1'' \partial_p, \quad X_2 = \phi_2 \partial_x + \phi_2' \partial_u - x\phi_2'' \partial_p,$$
$$Y_1 = \psi_1 \partial_y + \psi_1' \partial_v - y\psi_1'' \partial_p, \quad Y_2 = \psi_2 \partial_y + \psi_2' \partial_v - y\psi_2'' \partial_p.$$

The algebra conditions

$$[X, Y] \in L_4, \ \forall X, Y \in L_4$$

lead to

$$\phi_1\phi_2'' - \phi_1''\phi_2 = 0, \quad \psi_1\psi_2'' - \psi_1''\psi_2 = 0.$$

Since this algebra is a four-dimensional, this requires

$$\phi_1\phi_2' - \phi_1'\phi_2 \neq 0, \quad \psi_1\psi_2' - \psi_1'\psi_2 \neq 0.$$

The universal invariant of L_4 consists of the invariants

$$w, \ p + k(t)x^2 + l(t)y^2, \ z, \ t,$$

where $k = \phi_i''/(2\phi_i), \ l = \psi_i''/(2\psi_i)$. The rank of the Jacobi matrix of the universal invariant with respect to the dependent variables is equal to two. Hence, only partially invariant solutions with the defect $\delta \geq 2$ can be obtained.

A partially invariant solution with the minimum defect $\delta = 2$ is a regular partially invariant solution of $H(2, 2)$. This solution has the representation

$$w = 2f(z, t), \ p = h(z, t) - k(t)x^2 - l(t)y^2. \tag{5.75}$$

The functions $u = u(x, y, z, t), \ v = v(x, y, z, t)$ are superfluous. This class of solutions generalizes the class of partially invariant solutions studied in [115, 145, 100][20].

---

[19] Firstly the approach of partially invariant solutions to the Navier-Stokes equations was applied in [141]. Some applications of partially invariant solutions in hydrodynamics are considered in [4].

[20] The class of partially invariant solutions of the gas dynamics equations with $\phi_1 = 1, \psi_1 = 1, \phi_2 = t, \psi_2 = t$ is considered in [135].

8.1.2 Compatibility conditions

The analysis of compatibility is very cumbersome. A complete analysis is only given for the one particular case discussed here.

Introducing the functions $\hat{u}(x, y, z, t)$, $\hat{v}(x, y, z, t)$ by the formulae:

$$u = \hat{u} - x\frac{\partial f}{\partial z}, \quad v = \hat{v} - y\frac{\partial f}{\partial z},$$

the second equation of (5.74) becomes

$$\frac{\partial \hat{u}}{\partial x} + \frac{\partial \hat{v}}{\partial y} = 0.$$

The general solution of the last equation is

$$\hat{u} = \frac{\partial \psi}{\partial y}, \quad \hat{v} = -\frac{\partial \psi}{\partial x},$$

where $\psi = \psi(x, y, z, t)$. The first two scalar equations of (5.74) take the form

$$\begin{aligned} &\psi_{yt} + \psi_y\psi_{xy} - \psi_x\psi_{yy} + 2f\psi_{yz} - x(f_{zt} + f_z\psi_{xy} + 2ff_{zz} - f_z^2) \\ &\qquad -yf_z\psi_{yy} = \Delta\psi_y - xf_{zzz} + 2xk, \\ &-\psi_{xt} - \psi_y\psi_{xx} + \psi_x\psi_{xy} - 2f\psi_{xz} - y(f_{zt} - f_z\psi_{xy} + 2ff_{zz} - f_z^2) \\ &\qquad +xf_z\psi_{xx} = -\Delta\psi_x - yf_{zzz} + 2yl. \end{aligned} \tag{5.76}$$

The third coordinate of the motion equation (5.74) determines the function $h(z, t)$ through the function $f(z, t)$:

$$h_z + 2f_t - 2f_{zz} + 4ff_z = 0.$$

Differentiating the first equation of (5.76) with respect to x, the second equation with respect to y, and combining them, one obtains

$$\psi_{xy}^2 - \psi_{xx}\psi_{yy} = Lf_z + k + l, \tag{5.77}$$

where L is the linear operator

$$LF \equiv F_t + 2fF_z - F_{zz} - f_zF.$$

Equation (5.77) can be regarded as the Monge-Ampere equation with a constant (depending on the parameters z and t) right side. The method of solving the Monge-Ampere equation depends on the sign of the right side. If

$$Lf_z + k + l = \alpha^2 \geq 0,$$

then the Monge–Ampere equation can be integrated [57]. This helps to overcome all difficulties in analyzing compatibility[21] of the overdetermined system (5.76), (5.77).

Theorem 5.12. *Any solution of system (5.76) satisfies the Monge-Ampere equation (5.77). If the right side of the Monge-Ampere equation is non negative, $Lf_z + k + l = \alpha^2 \geq 0$, then the solution of the overdetermined system (5.76), (5.77) is either a solution of the system*

$$\begin{gathered} Lf_z + k + l = -c\gamma + \alpha^2, \; L\alpha = \alpha f_z + k - l, \\ L\gamma = f_z\gamma, \; L\lambda = \lambda\alpha - b\gamma, \; Lc = f_z c, \; Lb = \lambda c - \alpha b, \end{gathered} \tag{5.78}$$

or of the system

$$\begin{gathered} Lf_z + k + l = \alpha^2, \; L\alpha = \alpha f_z + k - l, \; L\lambda = \alpha\lambda, \\ L\varphi - \varphi_{yy} - (y(\alpha + f_z) + \lambda)\varphi_y + \alpha\varphi = 0, \end{gathered} \tag{5.79}$$

where $\varphi(y, z, t) = g_y(y, z, t)$.

In case (5.78)

$$\psi(x, y, z, t) = \frac{1}{2}(x^2\gamma(z, t) + y^2 c(z, t)) + x\lambda(z, t) + yb(z, t) + xy\alpha(z, t), \tag{5.80}$$

and in case (5.79)

$$\psi(x, y, z, t) = xy\alpha(z, t) + x\lambda(z, t) + g(y, z, t). \tag{5.81}$$

If $\alpha^2 - c\gamma < 0$, then the solution (5.80) gives an example where system (5.76) is also compatible in an elliptic case. Another example of a solution in an elliptic case is the solution in [145].

The solution (5.81) is separated into two parts. At first one has to find the functions $\alpha(z, t)$, $\lambda(z, t)$, $f(z, t)$, and after that one integrates a linear parabolic equations for the function $\varphi(y, z, t)$.

Thus, the representation (5.75) is very rich. The presence of two arbitrary functions $k(t)$ and $l(t)$ gives additional possibilities for satisfying boundary conditions. Systems (5.78) and (5.79) have infinite-dimensional admitted groups. Infinite-dimensionality is an obstacle for classification of such groups. To overcome these difficulties one can make a group stratification, which allows the splitting of the initial system into automorphic and resolving systems. Any solution of the automorphic system is obtained from a one fixed solution by a transformation belonging to the group. Therefore the problem of constructing solutions is reduced to finding solutions of the resolving systems. Admitted groups of the resolving systems are finite.

[21] Compatibility analysis is similar to the case studied in [115].

8.2 One class of irregular partially invariant solutions

Another example of a partially invariant solution is related with the two-dimensional gas dynamics equations

$$\begin{aligned}
&\rho(u_t + uu_x + vu_y) + p_x = 0,\\
&\rho(v_t + uv_x + vv_y) + p_y = 0,\\
&\rho_t + u\rho_x + v\rho_y + \rho(u_x + v_y) = 0,\\
&p_t + up_x + vp_y + A(p, \rho)(u_x + v_y) = 0.
\end{aligned} \tag{5.82}$$

Partially invariant solutions are considered with respect to the Lie algebra $\{t\partial_x + \partial_u, t\partial_y + \partial_v\}$. The universal invariant of this algebra is

$$u - xt^{-1},\ \ v - yt^{-1},\ p,\ \rho,\ t.$$

Before studying partially invariant solutions let us note that the rank of the Jacobi matrix of the universal invariant with respect to the dependent variables is equal to the number of the dependent variables. Hence, there is a representation of a solution invariant with respect to this algebra

$$u = \frac{x}{t} + U(t),\ v = \frac{y}{t} + V(t),\ p = P(t),\ \rho = R(t). \tag{5.83}$$

Substituting this representation into system (5.82) and solving it, one finds that this invariant solution is isentropic

$$RP' - A(P, R)R' = 0,$$

and that

$$U = \frac{c_1}{t},\ V = \frac{c_2}{t},\ R = \frac{c_3}{t^2}$$

with some constants c_1, c_2, c_3. Using Galilean transformations, one can set $c_1 = 0,\ c_2 = 0$.

Because a defect satisfies the inequality $0 < \delta \leq 1$, partially invariant solutions have the defect $\delta = 1$ and the rank $\sigma = 2$. One of the representations of an $H(2, 1)$-solution is

$$u = \frac{x}{t} + U(\rho, t),\ v = \frac{y}{t} + V(\rho, t),\ p = P(\rho, t). \tag{5.84}$$

Substituting (5.84) into (5.82) and taking linear combinations, one has

$$\begin{aligned}
D\rho + \rho(U_\rho\rho_x + V_\rho\rho_y + \frac{2}{t}) &= 0,\\
(P_\rho - (\rho U_\rho)^2)\rho_x - \rho^2 U_\rho V_\rho \rho_y &= -\frac{\rho}{t}((tU)_t - 2\rho U_\rho),\\
-\rho^2 U_\rho V_\rho \rho_x + (P_\rho - (\rho V_\rho)^2)\rho_y &= -\frac{\rho}{t}((tV)_t - 2\rho V_\rho),\\
P_t + (A - \rho P_\rho)(U_\rho\rho_x + V_\rho\rho_y + \frac{2}{t}) &= 0,
\end{aligned} \tag{5.85}$$

If the rank of the Jacobi matrix of (5.84) with respect to the derivatives ρ_t, ρ_x, ρ_y is equal to 3, then according to the Ovsiannikov reduction theorem [130] the solution is reduced to an invariant. Thus, for irreducible solutions,

$$P_\rho(P_\rho - (\rho U_\rho)^2 - (\rho V_\rho)^2) = 0,$$
$$P_\rho U_\rho(\rho P_\rho - A) = 0, \;\; P_\rho V_\rho(\rho P_\rho - A) = 0. \qquad (5.86)$$

Let us study nonisentropic solutions[22]. For the nonisentropic case $\rho P_\rho - A \neq 0$. Hence, from (5.85) $P_\rho = 0$.

If $U_\rho^2 + V_\rho^2 = 0$, then $tU = c_1, tV = c_2$ or without loss of generality one can set $c_1 = c_2 = 0$. The last equation (5.84) takes the form $P'(t) + \frac{2}{t}A(P(t), \rho) = 0$. If the state equation is such that $A_\rho \neq 0$, then from this equation $\rho = R(t)$. Hence, for $A_\rho \neq 0$ this partially invariant solution is reduced to the invariant solution (5.83). Therefore, for irreducible partially invariant solutions, $A_\rho = 0$. In this case

$$u = \frac{x}{t}, \;\; v = \frac{y}{t},$$

the pressure $p = P(t)$ is defined by the equation

$$P_t + \frac{2}{t}A(P) = 0,$$

and the density satisfies the equation

$$\rho_t + \frac{x}{t}\rho_x + \frac{y}{t}\rho_y + \frac{2}{t}\rho = 0.$$

In another case $U_\rho^2 + V_\rho^2 \neq 0$, equations (5.84) are rewritten

$$\rho(tU)_\rho P_t + A(tU)_t = 0, \;\; \rho(tV)_\rho P_t + A(tV)_t = 0, \qquad (5.87)$$

$$D\rho = \rho\frac{P_t}{A}, \;\; U_\rho\rho_x + V_\rho\rho_y = -\frac{P_t}{A} + \frac{2}{t}. \qquad (5.88)$$

System (5.88) is analyzed by using Poisson brackets for the function $R(\rho, x, y, z, t)$, which satisfies the condition $R(\rho(x, y, z, t), x, y, z, t) = const$ for any solution $\rho(x, y, z, t)$. Taking into account (5.87), one obtains

$$F = t^2 P_{tt} - 2tP_t - t^2 P_t^2 \frac{AA_p + \rho A_\rho + A}{A^2} - 2A = 0. \qquad (5.89)$$

If $F_\rho \neq 0$, then because of the inverse function theorem this solution is reduced to an invariant. Hence,

$$F_\rho = -(t^2 P_t^2 g_\rho + 2A_\rho) = 0, \qquad (5.90)$$

[22]Complete analysis of the isentropic case is given in [114].

where $g = A^{-2}\left(AA_p + \rho A_\rho + A\right)$. This equation is satisfied for[23] $A_\rho = 0$. If $A_\rho \neq 0$, then $g_\rho \neq 0$. Thus $t^2P_t^2 = -2A_\rho/g_\rho$. Similarly, if $(A_\rho/g_\rho)_\rho \neq 0$, then there is reduction to an invariant solution. The equation $(A_\rho/g_\rho)_\rho = 0$ is integrated to give

$$A = g\varphi + \psi,$$

with some functions $\varphi = \varphi(p) \neq 0$ and $\psi = \psi(p)$. With these functions equations (5.89) and (5.90) become

$$t^2P_{tt} - 2tP_t - 2\psi(P) = 0,$$
$$t^2P_t^2 = -2\varphi(P).$$

The compatibility of these equations gives the relationship between the functions $\psi = \psi(p)$ and $\varphi(p)$:

$$(2\psi + \varphi')^2 + 18\varphi(p) = 0.$$

9. Exact solutions constructed on the basis of partially invariant solutions

In this section the method, recently suggested by V.V.Pukhnachov [142][24], is considered. Originally this method was applied for systems of partial differential equations of the structure

$$L_1(u_1, u_2, \ldots, u_m) + w = 0,$$
$$L_i(u_1, u_2, \ldots, u_m) = 0, \ (i = 2, 3, \ldots, q), \tag{5.91}$$

where L_j, $(j = 1, 2, \ldots, q)$ are differential operators. Along with system (5.91) let us also consider the shortened system

$$L_k(u_1, u_2, \ldots, u_m) = 0, \ (k = 1, 2, \ldots, q), \tag{5.92}$$

Assume that the shortened system (5.92) admits a Lie group G such that there exists a partially invariant solution

$$u_i = \varphi(J_1, J_2, \ldots, J_k), \ (i = 1, 2, \ldots, k < m),$$
$$u_j = \psi(x_1, x_2, \ldots, x_n), \ (j = k+1, k+2, \ldots, m), \tag{5.93}$$

where $J_1, J_2, \ldots, J_k$ are invariants of the Lie group G. By virtue of the structure of the original system of equations (5.91), the functions (5.93) and $w = 0$

[23] For state equations with $A_\rho = 0$ these solutions belong to a class of double waves, which generalize simple waves [109].

[24] Further applications of the method can be found in [143, 117, 118].

form a solution of the original system. A new representation of a solution of the original system of equations (5.91) is assumed in the form

$$u_i = \Phi(J_1, J_2, \dots, J_k),\ (i = 1, 2, \dots, k < m),$$
$$u_j = \Psi(x_1, x_2, \dots, x_n),\ (j = k+1, k+2, \dots, m),$$

with unknown functions $\Phi(J_1, J_2, \dots, J_k)$, $(i = 1, 2, \dots, k < m)$, $\Psi(x_1, x_2, \dots, x_n)$, $(j = k+1, k+2, \dots, m)$. The representation of the function w is given by the first equation of (5.91). Since the functions (5.92) and $w = 0$ give a particular solution of the original system (5.91), this guarantees the consistency of the overdetermined system of equations for the functions (5.93).

The requirements on the structure of the original system when considering only partially invariant solutions of the shortened system can be weakened. The method allows the construction of solutions by sequentially solving systems with an increasing number of dependent variables.

In the next subsections two applications of the method are given. The first example deals with the original idea of the method. The second example shows its application in a weaker sense.

9.1 Rotationally symmetric motion of an ideal incompressible fluid

A rotationally symmetric motion of an ideal incompressible fluid is described by the equations[25]

$$\begin{gathered} r^3(u_t + uu_r + wu_z + p_r) - \Omega = 0, \\ w_t + uw_r + ww_z + p_z = 0, \\ u_r + r^{-1}u + w_z = 0, \\ \Omega_t + u\Omega_r + w\Omega_z = 0, \end{gathered} \tag{5.94}$$

where u, v, and w are the coordinates of the velocity in the cylindrical coordinate system (r, θ, z), p is the pressure, $\Omega = (rv)^2$. The corresponding shortened system of partial differential equations is formed by the equations:

$$\begin{gathered} u_t + uu_r + wu_z + \rho^{-1}p_r = 0, \\ w_t + uw_r + ww_z + \rho^{-1}p_z = 0, \\ u_r + r^{-1}u + w_z = 0. \end{gathered} \tag{5.95}$$

Let us consider the Lie group corresponding to the generators

$$\partial_z,\ t\partial_z + \partial_w,\ \partial_p.$$

[25] The method was applied to a rotationally symmetric motion of an ideal incompressible fluid in [142].

Since the invariants of this group are

$$J_1 = r, \ J_2 = t, \ J_3 = u,$$

a regular partially invariant solution has the representation

$$u = \varphi(t, r).$$

The functions

$$w = \psi_1(t, r, z), \ p = \psi_2(t, r, z)$$

form the set of superfluous functions. Substituting this representation into the last two equations of (5.95), and integrating them with respect to z, one finds

$$\psi_1 = zh + g, \ \psi_2 = z^2(h_t + \varphi h_r + h^2)/2 + z(g_t + \varphi g_r + hg) + \chi,$$

where

$$h(t, r) = -\varphi_r(t, r) - r^{-1}\varphi(t, r), \ g = g(t, r), \ \chi = \chi(t, r).$$

After substituting the results into the first equation of (5.95), and splitting it with respect to z, one arrives at the system of equations

$$(h_t + \varphi h_r + h^2)_r = 0, \ (g_t + \varphi g_r + hg)_r = 0, \ \varphi_t + \varphi\varphi_r + \chi_r = 0$$

for the functions $\varphi = \varphi(t, r), \ g = g(t, r), \ \chi = \chi(t, r)$.

According to the Pukhnachov method the constructed partially invariant solution can be used for finding a solution of the original system (5.94). For this solution one needs to assume

$$u = \Phi(t, r), \ w = \Psi_1(t, r, z), \ p = \Psi_2(t, r, z).$$

As for the shortened system, the functions $\Psi_1(t, r, z)$ and $\Psi_2(t, r, z)$ can be found from the second and third equations of system (5.94):

$$\Psi_1 = zH + G, \ \Psi_2 = z^2(H_t + \Phi H_r + H^2)/2 + z(G_t + \Phi G_r + HG) + F,$$

where

$$H(t, r) = -\Phi_r(t, r) - r^{-1}\Phi(t, r), \ G = G(t, r), \ F = F(t, r).$$

The first equation of (5.94) gives

$$\Omega = z^2\lambda + z\mu + \nu, \tag{5.96}$$

where

$$\lambda = (H_t + \Phi H_r + H^2)_r, \ \mu = (G_t + \Phi G_r + HG)_r, \ \nu = F_r - \Phi_t - \Phi\Phi_r. \tag{5.97}$$

The last equation of (5.94), after its splitting with respect to z, is reduced to the equations

$$\begin{gathered}\lambda_t + \Phi\lambda_r + (2H + 3r^{-1}\Phi)\lambda = 0,\\ \mu_t + \Phi\mu_r + (H + 3r^{-1}\Phi)\mu + G\lambda = 0,\\ \nu_t + \Phi\nu_r + 3r^{-1}\Phi\nu = 0.\end{gathered} \tag{5.98}$$

Thus one obtains a closed and consistent system of the six partial differential equations (5.97), (5.98) for the functions

$$\Phi(t,r),\ G(t,r),\ F(t,r),\ \lambda(t,r),\ \mu(t,r),\ \nu(t,r).$$

9.2 Application to a one dimensional gas flow

For the gas dynamic equations there is very well known the class of solutions with constant entropy. Isentropic flows can be considered as the set of solutions of the shortened system of equations for constructing nonisentropic solutions of the original system of equations. Let us consider a one dimensional gas flow

$$\begin{gathered}\rho_t + u\rho_t + \rho u_x = 0,\\ u_t + uu_t + \rho^{-1}P_\rho\rho_x + \rho^{-1}P_S S_x = 0,\\ S_t + uS_t = 0.\end{gathered} \tag{5.99}$$

Here u is the velocity of a gas, the pressure p is defined through the entropy S and the density ρ by the state equation $p = P(\rho, S)$. For example, for a polytropic gas one can use $P = \rho^\gamma S$. The class of isentropic solutions ($S = const$) satisfies the system of equations

$$\begin{gathered}\rho_t + u\rho_t + \rho u_x = 0,\\ u_t + uu_t + \rho^{-1}P_\rho\rho_x = 0.\end{gathered}$$

As a solution of the shortened system of equations one can take the set of simple waves. The simple waves are isentropic solutions and they are defined by the relation $u = u(\rho)$. According to the method, one assumes that the representation of a solution of the original system of equations is

$$u = U(\rho),\ S = S(x,t). \tag{5.100}$$

It is also assumed that $\rho_x \neq 0$.

Substituting this representation of a solution into system (5.99), one obtains

$$\begin{gathered}\rho_t + \rho_x(\rho u' + u) = 0,\\ PS_t = -\rho_x uS\left(\gamma\rho^{-1}P - \rho u'^2\right),\\ PS_x = \rho_x S\left(\gamma\rho^{-1}P - \rho u'^2\right).\end{gathered} \tag{5.101}$$

Equating the mixed derivatives for the function $S(x,t)$, one finds

$$u'(\rho^2 u'^2 - \gamma P)\rho_{xx} = \rho_x^2(u''(\gamma P - 3\rho^2 u'^2) + (\gamma - 2)\rho u'^3) \quad (5.102)$$

Then the equation $(\rho_t)_{xx} = (\rho_{xx})_t$ gives

$$F \equiv -u'''u'\rho(\rho^2 u'^2 - \gamma P)^2 + u''^2\rho(\gamma P - \rho^2 u'^2)(\gamma P - 3\rho^2 u'^2) - u''u'(\gamma P - \rho^2 u'^2)(\gamma P - (\gamma + 5)\rho^2 u'^2) + \rho u'^4(\gamma^2 - 4)(\gamma P - \rho^2 u'^2) = 0 \quad (5.103)$$

If $F_S \neq 0$, then from the last equation one gets $S = G(\rho)$. This leads to the trivial cases: either $S = const$ or $u = const$. Hence, one has to study the case $F_S = 0$:

$$\left(2\rho u'' - (\gamma - 2)u'\right)\left(\rho u'' + (\gamma + 2)u'\right) = 0.$$

Equation (5.103) is satisfied by virtue of the last equation. Thus the system of the equations (5.101) and (5.102) is compatible. The representation of a solution given for isentropic flow has led to a nonisentropic solution of the original system.

Remark 5.2. *Since all subgroups of the Lie group admitted by (5.99) contain the functions ρ and S among the invariants, the constructed solution cannot be obtained as a partially invariant solution with respect to any subgroup of the admitted group.*

10. Nonclassical, weak and conditional symmetries

The determining equations of a Lie group admitted by the system of equations (S) being studied and the differential constraints $\Phi(x,u,p) = 0$ are

$$XS_{|(S\Phi)} = 0, \quad X\Phi_{|(S\Phi)} = 0.$$

Hence, a solution of the determining equations

$$XS_{|(S)} = 0$$

is included in a solution of the previous equations. This provides an idea for constructing new symmetries of the system of equations for a particular class of solutions.

For example, let the generator

$$X = \xi^i(x,u)\partial_{x_i} + \eta^j(x,u)\partial_{u^j}$$

be admitted by a system of differential equations (S). An invariant solution $u = \phi(x)$ satisfies the conditions (differential constraints)

$$X(u^j - \phi^j(x))_{|u=\phi(x)} = -\xi^i(x,\phi(x))\phi^j_{x_i}(x) + \eta^j(x,\phi(x)) = 0, \quad (j = 1, 2, ..., m). \quad (5.104)$$

Thus, this solution satisfies the differential constraints

$$\Phi^j = \xi^i(x,u)p_i^j - \eta^j(x,u) = 0, \ (j = 1, 2, ..., m). \qquad (5.105)$$

10.1 Nonclassical symmetries

The nonclassical symmetries[26] were suggested in [8]. In this method the problem is to find an operator, which is admitted not only by the system (S) but also by equations (5.105). This means that the corresponding group of transformations leaves invariant the set of simultaneous solutions of the system (S) and the invariant conditions $\Phi = (\Phi^1, \Phi^2, ..., \Phi^m) = 0$. This approach is also called a nonclassical approach.

A method for obtaining the determining equations for the coefficients ξ^i and η^j of the generator X in the nonclassical approach is the same as in the classical approach. One has to apply the prolonged generator $\widetilde{X}$ to the equations of the system (S) and the differential constraints (5.105)

$$\widetilde{X}S_{|(S\Phi)} = 0, \ \widetilde{X}\Phi_{|(S\Phi)} = 0,$$

where the set $(S\Phi)$ is defined by the equations $S = 0$ and $\Phi = 0$. Notice that the prolonged generator $\widetilde{X}$ is a classical prolongation on the derivatives of the generator X.

It is directly checked that

$$\widetilde{X}\Phi^j = \Phi^\beta(\eta^j_{u^\beta} - p_i^j\xi^i_{u^\beta}), \ (j = 1, 2, ..., m).$$

Hence, $\widetilde{X}\Phi_{|(S\Phi)} \equiv 0$. Notice also that, opposite to the classical group analysis, if X is an admitted generator and $\lambda = \lambda(x,u)$ is an arbitrary function, then the prolongation formulae imply that

$$(\widetilde{\lambda X})_{|(\Phi)} = \lambda\widetilde{X}_{|(\Phi)}$$

This means that if X is an admitted nonclassical symmetry, then λX is also an admitted nonclassical symmetry. This property permits the normalization of any nonvanishing coefficient of the operator X by setting it equal to one.

The procedure of obtaining the equations $\widetilde{X}S_{|(S\Phi)} = 0$ yields an overdetermined, nonlinear system of equations for the coefficients ξ^i and η^j of the generator X, as opposed to a linear system in the classical case. Since $(S\Phi)$ is a submanifold of (S), the number of determining equations arising in the nonclassical method is fewer than in the classical method. All solutions of the classical determining equations necessarily satisfy the nonclassical determining equations. Hence, a set of nonclassical symmetries is larger than a set of

[26] The author uses the terminology, which is accepted in the literature.

classical symmetries. The vector fields associated with the nonclassical method do not form a vector space. For example, the sum of two nonclassical symmetry operators is not, in general, a symmetry operator. Similarly, the commutator of two nonclassical symmetry operators is not a nonclassical symmetry operator.

The next step, after obtaining an infinitesimal generator, is to study the compatibility of the overdetermined system $(S\Phi)$. Since the differential equations (Φ) are constraints for a solution, the method of differential constraints with m quasilinear differential constraints is more general.

10.1.1 Remark about involutive conditions

Let us denote by X^c a canonical operator [71] corresponding to the operator X:

$$X^c = (\eta^j - \xi^i p_i^j)\partial_{u^j}.$$

The prolongation of the canonical operator X^c is

$$\widetilde{X^c} = X^c + \zeta_\alpha^j \partial_{p_\alpha^j},$$

where

$$\zeta^j = \eta^j - \xi^i p_i^j,\ \zeta_\alpha^j = D^\alpha \zeta^j,\ (|\alpha| \geq 1,\ j = 1, 2, ..., m).$$

For any function $F(x, u, p)$ there is the identity

$$\widetilde{X^c}F = \widetilde{X}F - \xi^i D_i F.$$

Thus, for a nonclassical admitted operator X:

$$\widetilde{X^c} S_{|(\Phi)\cap(DS)} = \widetilde{X} S_{|(S\Phi)} = 0, \tag{5.106}$$

where (DS) denotes the manifold composed by the system (S) and its prolongations $D_i S = 0$:

$$(DS) = \{(x, u, p) \mid S = 0, D_i S = 0,\ (i = 1, 2, ..., n)\}.$$

Comparing this with the equations studied in section 6 one obtains sufficient conditions for the involutiveness of the system $(S\Phi)$.

10.2 Illustrative example of nonclassical symmetries

In this subsection the nonclassical method is used for symmetry reductions to the Boussinesq equation[27]

$$u_{tt} + uu_{xx} + u_x^2 + u_{xxxx} = 0. \tag{5.107}$$

[27] A complete application of the nonclassical method to the Boussinesq equation and comparisons with other methods can be found in [30].

The seeking generator is

$$X = \xi(x,t,u)\partial_x + \tau(x,t,u)\partial_t + \zeta(x,t,u)\partial_u.$$

Since, to obtain the determining equations one needs to transit onto the manifold defined by the system $(S\Phi)$, two cases have to be considered: (a) $\tau \neq 0$ and (b) $\tau = 0,\ \xi \neq 0$. The goal of this section is to demonstrate the idea of the method, therefore only case (a) is studied here. Without loss of generality, one can set $\tau = 1$. Hence, the derivative u_t is defined from the differential constraint

$$\Phi = \xi u_x + \tau u_t - \zeta = 0$$

Eliminating u_{tt} in the Boussinesq equation yields

$$\begin{gathered}\zeta_t + \zeta_u(\zeta - \xi u_x) - \xi_t u_x - \xi_u u_x(\zeta - \xi u_x) - \\ \xi(\zeta_x + \zeta_u u_x - \xi_x u_x + \xi_u u_x^2 + \xi u_{xx}) + u u_{xx} + u_x^2 + u_{xxxx} = 0.\end{gathered} \tag{5.108}$$

Classical group analysis with the generator

$$X = \xi\partial_x + \partial_t + \zeta\partial_u$$

was used to the last equation. In this case one only needs the following prolongation formulae

$$\begin{gathered}\zeta^{u_x} = D_x\zeta - u_x D_x\xi,\ \zeta^{u_{xx}} = D_x\zeta^{u_x} - u_{xx}D_x\xi,\\ \zeta^{u_{xxx}} = D_x\zeta^{u_{xx}} - u_{xxx}D_x\xi,\ \zeta^{u_{xxxx}} = D_x\zeta^{u_{xxx}} - u_{xxxx}D_x\xi.\end{gathered}$$

Substituting the derivative u_{xxxx} found from equation (5.108) and splitting with respect to u_x, u_{xx}, u_{xxx}, one obtains the following seven equations

$$\begin{gathered}\xi_u = 0,\ \zeta_{uu} = 0,\ 2\zeta_{xu} - 3\xi_{xx} = 0,\ \zeta_u + 2\xi_x = 0,\\ 6\zeta_{xxu} + \zeta + \xi\xi_t - 4\xi^2\xi_x + 2u\xi_x = 0,\\ \zeta_{tt} + 4\xi_x\zeta_t + \zeta_{xxxx} + u\zeta_{xx} - 2\xi_t\zeta_x - 2\xi_t\zeta_x\\ -4\xi\xi_x\zeta_x + 2\zeta\zeta_{xu} + 4\xi_u\zeta\zeta_u = 0,\\ 2u\zeta_{xu} - u\xi_{xx} + 2\zeta_x + 4\zeta_{xxxu} - 8\xi\xi_x\zeta_u\\ -2\xi\zeta_{tu} - \xi_{xxxx} + 4\xi\xi_x^2 - 2\xi_t\xi_x - \xi_{tt} = 0.\end{gathered}$$

The general solutions of these equations is

$$\begin{gathered}\xi = xf(t) + g(t),\ \zeta = -[2f(t)u + 2x^2 f(t)\left(\tfrac{df(t)}{dt} + 2f^2(t)\right) + \\ 2x\left(\tfrac{df(t)}{dt}g(t) + f(t)\tfrac{dg(t)}{dt} + 4f^2(t)g(t)\right) + 2g(t)\left(\tfrac{dg(t)}{dt} + 2f(t)g(t)\right)],\end{gathered} \tag{5.109}$$

where

$$f(t) = \frac{1}{2q(t)}\frac{dq(t)}{dt},\ g(t) = \frac{c_1}{2q(t)}\frac{dq(t)}{dt} + c_0\int\frac{q(s)}{(q'(s))^2}ds,$$

and $q(t)$ satisfies the equation

$$\left(\frac{dq(t)}{dt}\right)^2 = c_3 q^3(t) + c_2.$$

Here c_1, c_2, c_3 are arbitrary constants.

Remark 5.3. *Let us consider the differential constraint*

$$\Phi \equiv u_t - \phi(x, t, u, u_x) = 0 \tag{5.110}$$

compatible with the Boussinesq equation. Because all fourth derivatives of the function $u(x, t)$ can be found from the overdetermined system of equations (5.107), (5.110), and the prolongations

$$\begin{gathered} D_t\Phi = 0,\ D_x\Phi = 0, \\ D_t^2\Phi = 0,\ D_xD_t\Phi = 0,\ D_x^2\Phi = 0, \\ D_t^3\Phi = 0,\ D_xD_t^2\Phi = 0,\ D_x^2D_t\Phi = 0,\ D_x^3\Phi = 0, \end{gathered} \tag{5.111}$$

the general solution of this system is only defined up to arbitrary constants. Moreover, calculations show that for involutiveness the differential constraint (5.110) must be quasilinear

$$u_t = a(x, t, u)u_x + b(x, t, u).$$

Further calculations show that the overdetermined system (5.110), (5.110), (5.111) is involutive if, and only if,

$$a = -\xi,\ b = \zeta,$$

where the functions ξ, ζ are defined by the formulae (5.109). Thus, the main requirement for the system (5.110), (5.111) is that it is involutive.

10.3 Weak and conditional symmetries

10.3.1 Weak symmetries

Further generalization of the nonclassical method is given by weak symmetries [128]. If the overdetermined system $(S\Phi)$ is compatible, then the process of reducing it to an involutive system gives new partial differential equations (integrability conditions). Therefore one should compute the symmetry group not of just the system $(S\Phi)$, but also of any associated integrability conditions. A weak symmetry group of the system (S) is defined to be any symmetry group of the overdetermined system $(S\Phi)$ and all its integrability conditions. Notice that the first part of finding weak symmetries coincides with an application of the method of differential constraints with quasilinear first order m differential constraints.

Weak symmetry groups have some critical drawbacks. It can be shown that every solution of the system (S) can be derived from some weak symmetry group [128]. This obstacle is similar to the obstacle appearing in method of differential constraints, and it can be overcome by requiring the involutiveness of the system $(S\Phi)$ or some its prolongations.

10.3.2 Conditional symmetries

Let us consider a system of differential equations (S) and append to it the system of differential constraints

$$\Phi(x, u, p) = 0.$$

About the differential constraints it is only assumed that the overdetermined system $(S\Phi)$ is compatible. Classical symmetries of the overdetermined system $(S\Phi)$ are called conditional symmetries of the system (S).

10.4 B–symmetries

The method of B-symmetries [82] can be considered as a generalization of nonclassical symmetries. Let us consider a canonical generator $X^c = \Phi\partial_u$ of a Lie-Bäcklund group of transformations. The determining equation one can use is the same as for nonclassical symmetries (5.106):

$$\widetilde{X^c} S_{|(\Phi)\cap(DS)} = 0.$$

Considering only part of solutions of this equation one can simplify it to

$$\widetilde{X^c} S_{|(DS)} + B\Phi = 0 \tag{5.112}$$

with some matrix B. Equation (5.112) is called a determining equation of B-symmetry.

Let us consider the equation

$$u_{tx} = f(u).$$

The determining equation of B-symmetry is

$$D_t D_x \phi - f'\phi + B\phi = 0. \tag{5.113}$$

This equation has to be satisfied for any $u_{tx} = f(u)$ and all its prolongations. For example, assuming that

$$\phi = u_{xx} + g(u)u_x^2,$$

and substituting

$$u_{tx} = f,\ u_{txx} = f'u_x,\ u_{txxx} = f''u_x^2 + f'u_{xx},$$

into the determining equation, it is reduced to

$$u_{xx}(2g'u_tu_x + 2fg + B) + g''u_x^3u_t + u_x^2(f'' + 3fg' + f'g + Bg) = 0.$$

Supposing that the function B depends on u, u_t and u_x, the last equation can be split with respect to u_{xx}: $B = -2(g'u_tu_x + fg)$, and then with respect to u_x and u_t:

$$g'' - 2gg' = 0, \ f'' + 3fg' + f'g - 2fg^2 = 0.$$

One of solutions of these equations is

$$g = -u^{-1}, \ f = k_1u + k_2u\log(u).$$

The general solution of the overdetermined system

$$u_{xx} = -u^{-1}u_x^2, \ u_{tx} = k_1u + k_2u\log(u)$$

is $u = \psi(t)e^{x(k_2t+k_3)}$, where k_1, k_2 and k_3 are arbitrary constants and the function $\psi(t)$ satisfies the ordinary differential equation

$$(k_2t + k_3)\psi' = (k_1 - k_2)\psi + k_2\psi\log(\psi).$$

11. Group of tangent transformations

11.1 Lie groups of finite order tangency

A notion of a Lie group of point transformations is generalized to allow derivatives in the transformation[28]. Assume that the transformations

$$x'_k = f_k(x, u, p; a), \ u^{j'} = \phi^j(x, u, p; a), \ p_\alpha^{j'} = \psi_\alpha^j(x, u, p; a), \quad k = 1, 2, \dots, n; \ j = 1, 2, \dots, m; \ |\alpha| \le q \tag{5.114}$$

form a one parameter Lie group G^1, where the functions f_i, ϕ^j and ψ_α^j depend on the independent variables x, the dependent variables u, and the derivatives p_β^k, $(k = 1, 2, \dots, m; \ |\beta| \le q)$ of order up to q. The infinitesimal generator of the group G^1 is given by the vector field (ξ, η, ζ)

$$X = \xi^k\partial_{x_k} + \eta^j\partial_{u^j} + \zeta_\alpha^j\partial_{p_\alpha^j}$$

Conditions for the transformations (5.114) to be tangent transformations require preserving the tangent conditions

$$du^j - p_i^jdx_i = 0, \ dp_\alpha^j - p_{\alpha,k}^jdx_k = 0, \tag{5.115}$$

[28] Examples of such transformations were given in Chapter 1.

where the transformations of the differentials du^j, dx_i and dp^j_α are defined by the usual formulae for the differentials

$$
\begin{aligned}
dx_i' &= \tfrac{\partial f_i}{\partial x_l}dx_l + \tfrac{\partial f_i}{\partial u^s}du^s + \tfrac{\partial f_i}{\partial p^s_\beta}dp^s_\beta,\\
du^{k\prime} &= \tfrac{\partial \phi^k}{\partial x_l}dx_l + \tfrac{\partial \phi^k}{\partial u^s}du^s + \tfrac{\partial \phi^k}{\partial p^s_\beta}dp^s_\beta,\\
dp^{j\prime}_\alpha &= \tfrac{\partial \psi^j_\alpha}{\partial x_l}dx_l + \tfrac{\partial \psi^j_\alpha}{\partial u^s}du^s + \tfrac{\partial \psi^j_\alpha}{\partial p^s_\beta}dp^s_\beta,
\end{aligned}
\tag{5.116}
$$

The tangent conditions (5.115) are very strong, and they provide very strong restrictions for the transformations (5.114). Let us obtain them.

Substituting (5.116) into (5.115), one finds

$$
\begin{gathered}
\left(\tfrac{\partial \phi^k}{\partial x^i} + p^j_i\tfrac{\partial \phi^k}{\partial u^j} + \textstyle\sum_{|\beta|\le q-1}\tfrac{\partial \phi^k}{\partial p^j_\beta}p^j_{\beta,i} - \psi^k_l\left(\tfrac{\partial f_l}{\partial x^i} + p^j_i\tfrac{\partial f_l}{\partial u^j} + \textstyle\sum_{|\beta|\le q-1}\tfrac{\partial f_l}{\partial p^j_\beta}p^j_{\beta,i}\right)\right)\\
dx_i + \textstyle\sum_{|\beta|=q}\left(\tfrac{\partial \phi^k}{\partial p^j_\beta} - \psi^k_l\tfrac{\partial f_l}{\partial p^j_\beta}\right)dp^j_\beta = 0.\quad (k = 1, 2, ..., m).
\end{gathered}
\tag{5.117}
$$

$$
\begin{gathered}
\left(\tfrac{\partial \psi^j_\alpha}{\partial x_i} + p^j_i\tfrac{\partial \psi^j_\alpha}{\partial u^j} + \textstyle\sum_{|\beta|\le q-1}\tfrac{\partial \psi^j_\alpha}{\partial p^j_\beta}p^j_{\beta,i} - \psi^j_{\alpha,k}\left(\tfrac{\partial f_k}{\partial x^i} + p^s_i\tfrac{\partial f_k}{\partial u^s} + \textstyle\sum_{|\beta|\le q-1}\tfrac{\partial f_k}{\partial p^s_\beta}p^s_{\beta,i}\right)\right)\\
dx_i + \textstyle\sum_{|\beta|=q}\left(\tfrac{\partial \psi^j_\alpha}{\partial p^s_\beta} - \psi^j_{\alpha,k}\tfrac{\partial f_k}{\partial p^s_\beta}\right)dp^s_\beta = 0,\quad (j = 1, 2, ..., m,\ |\alpha| \le q-1).
\end{gathered}
\tag{5.118}
$$

Since the differentials dx_i and dp^j_β, $(i = 1, 2, ..., n;\ j = 1, 2, ..., m;\ |\beta| = q)$ are independent[29], one can split equations (5.117) and (5.118) with respect to these differentials:

$$
\begin{gathered}
\tfrac{\partial \phi^k}{\partial x^i} + p^j_i\tfrac{\partial \phi^k}{\partial u^j} + \textstyle\sum_{|\beta|\le q-1}\tfrac{\partial \phi^k}{\partial p^j_\beta}p^j_{\beta,i} - \psi^k_l\left(\tfrac{\partial f_l}{\partial x^i} + p^s_i\tfrac{\partial f_l}{\partial u^s} + \textstyle\sum_{|\beta|\le q-1}\tfrac{\partial f_l}{\partial p^s_\beta}p^s_{\beta,i}\right) = 0,\\
(i = 1, 2, ..., n;\ k = 1, 2, ..., m),\\
\tfrac{\partial \phi^k}{\partial p^j_\beta} - \psi^k_l\tfrac{\partial f_l}{\partial p^j_\beta} = 0,\ (|\beta| = q;\ k = 1, 2, ..., m),
\end{gathered}
$$

$$
\begin{gathered}
\tfrac{\partial \psi^j_\alpha}{\partial x_i} + p^j_i\tfrac{\partial \psi^j_\alpha}{\partial u^j} + \textstyle\sum_{|\beta|\le q-1}\tfrac{\partial \psi^j_\alpha}{\partial p^s_\beta}p^s_{\beta,i} - \psi^j_{\alpha,k}\left(\tfrac{\partial f_k}{\partial x^i} + p^s_i\tfrac{\partial f_k}{\partial u^s} + \textstyle\sum_{|\beta|\le q-1}\tfrac{\partial f_k}{\partial p^s_\beta}p^s_{\beta,i}\right) = 0,\\
(|\alpha| \le q-1;\ j = 1, 2, ..., m;\ i = 1, 2, ..., n),\\
\tfrac{\partial \psi^j_\alpha}{\partial p^s_\beta} - \psi^j_{\alpha,k}\tfrac{\partial f_k}{\partial p^s_\beta} = 0,\ (j, s = 1, 2, ..., m,\ |\alpha| \le q-1,\ |\beta| = q).
\end{gathered}
$$

Differentiating these relations with respect to the parameter a and substituting $a = 0$ into them, one obtains

$$
\begin{gathered}
\zeta^k_i = \tfrac{\partial \eta^k}{\partial x^i} + p^s_i\tfrac{\partial \eta^k}{\partial u^s} + \textstyle\sum_{|\beta|\le q-1}\tfrac{\partial \eta^k}{\partial p^s_\beta}p^s_{\beta,i} - p^k_l\left(\tfrac{\partial \xi_l}{\partial x^i} + p^s_i\tfrac{\partial \xi_l}{\partial u^s} + \textstyle\sum_{|\beta|\le q-1}\tfrac{\partial \xi_l}{\partial p^s_\beta}p^s_{\beta,i}\right),\\
(i = 1, 2, ..., n;\ k = 1, 2, ..., m),
\end{gathered}
\tag{5.119}
$$

[29] Notice that for admitted tangent transformations these differentials are not arbitrary.

$$\frac{\partial \eta^k}{\partial p^j_\beta} - p^k_l \frac{\partial \xi_l}{\partial p^j_\beta} = 0, \ (|\beta| = q; \ j, k = 1, 2, ..., m). \tag{5.120}$$

$$\zeta^j_{\alpha,i} = \frac{\partial \zeta^j_\alpha}{\partial x_i} + p^s_i \frac{\partial \zeta^j_\alpha}{\partial u^s} + \sum_{|\beta| \le q-1} \frac{\partial \zeta^j_\alpha}{\partial p^s_\beta} p^s_{\beta,i} - p^j_{\alpha,k} \left(\frac{\partial \xi_k}{\partial x^i} + p^s_i \frac{\partial \xi_k}{\partial u^s} + \sum_{|\beta| \le q-1} \frac{\partial \xi_k}{\partial p^s_\beta} p^s_{\beta,i} \right),$$
$$(|\alpha| \le q-1; \ j = 1, 2, ..., m; \ i = 1, 2, ..., n), \tag{5.121}$$

$$\frac{\partial \zeta^j_\alpha}{\partial p^s_\beta} - p^j_{\alpha,k} \frac{\partial \xi_k}{\partial p^s_\beta} = 0, \ (j, s = 1, 2, ..., m, \ |\alpha| \le q-1, \ |\beta| = q). \tag{5.122}$$

Using equations (5.120) and (5.122), equations (5.119) and (5.121) can be rewritten in the form

$$\zeta^k_i = D_i \eta^k - p^k_l D_i \xi_l, \ (i = 1, 2, ..., n; \ k = 1, 2, ..., m), \tag{5.123}$$

$$\zeta^j_{\alpha,i} = D_i \zeta^j_\alpha - p^j_{\alpha,k} D_i, \ (|\alpha| \le q-1; \ j = 1, 2, ..., m; \ i = 1, 2, ..., n), \tag{5.124}$$

For further study it is convenient to introduce the functions

$$W^k = \eta^k - \xi_i p^k_i, \ W^k_\beta = \zeta^k_\beta - \xi_i p^k_{\beta,i}, \ (k = 1, 2, ..., m; \ |\beta| \le q-1). \tag{5.125}$$

Let $q = 1$. Equations (5.120) become

$$\frac{\partial W^k}{\partial p^j_i} + \xi_i \delta_{jk} = 0, \ (i = 1, 2, ..., n; \ j; \ k = 1, 2, ..., m). \tag{5.126}$$

If $m > 1$, choosing $j \ne k$ and $j = k$ in equations (5.126), one obtains, respectively,

$$\xi_i = -\frac{\partial W^1}{\partial p^1_i} = -\frac{\partial W^2}{\partial p^2_i} = \ldots = -\frac{\partial W^m}{\partial p^m_i}, \qquad \frac{\partial W^k}{\partial p^j_i} = 0,$$
$$(i = 1, 2, ..., n; \ j, k = 1, 2, ..., m; \ j \ne k).$$

These equations give

$$\xi_i = V_i(x, u), W^j = U^j - \xi_k p^j_k = U^j - V_k p^j_k, (i = 1, 2, ..., n; j = 1, 2, ..., m),$$

where the functions $V_i = V_i(x, u)$ and $U^j = U^j(x, u)$ do not depend on the derivatives. Thus, the Lie group G^1 is a prolongation of a Lie group of point transformations.

If $m = 1$, then equations (5.123) and (5.126) give

$$\xi_i = -\frac{\partial W}{\partial p_i}, \quad \zeta_i = \frac{\partial W}{\partial x_i} + p_i \frac{\partial W}{\partial u}, \ (i = 1, 2, ..., n), \tag{5.127}$$

where $u = u^1$, $W = W^1$, $p_i = p_i^1$, $\zeta_i = \zeta_i^1$. Notice that from the definition of the function W and (5.127), one has

$$\eta = W - p_i \frac{\partial W}{\partial p_i}.$$

Hence, all coefficients of the generator X are expressed through the function W. The function W is called a characteristic function.

Let $q > 1$. Equations (5.120) and (5.122) become

$$\frac{\partial W^j}{\partial p_\beta^k} = 0, \quad \frac{\partial W_\alpha^j}{\partial p_\beta^k} + \xi_i \frac{\partial p_{\alpha,i}^j}{\partial p_\beta^k} = 0, \; (j, k = 1, 2, ..., m, \; |\alpha| \le q-1, \; |\beta| = q). \tag{5.128}$$

If $m > 1$, choosing $j \neq k$ in equations (5.128), one obtains

$$\frac{\partial W_\alpha^j}{\partial p_\beta^k} = 0, \quad (j, k = 1, 2, ..., m, \; |\alpha| \le q - 1, \; |\beta| = q). \tag{5.129}$$

Choosing $j = k$ and $\beta = \alpha, i$ in equations (5.128), one finds

$$\xi_i = -\frac{\partial W_\alpha^j}{\partial p_{\alpha,i}^j}, \quad (j = 1, 2, ..., m, \; |\alpha| \le q - 1). \tag{5.130}$$

Since in (5.128) the coordinates ξ_i do not depend on the derivatives p_β^j, $(|\beta| = q, \; j = 1, 2, ..., m)$. The general solution of (5.130) is

$$W_\alpha^j = U_\alpha^j - \xi_k p_{\alpha,k}^j, \; (j = 1, 2, ..., m, \; |\alpha| \le q - 1),$$

where the functions U_α^j do not depend on the derivatives of order q: p_β^s, $(s = 1, 2, ..., m; \; |\beta| = q)$. Because of (5.128) and (5.122) the coefficients η^k and ζ_α^j $(j = 1, 2, ..., m, \; |\alpha| \le q-1)$ are also independent of the derivatives of the order q. By induction on q one obtains (5.126). Hence, as in the case $q = 1$, the Lie group G^1 is a prolongation of a Lie group of point transformations.

If $m = 1$, equations (5.128) become

$$\frac{\partial W}{\partial p_\beta} = 0, \; (|\beta| = q), \quad \frac{\partial W_\alpha}{\partial p_\beta} + \xi_i \frac{\partial p_{\alpha,i}}{\partial p_\beta} = 0, \; (|\alpha \le q-1, \; |\beta| = q). \tag{5.131}$$

Notice that if for given β and α there is an i such that $\beta \neq \alpha, i$, then

$$\frac{\partial W_\alpha}{\partial p_\beta} = 0,$$

and, hence,

$$\frac{\partial}{\partial p_\beta}(\frac{\partial W_\alpha}{\partial p_{\alpha,i}}+\xi_i)=\frac{\partial}{\partial p_{\alpha,i}}(\frac{\partial W_\alpha}{\partial p_\beta})+\frac{\partial \xi_i}{\partial p_\beta}=\frac{\partial \xi_i}{\partial p_\beta}=0.$$

If $n>1$, then for any $1\leq i\leq n$ and β, $(|\beta|=q)$ there are α, $(|\alpha|=q-1)$ and k such that $k\neq i$, $\beta=\alpha,k$ and $\beta\neq\alpha,i$. Thus, the coefficients ξ_i, $(1\leq i\leq n)$ do not depend on p_β, $(|\beta|=q)$, and

$$W_\alpha=U_\alpha-\xi_i p_{\alpha,i},\ (|\alpha|=q-1),$$

where the functions U_α also do not depend on the derivatives p_β, $(|\beta|=q)$. By induction on q one finds that G^1 is a prolongation of a Lie group of contact transformations with the coefficients

$$\xi_i=-\frac{\partial W}{\partial p_i},\quad \zeta_\alpha=D^\alpha W-p_{\alpha,k}\frac{\partial W}{\partial p_k},\ (i=1,2,...,n;\ |\alpha|\geq 1).$$

If $n=1$, then equations (5.131) lead to

$$W=\eta-\xi p_1,\ W_1=\zeta_1-\xi p_2,\ W_2=\zeta_2-\xi p_3,...,\ W_{q-1}=\zeta_{q-1}-\xi p_q,$$

and

$$\frac{\partial W}{\partial p_q}=0,\ \frac{\partial W_1}{\partial p_q}=0,\ \frac{\partial W_2}{\partial p_q}=0,...,\ \frac{\partial W_{q-2}}{\partial p_q}=0,\ \xi=-\frac{\partial W_{q-1}}{\partial p_q}.\quad (5.132)$$

Because of (5.124), one finds

$$DW_{q-2}=D(\zeta_{q-2}-\xi p_{q-1})=D\zeta_{q-2}-p_{q-1}D\xi-\xi p_q=\zeta_{q-1}-\xi p_q=W_{q-1}.$$

Because of (5.132)

$$\xi=-\frac{\partial W_{q-1}}{\partial p_q}=-\frac{\partial W_{q-2}}{\partial p_{q-1}}.$$

Hence,

$$\eta=W-p_1\frac{\partial W_{q-2}}{\partial p_{q-1}},\ \zeta_1=W_1-p_2\frac{\partial W_{q-2}}{\partial p_{q-1}},...,\ \zeta_{q-2}=W_{q-2}-p_{q-1}\frac{\partial W_{q-2}}{\partial p_{q-1}},$$
$$\zeta_{q-1}=DW_{q-2}+\xi p_q=DW_{q-2}-p_q\frac{\partial W_{q-2}}{\partial p_{q-1}}=$$
$$\frac{\partial W_{q-2}}{\partial x}+p_1\frac{\partial W_{q-2}}{\partial u}+p_2\frac{\partial W_{q-2}}{\partial p_1}+...+p_{q-1}\frac{\partial W_{q-2}}{\partial p_{q-2}}.$$

Thus, ξ, η, ζ_1, ζ_2, ..., ζ_{q-1} only depend on x, u, p_1, p_2, ..., p_{q-1}. By induction on q one finds that the Lie group G^1 is a prolongation of a Lie group of contact transformations.

Therefore, any Lie group of transformations (5.114) is not a prolongation of a Lie group of point transformations only for $m=1$. For $m=1$, any Lie group (5.114) is a prolongation of a Lie group of contact transformations, and there exists a characteristic function $W = W(x_1, x_2, ..., x_n, u, p_1, p_2, ..., p_n)$ such that

$$\xi_i = -\frac{\partial W}{\partial p_i},\ \eta = W - p_i\frac{\partial W}{\partial p_i},\ \zeta_\alpha = D^\alpha W - p_{\alpha,k}\frac{\partial W}{\partial p_k},\quad (i = 1, 2, ..., n;\ |\alpha| \geq 1). \tag{5.133}$$

This statement is known as the Bäcklund theorem [71].

Remark 5.4. *The Bäcklund theorem was proven with the assumption that there are no restrictions on the derivatives* p^j_β, $(j = 1, 2, ..., m; |\beta| \leq q + 1)$ *of the order up to* $q + 1$.

11.2 An admitted Lie group of tangent transformations

A Lie group of tangent transformations with the generator X is said to be admitted by a system of partial differential equations

$$(S) \qquad F(x, u, p) = 0$$

if the determining equations

$$XF_{|(S)} = 0 \tag{5.134}$$

are satisfied. Here q is the order of the tangent Lie group, (D^qS) is the system of the equations formed by the prolongations of the equations of the system (S) up to the order q. For example, let a Lie group of tangent transformations with $m = 1$ be a prolongation of a Lie group of contact transformations with a characteristic function W. Because of (5.133)

$$\begin{aligned} XF = \xi_i F_{x_i} + \eta F_u + \zeta_\alpha F_{p_\alpha} &= -\frac{\partial W}{\partial p_i}F_{x_i} + (W - p_i\frac{\partial W}{\partial p_i})F_u \\ &+ (D^\alpha W - p_{\alpha,i}\frac{\partial W}{\partial p_i})F_{p_\alpha} \\ &= -\frac{\partial W}{\partial p_i}D_iF + \left(WF_u + (D^\alpha W)F_{p_\alpha}\right). \end{aligned}$$

Hence, the determining equations (5.134) can be rewritten in the form

$$XF_{|(S)} = XF_{|(S),(DS)} = \left(WF_u + (D^\alpha W)F_{p_\alpha}\right)_{|(S),(DS)} = 0. \tag{5.135}$$

As was proven in the general case[30] a Lie group of contact transformations which is not a prolongation of a Lie group of point transformations[31] is only possible for $m = 1$. An admitted Lie group of contact transformations is related with a system of differential equations (S). The differential equations provide restrictions for the derivatives. Hence, one has to take into account these restrictions, for example, when splitting equations (5.117), (5.118). This

[30] Without relating transformations with a studied system of equations.

[31] Such groups are called irreducible [73].

circumstance gives possibilities for the existence of admitted Lie groups of tangent transformations where the number of the independent variables is greater than one ($m > 1$). These transformations are called Bäcklund transformations. Here is the example[32]. Let us consider the differential equations[33]

$$U''' - UU'' + (U')^2 = 0, \quad V_t + UV_s - VU' = V_{ss}, \tag{5.136}$$

where

$$U = U(s), \ V = V(t, s).$$

Consider the transformation

$$\begin{gathered} \bar{t} = t, \ \ \bar{s} = s, \ \ \overline{U} = U, \ \ \overline{V} = V + a(tU' + 1), \\ \overline{V_t} = V_t + aU', \ \ \overline{V_s} = V_s + atU'', \ \ \overline{V_{ss}} = V_{ss} + at\left(UU'' - (U')^2\right), \end{gathered} \tag{5.137}$$

where the other derivatives of first and second order are mapped identically. The infinitesimal operator[34] of this Lie group is

$$X = (tU' + 1)\partial_V + U'\partial_{V_t} + tU''\partial_{V_s} + t\left(UU'' - (U')^2\right)\partial_{V_{ss}}.$$

According to the construction, the tangent conditions are satisfied for any solution of system (5.136). By direct calculations one finds that the determining equations (5.134) are also satisfied. Thus, the transformations (5.137) are Bäcklund transformations.

Notice that system of equations (5.136) has another Lie group of tangent transformations corresponding to the generator $U'\partial_V$.

Remark 5.5. *The Lie group of transformations (5.137) was originally found by seeking an admitted Lie group of point transformations for the equivalent system*

$$U' = W, \ \ W'' - UW' + W^2 = 0, \ \ V_t + UV_s - VU' = V_{ss}. \tag{5.138}$$

The same result is obtained if one chooses the set of the dependent variables

$$u^1 = U, \ \ u^2 = V, \ \ u^3 = U', \ \ u^4 = U''.$$

Although the search for a Lie group of contact transformations of system (5.138) by the usually accepted technique where this group is seeking a Lie group of point transformations with the dependent variables

$$u^1 = U, \, u^2 = V, \, u^3 = U', \, u^4 = V_t, \, u^5 = V_s,$$

does not give the transformation (5.137).

---

[32]The group was constructed by K.Thailert.

[33]These equations were obtained in the process of studying partially invariant solutions of the Navier-Stokes equations.

[34]This operator is uniquely defined by the first term $(tU' + 1)\partial_V$.

11.3 Contact transformations of the Monge-Ampere equation

Let us apply the theory of contact transformations to the Monge-Ampere equation[35] (1.93)

$$u_{qq}u_{pp} - u_{pq}^2 + a(p,q) = 0, \tag{5.139}$$

where[36] $a(p,q) = H(q)p^{\kappa} > 0,\ -2 < \kappa < -1$, and $H' \neq 0$.

Notice that the Lie group of equivalent transformations corresponds to the generators

$$\begin{gathered}\partial_q,\ \partial_u,\ p\partial_u,\ q\partial_u,\ p\partial_p - u(\kappa+2)\partial_u,\ u\partial_u + q\partial_q,\\ q\left(p\partial_p + q\partial_q + u\partial_u - (\kappa+4)H\partial_H\right),\ u\partial_u + 2H\partial_H.\end{gathered} \tag{5.140}$$

It is possible to prove that a Lie group of contact transformations admitted by the Monge-Ampere equation (5.139) is defined by the characteristic function[37] $W = W(p,q,u,u_p,u_q)$ with the coefficients of the infinitesimal generator (5.133). Thus, the determining equations (5.135) are

$$u_{qq}D_p^2W + u_{pp}D_q^2W - 2u_{pq}D_pD_qW = 0. \tag{5.141}$$

Splitting the determining equations gives the following four partial differential equations:

$$\begin{gathered}S_1 = 2W_{qu}u_q + W_{qq} + W_{uu}u_q^2 - W_{u_pu_p}Hp^{\kappa} = 0,\\ S_2 = W_{pq} + W_{pu}u_q + W_{qu}u_p + W_{uu}u_pu_q + W_{u_pu_q}Hp^{\kappa} = 0,\\ S_3 = 2W_{pu}u_p + W_{pp} + W_{uu}u_p^2 - W_{u_qu_q}Hp^{\kappa} = 0,\\ S_4 = H'W_{u_q}p + 2pH(W_{pu_p} + W_{qu_q} + W_{uu_p}u_p + W_{uu_q}u_q + W_u)\\ +W_{u_p}\kappa H = 0.\end{gathered} \tag{5.142}$$

To find transformations which are admitted for any function $H(q)$, one has to split equations (5.142) with respect to H and H'. Splitting the last equation of (5.142) with respect to H', one has

$$W_{u_q} = 0.$$

Then splitting the remaining equations with respect to H and u_q, one finds

$$\begin{gathered}W_{qu} = 0,\ W_{qq} = 0,\ W_{uu} = 0,\ W_{u_pu_p} = 0,\ W_{pq} = 0,\ W_{pu} = 0,\ W_{pp} = 0,\\ 2p(W_{pu_p} + W_{uu_p}u_p + W_u) + W_{u_p}\kappa = 0.\end{gathered}$$

[35] Complete classification of the Monge-Ampere equation with respect to Lie groups of contact transformations is given in [83].

[36] For a polytropic gas $\kappa = -\gamma^{-1}(\gamma+1)$, where γ is a polytropic exponent ($\gamma > 1$).

[37] The Monge-Ampere equation does not effect the study of the tangent conditions for a contact Lie group.

Thus, the kernel of admitted Lie groups is formed by the characteristic function

$$W = c_1 p + c_2 q + c_3((\kappa + 2)u - pu_p) + c_4, \tag{5.143}$$

where c_i, $(i = 1, 2, 3, 4)$ are constant. This characteristic function corresponds to a Lie group of point transformations with the generators

$$X_1 = \partial_u,\ X_2 = p\partial_u,\ X_3 = q\partial_u,\ X_4 = p\partial_p + (\kappa + 2)u\partial_u. \tag{5.144}$$

A nontrivial admitted Lie group of contact transformations can be only obtained for special functions $H(q)$. Let us find these functions.

From equations (5.142) one can find W_{qq}, W_{pq}, W_{pp}, W_{pu_p}. Introducing the function

$$W_1 = 4pHW_u - pH'W_{u_q} - W_{u_p}\kappa H, \tag{5.145}$$

from the equations $(W_{pp})_q - (W_{pq})_p = 0$, $(W_{qq})_p - (W_{pq})_q = 0$, and $(W_{pu_p})_q - (W_{pq})_{u_p} = 0$, one obtains, respectively,

$$W_{1u_q} = 0,\ W_{1u_p} = 0,\ -W_1 H' + W_{1q} H + W_{1u} H u_q = 0.$$

Since $W_{1u_q} = 0$ from the last equation one has $W_{1u} = 0$, and

$$W_1 = 4p\lambda(p)H(q)$$

with some function[38] $\lambda(p)$. Integrating (5.145) with the obtained function W_1, one finds

$$W = \Phi(p, q, \xi, \eta) + u\lambda(p), \tag{5.146}$$

where

$$\xi = u_p + \frac{\kappa}{4p}u,\ \ \eta = u_q + \frac{H}{4H'}u.$$

Substituting the representation (5.146) into (5.142), one obtains

$$S_i = F_{i0} + F_{i1}u + F_{i2}u^2 = 0,\ (i = 1, 2, 3, 4),$$

where the coefficients F_{ij}, $(i = 1, 2, 3, 4;\ j = 0, 1, 2)$ do not depend on the variable u. Hence, these equations can be split with respect to u

$$F_{ij} = 0,\ (i = 1, 2, 3, 4;\ j = 0, 1, 2). \tag{5.147}$$

Equations (5.147) are equations for the function $\Phi(p, q, \xi, \eta)$. For example, the equations $F_{4j} = 0$, $(j = 0, 1)$ give

$$H'p(\Phi_{\xi\eta}\xi + \Phi_{\eta\eta}\eta + 3\Phi_\eta) +$$
$$H(4p(\Phi_{p\xi} + \Phi_{q\eta} + \lambda) + \kappa(\Phi_{\xi\eta}\eta + \Phi_{\xi\xi}\xi + 3\Phi_\xi)) = 0,$$
$$p^2\Phi_{\eta\eta}(-4H''H + 5(H')^2) + 2H'\Phi_{\xi\eta}H\kappa p + \Phi_{\xi\xi}H^2\kappa(\kappa + 4)\Phi_{\xi\xi} = 0.$$

[38]The value $4p$ is chosen for convenience.

Finding from the last equation $\Phi_{\xi\xi}$ and substituting it into the equations $F_{i2} = 0,\ (i = 2, 3)$, one finds

$$\Phi_{\xi\eta}\left((\kappa+4)HH'' - (\kappa+5)(H')^2\right) = 0,$$
$$\Phi_{\xi\xi}\left((\kappa+4)HH'' - (\kappa+5)(H')^2\right) = 0.$$

Further analysis depends on the function $H(q)$.

If

$$(\kappa+4)HH'' \neq (\kappa+5)(H')^2,$$

the general solution of equations (5.147) corresponds to the characteristic function[39] (5.143).

Let

$$(\kappa+4)HH'' = (\kappa+5)(H')^2.$$

Because of the equivalence group (5.140), without loss of generality, one can assume that

$$H = q^{-(\kappa+4)}.$$

In this case the general solution of equations (5.147) is

$$W = q\omega(\xi_1, \xi_2) + k_1\left(pu_p + qu_q\right) +$$
$$k_2\left(\xi_1\left(2(\kappa+3)pu_p + 2(\kappa+1)qu_q - \tfrac{\kappa-2}{2}u\right) - \tfrac{2}{(\kappa+2)}\xi_2^{\kappa+2}\right),$$

where $k_1,\ k_2$ are constant, $\omega(\xi_1, \xi_2)$ is the function with the arguments

$$\xi_1 = u - pu_p - qu_q,\quad \xi_2 = p/q,$$

satisfying the equation

$$\omega_{\xi_2\xi_2} - \xi_2^{\kappa}\omega_{\xi_1\xi_1} = 0. \tag{5.148}$$

Let us consider transformations corresponding $k_1 = 0$ and $k_2 = 0$. The generator of these Lie groups is

$$X_\omega = pq\omega_{\xi_1}\partial_p + q^2\omega_{\xi_1}\partial_q + q(\omega + (u-\xi_1)\omega_{\xi_1})\partial_u +$$
$$\omega_{\xi_2}\partial_{u_p} + (\omega - \xi_2\omega_{\xi_2})\partial_{u_q}.$$

Invariants of these Lie groups[40] are ξ_1 and ξ_2. Hence, the representation of an invariant solution is $\xi_1 = f(\xi_2)$, or

$$u - pu_p - qu_q = \varphi(p/q).$$

This is an intermediate[41] integral of equation (5.139).

---

[39] Complete study is given in [83].

[40] The function $\omega(\xi_1, \xi_2)$ is assumed to be an arbitrary function satisfying equation (5.148).

[41] Equation (1.98) of Section 1.9.

11.4 Lie-Bäcklund operators

The limitations dictated by the Bäcklund theorem can be overcome by extending the space of the derivatives involved in the transformations up to infinity ($q = \infty$). Although in this case there is a fundamental difficulty of relating the infinite-order tangency with the group property. This difficulty consists in providing a justification for the functions of the transformations f, ϕ, ψ. This problem is overcome in [70] by using formal power series[42]. Here only operator theory, without relating operators with transformations, is considered.

Definition 5.19. *A differential operator of the form*

$$X = \xi_i \partial_{x_i} + \eta^j \partial_{u^j} + \zeta^j_\alpha \partial_{p^j_\alpha},$$

where ξ_i, η^j *are infinitely many times continuously differentiable,* $\zeta^j_0 = \eta^j$, *and*

$$\zeta^j_{\beta,i} = D_i \zeta^j_\beta - p^j_\beta D_k \xi^k,$$

is called a Lie-Bäcklund operator.

It is assumed that there is an integer number q such that the functions ξ^i, η^j and ζ^j_β, $(|\beta| \leq q)$ are defined in some neighborhood of the space J_q.

A Lie-Bäcklund operator is a formal infinite sum. It truncates when acting on any differential equation.

Let

$$X_k = \xi^k_i \partial_{x_i} + \eta^j_k \partial_{u^j} + \ldots, \quad (k = 1, 2)$$

be two Lie-Bäcklund operators. Their commutator, defined by the formula

$$[X_1, X_2] = X_1X_2 - X_2X_1 = (X_1(\xi^i_2) - X_2(\xi^i_1))\partial_{x_i} + (X_1(\eta^j_2) - X_2(\eta^j_1))\partial_{u^j} + \ldots,$$

is a Lie-Bäcklund operator. The set of all Lie-Bäcklund operators is a Lie algebra with respect to the commutation operation. This algebra is called a Lie-Bäcklund algebra.

Lie-Bäcklund operators X satisfy the property

$$XD_i - D_iX = -D_i(\xi^j)D_j.$$

This property forms conditions of equivalence for two Lie-Bäcklund operators. Two Lie-Bäcklund operators

$$X_k = \xi^i_k \partial_{x_i} + \eta^j_k \partial_{u^j}, \quad (k = 1, 2)$$

are equivalent if there exists a set of functions ξ^i_* such that

$$X_2 - X_1 = \xi^i_* D_i.$$

[42] Detail discussion of the theory can be found in [71, 72].

For example, the operator

$$\widetilde{X} = X - \xi^i D_i \equiv (\eta^j - \xi^i p_i^j)\partial_{u^j} + ... = \widetilde{\eta}^j \partial_{u^j} + ...$$

is equivalent to the operator X. A Lie-Bäcklund operator with $\xi^i = 0$, $(i = 1, 2, ..., n)$ is called a canonical Lie-Bäcklund operator.

Let

$$F(x, u, p) = 0 \tag{5.149}$$

be the s-th order system of partial differential equations. Denote by $[F]$ the set of values x_k, u^j, p_α^j, $(k = 1, 2, ..., n;\ j = 1, 2, ..., m;\ |\alpha| = 1, 2, ..., \infty)$ defined by the infinite system[43] of equations

$$[F]: \qquad F = 0,\ D_i F = 0,\ D^\alpha F = 0,\ (2 \le |\alpha|).$$

The set $[F]$ is called the extended frame of the system of differential equations $F = 0$. The following criterion relates differential equations and Lie-Bäcklund operators.

Definition 5.20. *A Lie-Bäcklund operator*

$$X = \xi_i \partial_{x_i} \partial_{x_i} + \eta^j \partial_{u^j} + \zeta_\alpha^j \partial_{p_\alpha^j}$$

is said to be admitted by a system of equations (5.149) (or a system (5.149) admits the operator X) if

$$XF_{|[F]} = 0.$$

In the last formula the notation $|[F]$ means that the expression XF is evaluated on the extended frame[44].

The theory of Lie-Bäcklund operators gives new ways for constructing solutions[45].

Let us consider a canonical Lie-Bäcklund operator $X = \eta^j \partial_{u^j} + ...$, where the functions $\eta^j = \eta^j(x, u, p)$ are defined on the space J_q. A function $u = u(x)$ is said to be invariant with respect to the Lie-Bäcklund operator X if it satisfies the differential constraints $\eta^j(x, u(x), p(x)) = 0$. Therefore, any invariant with respect to the Lie-Bäcklund operator X solution of system (5.149) satisfies the overdetermined system of partial differential equations

$$F(x, u, p) = 0,\ \eta^j(x, u, p) = 0,\ (j = 1, 2, ..., m).$$

---

[43] For involutive systems of equations it is enough to have a finite prolongation of the system ($|\alpha| \le q$).

[44] Compare with (5.135).

[45] There are other applications of Lie-Bäcklund operators which are outside of the content of the book. For example, applications of Lie-Bäcklund operators for constructing conservation laws and Bäcklund transformations.

11.4.1 Boussinesq equation

The Lie algebra admitted by the Boussinesq equation

$$u_{tt} + uu_{xx} + u_x^2 + u_{xxxx} = 0.$$

consists of the generators [30]

$$\partial_x, \ \partial_t, \ x\partial_x + 2t\partial_t - 2u\partial_u.$$

The coefficient η of the canonical Lie-Bäcklund operator $\eta\partial_x$ has to satisfy the equation

$$D_t^2\eta + uD_x^2\eta + u_{xx}\eta + 2u_x D_x\eta + D_x^4\eta = 0.$$

Calculations show that if the order of the derivatives in the function η is restricted to 3, then the admitted Lie-Bäcklund operator $\eta\partial_x$ is reduced to the canonical Lie-Bäcklund operator equivalent to the generator of the admitted Lie group

$$(c_1(2u + xu_x + 2tu_t) + c_2u_x + c_3u_t)\,\partial_u,$$

where c_1, c_2, c_3 are arbitrary constants.

This example of the Boussinesq equation shows the problem of finding nontrivial Lie-Bäcklund operators.

11.4.2 Nontrivial Lie-Bäcklund operators

Let us consider an isentropic flow of an ideal gas

$$r_t + \lambda_1 r_x = 0, \ \ s_t + \lambda_2 s_x = 0. \qquad (5.150)$$

Here r and s are Riemann invariants,

$$\lambda_1 = \frac{r+s}{2} + \psi, \ \ \lambda_2 = \frac{r+s}{2} - \psi,$$

the function $\psi = \psi(r - s)$ depends on the state equation. For example, for a polytropic gas with the polytropic exponent γ the function $\psi = \frac{\gamma-1}{4}(r - s)$, and

$$\lambda_1 = \alpha r + \beta s, \ \ \lambda_2 = \alpha s + \beta r,$$

where

$$\alpha = \frac{1}{2} + \frac{\gamma - 1}{4}, \ \beta = \frac{1}{2} - \frac{\gamma - 1}{4}.$$

The coefficients of a Lie-Bäcklund operator $X = \zeta^1\partial_r + \zeta^2\partial_s$ admitted by system (5.150) have to satisfy the determining equations

$$2D_t\zeta^1 + 2\lambda_1 D_x\zeta^1 + r_x((1 + 2\psi')\zeta^1 + (1 - 2\psi')\zeta^2) = 0,$$
$$2D_t\zeta^2 + 2\lambda_2 D_x\zeta^2 + s_x((1 - 2\psi')\zeta^1 + (1 + 2\psi')\zeta^2) = 0.$$

These equations have to be considered when the derivatives substituted into them

$$\frac{\partial^k r}{\partial t^{k-i}\partial x^i},\ \frac{\partial^k s}{\partial t^{k-i}\partial x^i},\ (0 \le i \le (k-1),\ 1 \le k \le q+1),$$

found from the prolongations of equations (5.150):

$$\frac{\partial^k r}{\partial t^{k-i}\partial x^i} + D_t^{k-i-1}D_x^i(\lambda_1 r_x) = 0,\ \frac{\partial^k s}{\partial t^{k-i}\partial x^i} + D_t^{k-i-1}D_x^i(\lambda_2 s_x) = 0.$$

Assume that $q = 2$. The determining equations become

$$\begin{gathered}
2\zeta_t^1 + 4\zeta_s^1\psi s_x + \zeta_x^1(2\psi + r + s) + \zeta^1(2\psi' + 1)r_x - \zeta^2(2\psi' - 1)r_x + \\
\zeta_{r_x}^1 r_x((2\psi' - 1)s_x - (2\psi' + 1)r_x) + 4\zeta_{s_{xx}}^1 s_{xxx}\psi + \\
\zeta_{s_x}^1(s_x((2\psi' - 1)r_x - (2\psi' + 1)s_x) + 4\psi s_{xx}) + \\
\zeta_{r_{xx}}^1(s_{xx}r_x(2\psi' - 1) - r_{xx}(3r_x(2\psi' + 1) - 2s_x(2\psi' - 1)) \\
-2\psi'' r_x(r_x - s_x)^2) + \zeta_{s_{xx}}^1(2\psi'' s_x(r_x - s_x)^2 + s_{xx}(2r_x(2\psi' - 1) \\
-3s_x(2\psi' + 1)) + r_{xx}s_x(2\psi' - 1)) = 0,
\end{gathered} \tag{5.151}$$

$$\begin{gathered}
2\zeta_t^2 - 4\zeta_r^2\psi r_x + \zeta_x^2(-2\psi + r + s) + \zeta^2(2\psi' + 1)s_x - \zeta^1(2\psi' - 1)s_x + \\
\zeta_{r_x}^2(r_x((2\psi' - 1)s_x - (2\psi' + 1)r_x) - 4\psi r_{xx}) - 4\zeta_{r_{xx}}^2 r_{xxx}\psi + \\
\zeta_{s_x}^2 s_x((2\psi' - 1)r_x - (2\psi' + 1)s_x) + \\
\zeta_{r_{xx}}^2(-2\psi'' r_x(r_x - s_x)^2 + (2\psi' - 1)r_x s_{xx} + (2(2\psi' - 1)s_x \\
-3(2\psi' + 1)r_x)r_{xx}) + \zeta_{s_{xx}}^2(2\psi'' s_x(r_x - s_x)^2 + s_{xx}(2(2\psi' - 1)r_x \\
-3(2\psi' + 1)s_x) + (2\psi' - 1)r_{xx}s_x) = 0.
\end{gathered} \tag{5.152}$$

Since the functions

$$\zeta^1 = \zeta^1(x, t, r, s, r_x, s_x, r_{xx}, s_{xx}),\ \zeta^2 = \zeta^2(x, t, r, s, r_x, s_x, r_{xx}, s_{xx})$$

do not depend on the derivatives r_{xxx} and s_{xxx}, the determining equations can be split with respect to these derivatives. This gives an overdetermined system of four partial differential equations which have to be studied for compatibility.

Remark 5.6. *The case where the functions ζ^1 and ζ^2 do not depend on the independent variables x and t is studied in [72, 108, 153]*[46]. *Notice also that*

[46] In [153] group classification with respect to the state equation is considered. The operators of first order corresponding to an infinite set of Lie-Bäcklund operators were found in [108].

system of equations (5.150) belongs to the class of hydrodynamic-type systems. A review of the results for hydrodynamic-type systems can be found in [72].

Here a solution of the determining equations in the case of a polytropic gas[47] where the functions ζ^1 and ζ^2 depend on all variables $(x, t, r, s, r_x, s_x, r_{xx}, s_{xx})$ is given.

Splitting the determining equations (5.151) and (5.152) with respect to r_{xxx} and s_{xxx} gives $\zeta^1_{s_{xx}} = 0$ and $\zeta^2_{r_{xx}} = 0$, and then differentiating equation (5.151) with respect to s_{xx} and equation (5.152) with respect to r_{xx}, one finds the derivatives of the functions ζ^1 and ζ^2 with respect to r_{xx} and s_{xx}:

$$\begin{gathered}\zeta^1_{r_{xx}}(2\psi' - 1)s_x = \zeta^2_{s_{xx}}(2\psi' - 1)s_x - 4\zeta^2_{r_x}\psi, \quad \zeta^1_{s_{xx}} = 0,\\ \zeta^2_{r_{xx}} = 0, \quad \zeta^2_{s_{xx}}(2\psi' - 1)r_x = \zeta^1_{r_{xx}}(2\psi' - 1)r_x + 4\zeta^1_{s_x}\psi.\end{gathered} \tag{5.153}$$

The remanding equations are

$$\begin{gathered}2\zeta^1_t + 4\zeta^1_s\psi s_x + \zeta^1_x(2\psi + r + s) + \zeta^1(2\psi' + 1)r_x +\\ \zeta^1_{r_x}r_x((2\psi' - 1)s_x - (2\psi' + 1)r_x) + \zeta^1_{s_x}s_x((2\psi' - 1)r_x - (2\psi' + 1)s_x) -\\ \zeta^1_{r_{xx}}(r_{xx}(3(2\psi' + 1)r_x + 2(2\psi' - 1)s_x)) -\\ \zeta^2(2\psi' - 1)r_x + (4\zeta^1_{s_x}\psi + \zeta^1_{r_{xx}}r_x(2\psi' - 1))s_{xx} = 0,\end{gathered} \tag{5.154}$$

$$\begin{gathered}2\zeta^2_t - 4\zeta^2_r\psi r_x + \zeta^2_x(-2\psi + r + s) + \zeta^2(2\psi' + 1)s_x +\\ \zeta^2_{r_x}r_x((2\psi' - 1)s_x - (2\psi' + 1)r_x) + \zeta^2_{s_x}s_x((2\psi' - 1)r_x - (2\psi' + 1)s_x) +\\ \zeta^2_{s_{xx}}(s_{xx}(2(2\psi' - 1)r_x - 3(2\psi' + 1)s_x)) -\\ \zeta^1(2\psi' - 1)s_x - (4\zeta^2_{r_x}\psi - \zeta^2_{s_{xx}}(2\psi' - 1)s_x)r_{xx} = 0.\end{gathered} \tag{5.155}$$

Further analysis of the overdetermined system (5.153)-(5.155) is cumbersome. This analysis was made on a computer by using the symbolic calculation system Reduce [69]. The result of the calculations shows that the Lie-Bäcklund operator is

$$X = X_\infty + \sum_{j=1}^{9} c_j X_j,$$

where the coefficients of the operators $X_\infty = \zeta^1_\infty\partial_r + \zeta^2_\infty\partial_s$, $X_j = \zeta^1_j\partial_r + \zeta^2_j\partial_s$, $(j = 1, 2, ..., 9)$ are defined by the formulae:

$$\zeta^1_1 = \frac{r_{xx}}{r_x^2} - \frac{(\gamma-3)}{2(\gamma-1)}\frac{(r_x - s_x)^2}{r_x s_x(r-s)}, \quad \zeta^2_1 = \frac{s_{xx}}{s_x^2} - \frac{(\gamma-3)}{2(\gamma-1)}\frac{(r_x - s_x)^2}{r_x s_x(r-s)},$$

[47] The cases of the polytropic exponent $\gamma = 3$ and $\gamma = 5/3$ are excluded from the study. Notice that in these cases the Darboux equation, which is obtained from the gas dynamic equations (5.150), is integrable in quadratures.

$$\zeta_2^1 = r\frac{r_{xx}}{r_x^2} - \frac{(\gamma-3)}{2(\gamma-1)}\frac{(rs_x - r_x s)(r_x - s_x)}{(r-s)r_x s_x},$$

$$\zeta_2^2 = s\frac{s_{xx}}{s_x^2} - \frac{(\gamma-3)}{2(\gamma-1)}\frac{(rs_x - r_x s)(r_x - s_x)}{(r-s)r_x s_x},$$

$$\zeta_3^1 = r^2\frac{r_{xx}}{r_x^2} - \frac{(\gamma-3)}{2(\gamma-1)}\frac{(rs_x - r_x s)^2}{(r-s)r_x s_x},\quad \zeta_3^2 = s^2\frac{s_{xx}}{s_x^2} + \frac{(\gamma-3)}{2(\gamma-1)}\frac{(rs_x - r_x s)^2}{(r-s)r_x s_x},$$

$$\begin{aligned}\zeta_4^1 &= 2(\gamma-1)r^3 r_x^{-2} r_{xx} + xr_x\left((\gamma+1)r + (3-\gamma)s\right) - \\ &(r-s)^{-1}\left((\gamma-3)(r^3 s_x r_x^{-1} + s^3 r_x s_x^{-1}) + r^2\left(r(\gamma+1) + s(5-3\gamma)\right)\right)\\ \zeta_4^2 &= 2(\gamma-1)s^3 s_x^{-2} s_{xx} + xs_x\left((3-\gamma)r + (\gamma+1)s\right) + \\ &(r-s)^{-1}\left((\gamma-3)(s^3 r_x s_x^{-1} + r^3 s_x r_x^{-1}) + s^2((5-3\gamma)r + (\gamma+1)s)\right),\end{aligned}$$

$$\begin{aligned}\zeta_5^1 &= 4(\gamma-1)^2 r^4 r_x^{-2} r_{xx} + xr_x(\gamma+1)(-r^2(3\gamma-1) + (\gamma-3)(2r+s)s) + \\ &4^{-1}tr_x(\gamma+1)(r^3(5\gamma-3)(3\gamma-1) - (\gamma-3)s((3\gamma-1)(3r^2+s^2)\\ &+3rs(\gamma+1)))2(\gamma-1)(r-s)^{-1}((3-\gamma)(s^4 r_x s_x^{-1} + r^4 s_x r_x^{-1})\\ &-2r^3((3\gamma-1)r - 4s(\gamma-1))),\end{aligned}$$

$$\begin{aligned}\zeta_5^2 &= 4(\gamma-1)^2 s^4 s_x^{-2} s_{xx} + xs_x(\gamma+1)(-s^2(3\gamma-1) + (\gamma-3)(r+2s)r) + \\ &4^{-1}ts_x(\gamma+1)(s^3(5\gamma-3)(3\gamma-1) - (\gamma-3)r((3\gamma-1)(r^2+3s^2)\\ &+3rs(\gamma+1))) + -2(\gamma-1)(r-s)^{-1}((3-\gamma)(s^4 r_x s_x^{-1} + r^4 s_x r_x^{-1})\\ &-2s^3((3\gamma-1)s - 4r(\gamma-1))),\end{aligned}$$

$$\zeta_6^1 = 4^{-1}tr_x(\gamma+1)(r^2(3\gamma-1) - (\gamma-3)s(2r+s)) - xr_x(r(\gamma+1) - s(\gamma-3))$$
$$-2(\gamma-1)r^2,$$

$$\zeta_6^2 = 4^{-1}ts_x(\gamma+1)(s^2(3\gamma-1) - (\gamma-3)r(r+2s)) - xs_x(s(\gamma+1) - r(\gamma-3))$$
$$-2(\gamma-1)s^2,$$

$$\zeta_7^1 = tr_x((\gamma+1)r - (\gamma-3)s) - 4xr_x,\quad \zeta_7^2 = ts_x((\gamma+1)s - (\gamma-3)r) - 4xs_x,$$

$$\zeta_8^1 = tr_x - 1,\quad \zeta_8^2 = ts_x - 1.$$

$$\zeta_9^1 = xr_x - r,\quad \zeta_9^2 = xs_x - s.$$

$$\zeta_\infty^1 = r_x\mu_r,\quad \zeta_\infty^2 = s_x\mu_s,$$

where the function $\mu(r, s)$ satisfies the Darboux equation

$$\mu_{rs} + \frac{(\gamma-3)}{2(\gamma-1)(r-s)}(\mu_r - \mu_s) = 0. \tag{5.156}$$

Invariant solutions with respect to the operators X_1, X_2 and X_3 are discussed in[48] [153]. The operators X_7, X_8 and X_9 are equivalent to the generators of the admitted Lie group

$$X_7 \sim t\partial_t + x\partial_x, \quad X_8 \sim \partial_r + \partial_s + t\partial_x, \quad X_9 \sim r\partial_r + s\partial_s + x\partial_x.$$

The operator X_∞ is defined by an arbitrary function $\mu(r,s)$ satisfying the Darboux equation[49] (5.156). The operators X_4, X_5 and X_6 include the independent variables in their coefficients[50].

Solutions which are invariant with respect to the admitted Lie-Bäcklund operators have to satisfy the differential constraints

$$\zeta^1(x,t,r,s,r_x,s_x,r_{xx},s_{xx}) = 0, \quad \zeta^2(x,t,r,s,r_x,s_x,r_{xx},s_{xx}) = 0. \quad (5.157)$$

Hence, when finding invariant solutions one has to solve the overdetermined system of equations(5.150), (5.157). These systems are compatible [108][51].

[48] See also review in [72].

[49] For an arbitrary state equation the function $\mu(r,s)$ has to satisfy the Darboux equation [108]

$$\mu_{rs} + \frac{2\psi' - 1}{4\psi}(\mu_r - \mu_s) = 0.$$

[50] These operators are new, and they were found during the preparation of this book. Their invariant solutions have not been studied. Notice that the operator X_6 is an operator of first order.

[51] Compatibility and arbitrariness of solutions of these systems were studied in Section 4.6.

Chapter 6

SYMMETRIES OF EQUATIONS WITH NONLOCAL OPERATORS

Equations describing real phenomena in mathematical modelling take various forms, such as ordinary differential equations, partial differential equations, integro-differential equations, functional differential equations and many others. The algorithmic approach of group analysis was developed especially for differential equations. Applying it to equations having nonlocal terms presents some difficulties. The main ones of these arise from the nonlocal terms. There are several ways[1] of overcoming these difficulties by using classical group analysis developed for partial differential equations. All of these approaches are heuristic and can be applied to specific equations.

In applications of group analysis to equations with nonlocal operators it is necessary to use successive steps, as for partial differential equations. The first step involves constructing an admitted Lie group. Since the definition of an admitted Lie group given for partial differential equations cannot be applied to equations with nonlocal terms, this concept requires further investigation. Notice that even for partial differential equations the notion of an admitted Lie group needs to be discussed: there are three definitions of the admitted Lie group[2]. The first section of the chapter is devoted to a discussion of these definitions. This discussion assists in establishing a definition of an admitted Lie group for integro-differential and functional differential equations.

As for partial differential equations, an admitted Lie group of equations with nonlocal terms is a Lie group satisfying determining equations. In contrast to partial differential equations the admitted Lie group does not have the property of mapping any solution into a solution of the same equations, although the

[1]A short review of these methods is presented later.

[2]Relations between these definitions are discussed in [96], chapter 6, section 1, [127], section 2.6, [72], section 1.3, [73], section 9.2[3], [150] and references therein.

method developed for constructing the determining equations used this property. In practice the algorithm for obtaining determining equations is no more difficult than for partial differential equations. The main difficulty consists of solving the determining equations. Since this depends on the properties of the Cauchy problem, the method of solving determining equations also depends on the nonlocal equations studied. For example, symmetries of the system of Vlasov-Maxwell kinetic equations, of the Boltzmann equation and of equations describing a one-dimensional motion of a viscoelastic medium are found using different approaches for splitting determining equations.

Finally, the group analysis method is applied to functional differential equations. One class of functional differential equations is the class of delay differential equations. By virtue of the existence theorem of the Cauchy problem for delay differential equations, the process of splitting their determining equations is similar to partial differential equations. Thus the method for solving determining equations for delay differential equations is also similar to partial differential equations. This is demonstrated by examples.

1. Definitions of an admitted Lie group

The first definition of a Lie group admitted by a system of partial differential equations (S) is based on a knowledge of the solutions: a Lie group is admitted by the system (S) if any solution of this system is mapped into a solution of the same system. Two other definitions are based on the geometrical approach: equations are considered as manifolds. One of these definitions deals with the manifold defined by the system (S). Another definition works with the extended frame of the system (S): system (S) and all its prolongations[4]. Notice that the definitions based on the geometrical approach have the following inadequacy. There are equations which have no solutions, however they have an admitted (in this meaning) Lie group. Although the geometrical approach has the advantage that it is simple in applications.

Here it should be also mentioned that different approaches have been developed for finite-difference equations. Review of these approaches can be found in [34] and in references therein.

1.1 The geometrical approach

The classical geometrical definition of an admitted Lie group deals with invariant manifolds: the group is admitted by the system of equations

$$(S) \qquad S(x, u, p) = 0 \tag{6.1}$$

[4]According to the Cartan-Kähler theorem, after a finite number of prolongations the system (S) becomes either involutive or incompatible. Therefore, from the theory of compatibility point of view, there is no necessity for infinite prolongations of the system (S).

if the manifold defined by these equations is invariant with respect to this group. All functions are assumed enough times continuously differentiable, for example, of the class C^∞. The manifold $(S) = \{(x, u, p) \mid S(x, u, p) = 0\}$, defined by equations (6.1), is considered in the space J^l of the variables

$$x = (x_1, x_2, \ldots, x_n),\ u = (u^1, u^2, \ldots, u^m),\ p = (p^j_\alpha),$$
$$(j = 1, 2, \ldots, m;\ \ |\alpha| \le l).$$

Here the following notations are used: $|\alpha| \equiv \alpha_1 + \alpha_2 + \ldots + \alpha_n \le l$, $\alpha, i \equiv (\alpha_1, \alpha_2, \ldots, \alpha_{i-1}, \alpha_i + 1, \alpha_{i+1}, \ldots, \alpha_n)$ and $p^j_\alpha = \partial^{|\alpha|} u^j / \partial x^\alpha$. Any local Lie group of point transformations

$$\overline{x}_i = f^i(x, u; a),\ \ \overline{u}^j = \varphi^j(x, u; a), \tag{6.2}$$

is defined by the transformations of the independent and dependent variables[5]. Here a is the group parameter. The transformations of the derivatives $\overline{p}^j_\alpha = \varphi^j_\alpha(x, u, p; a)$ in the space J^l are defined by the formulae of prolongations. These formulae are obtained by requiring the tangent conditions[6]

$$du^j - p^j_k dx_k = 0,\ \ dp^j_\alpha - p^j_{\alpha,k} dx_k = 0, \tag{6.3}$$

to be invariant. For example, for the first order derivatives

$$d\overline{u}^j - \overline{p}^j_k d\overline{x}_k = \Big((\varphi^j_{x_k} + \varphi^j_{u^i} p^i_k) - \overline{p}^j_s (f^s_{x_k} + f^s_{u^i} p^i_k) \Big)\, dx_k = 0$$

or

$$\overline{p}^j_s = \left(f^s_{x_k} + f^s_{u^i} p^i_k \right)^{-1} (\varphi^j_{x_k} + \varphi^j_{u^i} p^i_k).$$

For higher order derivatives the prolongation formulae are obtained recurrently. According to the Lie theorem there is one-to-one correspondence between a Lie group and the infinitesimal generator

$$X = \xi^i(x, u)\partial_{x_i} + \eta^j(x, u)\partial_{u^j},$$

where

$$\xi^i(x, u) = \frac{df^i}{da}(x, u; 0),\ \ \eta^j(x, u) = \frac{d\varphi^j}{da}(x, u; 0).$$

For dealing with the derivatives one has to use the prolonged generator

$$\underset{l}{X} = X + \eta^j_\alpha \partial_{p^j_\alpha},\ \ |\alpha| \le l,$$

---

[5]For the sake of simplicity only a Lie group of point transformations is discussed. For tangent transformations the study is similar.

[6]How to obtain these formulae was discussed in the previous section.

where the coefficients η^j_α are defined recurrently

$$\eta^j_{\alpha,k} = D_k\eta^j_\alpha - p^j_{\alpha,i}D_k\xi^i$$

and the operators

$$D_k = \frac{\partial}{\partial x_k} + p^j_{\alpha,k}\frac{\partial}{\partial p^j_\alpha}$$

are the operators of the total derivatives with respect to x_k, $(k = 1, 2, \ldots, n)$. For the sake of simplicity, for $|\alpha| = 0$ it is denoted $u^j = p^j_\alpha$ and $\eta^j = \eta^j_\alpha$. Notice that[7]

$$\begin{aligned} D_k(\underset{l}{X}S) &= (D_k\xi^i)S_{x_i} + (D_k\eta^j_\alpha)S_{p^j_\alpha} + \eta^j_\alpha(D_kS_{p^j_\alpha}) + \xi^i(D_kS_{x_i}) = \\ &= (D_k\xi^i)S_{x_i} + (\eta^j_{\alpha,k} + p^j_{\alpha,i}D_k\xi^i)S_{p^j_\alpha} + \eta^j_\alpha(D_kS_{p^j_\alpha}) + \xi^i(D_kS_{x_i}) = \\ &= (D_k\xi^i)\left(S_{x_i} + p^j_{\alpha,i}S_{p^j_\alpha}\right) + \eta^j_{\alpha,k}S_{p^j_\alpha} + \xi^i(D_kS_{x_i}) + \eta^j_\alpha(D_kS_{p^j_\alpha}) = \\ &= (D_k\xi^i)\left(S_{x_i} + p^j_{\alpha,i}S_{p^j_\alpha}\right) + \underset{l+1}{X}(D_kS) = \\ &= (D_k\xi^i)(D_iS) + \underset{l+1}{X}(D_kS). \end{aligned} \tag{6.4}$$

The classical geometrical definition of the admitted Lie group requires that the manifold (6.1) is invariant with respect to this group. This requirement is equivalent to

$$\left(\underset{l}{X}S\right)_{|(S)} = 0. \tag{6.5}$$

For the admitted Lie group equalities (6.4) imply

$$\left(\underset{l+s}{X}(D^\alpha S)\right)_{|(DS)} = \left(D^\alpha\left(\underset{l}{X}S\right)\right)_{|(DS)} = 0, \ \ (|\alpha| \le s), \tag{6.6}$$

where the manifold (DS) is defined by the equations

$$(DS) \qquad\qquad D^\alpha S = 0, \ \ (0 \le |\alpha| \le s).$$

In particular, equations (6.6) mean that the Lie group admitted by the system of equations (S), is also admitted by the system of equations (DS). The inverse, in the general case, is not correct. In fact, by virtue of equations (6.6) to find the admitted Lie group of the system (DS), one needs to solve the determining equations

$$\left(\underset{l}{X}S\right)_{|(DS)} = 0. \tag{6.7}$$

Since the manifold defined by the equations (S), includes the manifold, defined by the equations (DS), the set of solutions of equations (6.7) includes[8] the set of solutions of equations (6.5).

---

[7] It is a generalization of formula (5.9).

[8] There are transformations, which are admitted by the system (DS), but not admitted by the system (S). One of such examples can be found in [127] or in [72] (vol. 2).

In the geometrical approaches the problem of the coincidence of the Lie group admitted by a system (S), and the Lie group admitted by the system (DS), can be solved in the following way. If the system (S) is involutive, then the Lie group admitted by the system (DS) is admitted by the system (S). The proof of this statement uses the main property of involutive systems to be passive: involutive system does not produce new equations of the same order as the initial system (S). If the system (S) is not involutive, then the Lie group admitted by the system (S) is included into the Lie group admitted by the system (DS).

As an example of the first situation, where the admitted Lie groups coincide although the system is not involutive, one can consider the Navier-Stokes equations:

$$u^{\alpha}_{x_\alpha} = 0, \quad S_i = u^i_t + u^\alpha u^i_{x_\alpha} + p_{x_i} = \Delta u^i, \quad (i = 1, 2, 3). \tag{6.8}$$

Here the independent variables are (t, x_1, x_2, x_3) and the dependent variables are (p, u_1, u_2, u_3). Since this system is a second order system of partial differential equations, one needs to prolong the first order equations

$$u^{\alpha}_{x_\alpha t} = 0, \quad u^{\alpha}_{x_\alpha x_i} = 0, \quad (i = 1, 2, 3). \tag{6.9}$$

and append them to the Navier-Stokes equations. The overdetermined system (6.8), (6.9) is still not involutive. It becomes involutive after appending the equation

$$(S_i)_{x_i} = \Delta p + u^{\alpha}_{x_i} u^i_{x_\alpha} = 0. \tag{6.10}$$

Direct calculations show that the Lie group admitted by the Navier-Stokes equations (6.8) and the Lie group admitted by the overdetermined system of equations (6.8), (6.9), (6.10) coincide[9].

As an example of the second situation, where the admitted Lie groups are different one can consider the equations [127]:

$$S_1 = u_z + y u_x = 0, \quad S_2 = u_y = 0. \tag{6.11}$$

This system becomes involutive after appending the equation

$$(S_1)_y = u_x = 0. \tag{6.12}$$

The system (6.11), (6.12) admits the shift

$$y \to y + a,$$

which is not admitted by the system (6.11).

[9]Fortunately, this equation does not have influence on the admitted Lie group found in [20].

The geometrical definitions of an admitted Lie group do not require the existence of a solution. For example, the equation[10]

$$u_x^2 + u_y^2 = -1$$

has no solution, but it admits the scaling

$$x \to xe^a, \ y \to ye^a, \ u \to ue^a.$$

Another example is the overdetermined system of equations

$$u_x = y^3, \ u_y = -x^3$$

which admits, for example, the scaling:

$$x \to xe^a, \ y \to ye^a, \ u \to ue^{4a}.$$

The prolonged system (DS) includes the equation

$$(u_x)_y - (u_y)_x = 3(y^2 + x^2) = 0,$$

which is degenerate.

1.2 The approach based on a solution

Lie groups admitted in the sense of the geometrical approach have the property to transform any solution of the system of equations (S) (or (S), (DS)) into a solution of the same system. This property can be taken as a definition of the admitted Lie group[11].

Definition 6.1. *A Lie group (6.2) is admitted by the system* (S) *if it maps any solution of the system* (S) *into a solution of the same system.*

This definition supposes that the system (S) has at least one solution.

The determining equations for the admitted group are obtained as follows. Let a function $u = u_o(x)$ be given. Substituting it into the first part of transformation (6.2) and using the inverse function theorem one finds

$$x = g^x(\overline{x}, a). \tag{6.13}$$

The transformed function $u_a(\overline{x})$ is given by the formula

$$u_a(\overline{x}) = f^u(g^x(\overline{x}, a), u_o(g^x(\overline{x}, a)); a).$$

Because the tangent conditions are invariant with respect to any transformation of the admitted Lie group, the transformed derivatives are $\overline{p}^j_\alpha(\overline{x}, a) =$

[10]This example is given in [130].

[11]The author thinks that the determining equations of an admitted Lie group were originally obtained on the basis of this property.

$\varphi^j_\alpha(x, u_o(x), p(x); a)$, where $p(x)$ are derivatives of the function $u_o(x)$, and x is defined by (6.13). Let the function $u_o(x)$ be a solution of a system (S). Since any transformation of the admitted Lie group transforms any solution to a solution of the same system, the function $u_a(\overline{x})$ is also a solution of the system (S):

$$\overline{S}(\overline{x}, a) = S(\overline{x}, u_a(\overline{x}), \overline{p}^j_\alpha(\overline{x}, a)) = 0.$$

In the last equations, instead of the independent variables $\overline{x},\ a$ one can consider the independent variables $x,\ a$:

$$\overline{\overline{S}}(x, a) = \overline{S}(f^x(x, u_o(x); a), a).$$

Differentiating the functions $\overline{\overline{S}}(x, a)$ or $\overline{S}(\overline{x}, a)$ with respect to the group parameter a and setting $a = 0$, one obtains the determining equations

$$\left(\frac{\partial}{\partial a}\overline{\overline{S}}(x, a)\right)(x, 0) = (XS)\,(x, u_o(x), p(x)) = 0 \tag{6.14}$$

or

$$\left(\frac{\partial}{\partial a}\overline{S}(\overline{x}, a)\right)(\overline{x}, 0) = \left(\widetilde{X}S\right)(x, u_o(x), p(x)) = 0. \tag{6.15}$$

The operator $\widetilde{X}$ is the canonical Lie-Bäcklund operator [71]

$$\widetilde{X} = \overline{\eta}^j \partial_{u^j} + D^\alpha \overline{\eta}^j \partial_{p^j_\alpha}$$

equivalent to the generator X. Here $\overline{\eta}^j = \eta^j(x, u) - \xi^\beta(x, u) p^j_\beta$.

Since the function $u_o(x)$ is a solution of the system (S), the solutions of the determining equations (6.14) and (6.15) coincide. Notice also that the equations (6.5) and (6.7) with substituted into them $u = u_o(x)$ and $p = p(x)$ coincide with (6.14).

For solving the determining equations one needs to know arbitrary elements. In the geometrical definitions the arbitrary elements are coordinates of the manifolds. In the case of the determining equations (6.14) or (6.15) for establishing the arbitrary elements one can use, for example, a knowledge of the existence of a solution of the Cauchy problem.

Definition 6.2. *A system of partial differential equations (6.1) is locally solvable at the point* $(x_0, u_0, p_0) \in (S)$ *if there exists a smooth solution* $u = v(x)$ *of the system* (S) *defined in a neighborhood of the point* x_0 *such that* $v(x_0) = u_0$, $\dfrac{\partial^{|\alpha|} v}{\partial x^\alpha}(x_0) = (p^j_\alpha)_0$. *The system (6.1) is locally solvable if it is locally solvable at any point* $(x_0, u_0, p_0) \in (S)$.

Under conditions of solvability, the geometrical definition of an admitted Lie group of involutive system and the definition based on a solution are equivalent.

Notice that for analytical systems (S) involutiveness guarantees the property of local solvability.

From one point of view the last definition (related to a solution) is more difficult for applications than the geometrical definitions. Although, from another point of view, this definition allows the construction of the determining equations for more general objects than differential equations: integro-differential equations, functional differential equations or even for more general type of equations.

2. Symmetry groups for integro–differential equations

2.1 Short review of the methods

As was mentioned the main difficulty when applying group analysis to integro–differential equations arises from the integral (nonlocal) terms present in these equations. There are several heuristic ways for overcoming this difficulty. Among these ways the following are pointed out[12]:

(1) finding a representation of an admitted group or a solution (on the basis of a priory assumptions);

(2) studying a system of moments – the method of moments;

(3) transforming the original integro-differential equation into a differential equation.

The first approach supposes an a priori choice of the form of symmetries or solutions on the basis of some assumptions. This approach is the simplest and the most efficient. For example, the well–known BKW–solution of the Boltzmann equation was found in this way. For the Boltzmann equation this approach was also applied in [60, 12]. The methods used in [123, 124, 125, 9, 11] can be related with this approach. The main problem in this approach is to discover a representation of an admitted group (or solution).

In the second approach (the method of moments) the original system of integro-differential equations is reduced to an infinite system of differential equations (system of moment equations). For a finite number N of equations of this system, containing a finite number of terms, the classical group analysis (for differential equations) is applied. Then the process of taking a limit[13] $N \to \infty$ is carried out. The first application of this approach for finding an admitted Lie group was done in [166], and then it was used for one of the models of the Boltzmann equation in [17, 18]. There are some problems in the application of this approach. One of them is that, for some equations, the construction of the moment system is impossible. Another problem with the moment system is the infinite number of equations that are involved.

[12] Some discussions of applications of group analysis to integro–differential equations can be found in [72].

[13] The problem of obtaining a complete admitted Lie group by the system of moments is discussed in [76].

In the third approach, as previously, initial integro–differential equations are transformed into differential equations. After that a classical algorithm of group analysis is applied to the differential equations. The results of works [91, 38, 46, 126] were obtained by this approach. In this way there are the same problems as in the previous approaches.

For a complete description of group properties of integro-differential equations it is necessary to use successive approaches of group analysis: constructing determining equations and finding their solutions. Such an approach for integro-differential equations was developed in [60, 61]. Later it was applied to the Boltzmann equation and its models [112, 62, 63, 64]. For other models this approach was used in [151, 88, 89].

2.2 Definition of a Lie group admitted by integro–differential equations

Let us consider an integro-differential equation:

$$\Phi(x, u) = 0. \tag{6.16}$$

Here u is the vector of the dependent variables, x is the vector of the independent variables. Assume that a one–parameter Lie group $G^1(X)$ of transformations

$$\overline{x} = f^x(x, u; a), \quad \overline{u} = f^u(x, u; a) \tag{6.17}$$

with the generator

$$X = \zeta^u(x, u)\partial_u + \xi^x(x, u)\partial_x,$$

transforms a solution $u_0(x)$ of equations (6.16) into the solution $u_a(x)$ of the same equations. The transformed function $u_a(x)$ is

$$u_a(\overline{x}) = f^u(x, u(x); a),$$

where $x = \psi^x(\overline{x}; a)$ is substituted into this expression. The function $\psi^x(\overline{x}; a)$ is found from the relation $\overline{x} = f^x(x, u(x); a)$. Differentiating the equations $\Phi(x, u_a(x))$ with respect to the group parameter a and considering these equations for the value $a = 0$, one obtains the equations

$$\left(\frac{\partial}{\partial a}\Phi(x, u_a(x))\right)_{|a=0} = 0. \tag{6.18}$$

The obstacle for applying to integro-differential equations the definition of an admitted Lie group, based on a solution, is the localness of the inverse function theorem. However, notice that equations (6.18) coincide with the equations

$$(\bar{X}\Phi)(x, u_0(x)) = 0. \tag{6.19}$$

obtained by the action of the canonical Lie–Bäcklund operator $\bar{X}$, which is equivalent to the generator X:

$$\bar{X} = \bar{\zeta}^j \partial_{u^j},$$

where $\bar{\zeta}^j = \zeta^{u^j}(x,u) - \xi^{x_i}(x,u)u^j_{x_i}$. The actions of the derivatives ∂_{u^j} and $\partial_{p^j_\alpha}$ are considered in terms of the Frechet derivatives. Equations (6.19) can be constructed without requiring the property that the Lie group should transform a solution into a solution. This allows the following definition of an admitted Lie group.

Definition 6.3. *A one–parameter Lie group G^1 of transformations (6.17) is a symmetry group admitted by equation (6.16) if G^1 satisfies equations (6.19) for any solution $u_0(x)$ of (6.16). Equations (6.19) are called the determining equations.*

Remark 6.1. *For a system of differential equations (without integral terms) the determining equations (6.19) coincide with the determining equations (6.15).*

The way of obtaining determining equations for integro-differential equations is similar (and not more difficult) to the way used for differential equations. Notice that the determining equations of integro-differential equations are integro-differential. The main difficulty is in solving the determining equations. Introducing new variables, denoting the integral terms, allows the simplification of some part of the equations [88]. The main simplification is related to the splitting of the determining equations. It should be noted that, contrary to differential equations, the splitting of integro-differential equations depends on the studied equations[14]. Since the determining equations (6.19) have to be satisfied for any solution of the original equations (6.16), the arbitrariness of the solution $u_0(x)$ plays a key role in the process of solving the determining equations. The important circumstance in this process is the knowledge of the properties of solutions of the original equations. One of these properties is a theorem of the existence of a solution of the Cauchy problem.

An advantage of the given definition of an admitted Lie group is that it provides a constructive method for obtaining the admitted group. Another advantage of this definition is a possibility to apply it when seeking Lie–Bäcklund transformations[15], conditional symmetries and other types of symmetries for integro–differential equations.

Remark 6.2. *A geometrical approach for constructing an admitted Lie group for integro–differential equations is applied in [26, 27].*

[14]In this and the next sections the process of splitting is demonstrated by three different types of integro-differential equations.

[15]There are some trivial examples of such applications for integro-differential equations.

2.3 The kinetic Vlasov equation

The one–dimensional high–frequency vibrations of collisionless plasma are described by the system of equations

$$f_t + vf_x - Ef_v = 0,\ E_v = 0,\ E_t = \int vf\,dv,\ E_x = 1 - \int f\,dv, \quad (6.20)$$

where $f = f(t, x, v)$, $E = E(t, x, v)$, and the integration is carried out over the one–dimensional velocity space R^1. The Cauchy problem is defined as follows[16]

$$f(t_0, x, v) = f_0(x, v).$$

A first application of group analysis to these equations was done by V.B.Taranov [166]. In [166] system (6.20) was reduced to an infinite system of differential equations: a system of moments. For any finite subsystem of N equations of this infinite system the classical group analysis was applied. As a result the admitted generators were found. Then an intersection of all these generators is defined and the limit $N \to \infty$ is taken. The resultant algebra of generators thus obtained is used to construct the generators admitted by the original system of equations

$$\begin{gathered} X_1 = \partial_t,\ X_2 = \partial_x,\ X_3 = x\partial_x + v\partial_v + E\partial_E - f\partial_f, \\ X_4 = \cos t\,\partial_x - \sin t\,\partial_v + \cos t\,\partial_E, \\ X_5 = \sin t\,\partial_x + \cos t\,\partial_v + \sin t\,\partial_E, \end{gathered} \quad (6.21)$$

In this section it is proven that this group is complete with respect to the definition of an admitted Lie group given in this section: there are no other symmetries except (6.21).

Let a one-parameter Lie group of transformations admitted by (6.20) has the infinitesimal generator

$$X = \zeta_1\partial_E + \zeta_2\partial_f + \xi\partial_x + \eta\partial_t + z\partial_t,$$

where the coefficients $\zeta_1,\ \zeta_2,\ \xi,\ \eta$ and z are functions of the variables t, x, v, E, f. The action of the generator X on the first two equations of system (6.20) is defined by the usual procedure (as for differential equations)

$$\left(D_t\tilde{\zeta}_2 + vD_x\tilde{\zeta}_2 - ED_v\tilde{\zeta}_2 - f_v\tilde{\zeta}_1\right)_{|(S)} = 0,\ \left(D_v\tilde{\zeta}_1\right)_{|(S)} = 0, \quad (6.22)$$

where D_t, D_x, D_v denote the total derivatives with respect to the independent variables t, x, v, the subscript $|(S)$ means that the expression is satisfied for any solution of system (6.20), and

$$\tilde{\zeta}_1 = \zeta_1 - \xi E_x - \eta E_t - zE_v,\ \tilde{\zeta}_2 = \zeta_2 - \xi f_x - \eta f_t - zf_v.$$

[16] See, for example, [56].

According to the given definition of an admitted Lie group, the determining equations related with the last two equations of (6.20) are

$$\left(D_t\tilde{\zeta}_1 - \int v\tilde{\zeta}_2\,dv\right)_{|(S)} = 0, \quad \left(D_x\tilde{\zeta}_1 - \int \tilde{\zeta}_2\,dv\right)_{|(S)} = 0. \tag{6.23}$$

Since the initial function $f_0(x, v)$ can be chosen arbitrary, the derivatives E_t, E_x, f_v, f_x at the fixed (but arbitrary) point (t_0, x_0, v_0) also can take arbitrary values. This allows the splitting of equations (6.22) with respect to the derivatives E_t, E_x, f_v, f_x. The general solution of these split equations is

$$\begin{gathered} \zeta_1 = E(c_1 - \eta'(t)) - \varphi''(t) - x\eta''(t),\ \zeta_2 = \psi(f), \\ z = x\eta'(t) + \varphi'(t) + c_1 v,\ \xi = x(c_1 + \eta'(t)) + \varphi(t), \end{gathered} \tag{6.24}$$

where $\eta(t)$, $\varphi(t)$, $\psi(f)$ are arbitrary functions, c_i are arbitrary constants. Substituting this solution (6.24) into (6.23), one obtains

$$\begin{gathered} -E_x(c_1 + 2\eta') + c_1 + \int \psi(f)\,dv = 0, \\ -E_t(c_1 + 2\eta') - 3E\eta'' - \varphi''' - x\eta'' - \varphi' - \int v\psi(f)\,dv = 0. \end{gathered} \tag{6.25}$$

Because $f = 0$, $E = x$ is a solution of (6.20), the integral $\int \psi\ (0)\ dv$ has to be bounded. This is only possible if $\psi(0) = 0$. Hence, equations (6.20), when substituted into them the solution $f = 0$, $E = x$, give

$$\eta' = 0,\ \varphi''' + \varphi' = 0.$$

Thus

$$\eta = c_2,\ \varphi = c_3 + c_4 \sin t + c_5 \cos t, \tag{6.26}$$

and equations (6.25) become

$$\int (c_1 f + \psi(f))\,dv = 0, \quad \int v(c_1 f + \psi(f))\,dv = 0.$$

Notice that these equations have to be satisfied for any solution of equations (6.20). In particular, for arbitrary initial function $f(t, x, v) = f_0(v)$. This is only possible if

$$\psi(f) = -c_1 f.$$

Therefore the admitted generator is

$$X = c_1 X_3 + c_2 X_1 + c_3 X_2 + c_4 X_5 + c_5 X_4$$

and the Lie group G^5 is complete, i.e., there are no additional admitted transformations except the transformations corresponding to the generators (6.21).

3. Invariant solutions of the Fourier–image of the spatially homogeneous isotropic Boltzmann equation

This section is devoted to the application of group analysis to the Fourier–image of the spatially homogeneous and isotropic Boltzmann equation

$$\Phi \equiv \varphi_t(x,t) + \varphi(x,t)\varphi(0,t) - \int_0^1 \varphi(xs,t)\varphi(x(1-s),t)\,ds = 0. \quad (6.27)$$

Existence of a solution is guaranteed by the theorem of existence for the Cauchy problem (6.27) with the initial conditions[17]

$$\varphi(x,t_0) = \varphi_0(x).$$

Using equation (6.27) one can find the derivatives of the function $\varphi(x,t)$ at the time $t = t_0$:

$$\begin{array}{c}
\varphi_t(x,t_0) = -\varphi_0(0)\varphi_0(x) + \int_0^1 \varphi_0(sx)\varphi_0((1-s)x)\,ds,\\
\varphi_{xt}(x,t_0) = -\varphi_0(0)\varphi_0'(x) + 2\int_0^1 s\varphi_0'(sx)\varphi_0((1-s)x)\,ds,\\
\varphi_{tt}(x,t_0) = -\varphi_0^2(0)\varphi_0(x) - 3\varphi_0(0)\int_0^1 \varphi_0(sx)\varphi_0((1-s)x)\,ds +\\
2\int_0^1\int_0^1 \varphi_0((1-s)x)\varphi_0(ss'x)\varphi_0(s(1-s')x)\varphi_0((1-s)x)\,ds\,ds'.
\end{array}$$

Notice also that multiplying any solution of equation (6.27) by $e^{\lambda x}$, one obtains a solution.

3.1 Admitted Lie group

Let the generator of the admitted Lie group has the form

$$X = \xi(x,t,\varphi)\partial_x + \eta(x,t,\varphi)\partial_t + \zeta(x,t,\varphi)\partial_\varphi.$$

The coefficients of the infinitesimal generator X are assumed to be represented by the formal Taylor series with respect to φ:

$$\begin{array}{rcl}
\xi(x,t,\varphi) = \sum\limits_{l\ge 0} q_l(x,t)\varphi^l, \quad \eta(x,t,\varphi) & = & \sum\limits_{l\ge 0} r_l(x,t)\varphi^l, \quad \zeta(x,t,\varphi)\\
& = & \sum\limits_{l\ge 0} p_l(x,t)\varphi^l.
\end{array}$$

According to the definition of the admitted Lie group, the determining equation (6.19) for equation (6.27) is

$$\begin{array}{c}
D_t\psi(x,t) + \psi(0,t)\varphi(x,t) + \psi(x,t)\varphi(0,t) -\\
-2\int_0^1 \varphi(x(1-s)s,t)\psi(xs,t)\,ds = 0,
\end{array} \quad (6.28)$$

[17] See, for example, [10]

where $\varphi(x,t)$ is an arbitrary solution of (6.27), D_t is the total derivative with respect to t, and the function $\psi(x,t)$ is

$$\psi(x,t) = \zeta(x,t,\varphi(x,t)) - \xi(x,t,\varphi(x,t))\varphi_x(x,t) - \eta(x,t,\varphi(x,t))\varphi_t(x,t).$$

The method of solving the determining equation (6.28) consists of studying the properties of the functions $\xi(x,t,\varphi)$, $\eta(x,t,\varphi)$ and $\zeta(x,t,\varphi)$. These properties are obtained by sequentially considering the determining equation on a particular class of solutions of equation (6.27). This class of solutions is defined by the initial conditions

$$\varphi_0(x) = bx^n \tag{6.29}$$

at the given (arbitrary) time $t = t_0$. Here n is a positive integer. The determining equation is considered at any arbitrary initial time t_0. Equation (6.28) is studied by setting $n = 0, 1, 2, \ldots$, and varying the parameter b.

Let $n = 0$, then in this case the determining equation (6.28) becomes

$$\hat{\zeta}(x,t) + b(\hat{\zeta}(0,t) + \hat{\zeta}(x,t)) - 2b\int_0^1 \hat{\zeta}(xs,t)\,ds = 0.$$

From this equation one obtains

$$\frac{\partial p_0}{\partial t} = 0, \quad \frac{\partial p_{l+1}}{\partial t}(x,t) + p_l(x,t) + p_l(0,t) - 2\int_0^1 p_l(xs,t)\,ds = 0,$$
$$(l = 0, 1, \ldots). \tag{6.30}$$

Let $n \geq 1$ in equation (6.29). In terms of the beta function

$$\int_0^1 s^m(1-s)^n\,ds = \frac{m!n!}{(m+n+1)!},$$

one finds

$$\varphi_t(x,t_0) = P_n b^2 x^{2n}, \quad \varphi_x(x,t_0) = nbx^{n-1}, \quad \varphi_{tt}(x,t_0) = Q_n b^3 x^{3n},$$
$$\varphi_{tx}(x,t_0) = 2nP_n b^2 x^{2n-1},$$

where

$$P_n = \frac{(n!)^2}{(2n+1)!}, \quad Q_n = 2P_n\frac{(2n)!n!}{(3n+1)!}.$$

and the determining equation (6.28) becomes

$$\begin{gathered}
\hat{\zeta}_t + b\left(-nx^{n-1}\hat{\xi}_t + x^n\hat{\zeta}(0,t) - 2x^n\textstyle\int_0^1(1-s)^n\hat{\zeta}(xs,t)\,ds\right) + \\
+b^2\left(-P_n x^{2n}\hat{\eta}_t + P_n x^{2n}\hat{\zeta}_\varphi - 2nP_n x^{2n-1}\hat{\xi} - \delta_{n1}\hat{\xi}(0,t) + \right. \\
\left. 2nx^{2n-1}\textstyle\int_0^1(1-s)^n s^{n-1}\hat{\xi}(xs,t)\,ds\right) + \\
+b^3\left(-nP_n x^{2n-1}\hat{\xi}_\varphi - Q_n x^{3n}\hat{\eta} + 2P_n x^{3n}\textstyle\int_0^1(1-s)^n s^{2n}\hat{\eta}(xs,t)\,ds\right) - \\
-b^4\left(P_n^2 x^{4n}\hat{\eta}_\varphi\right) = 0,
\end{gathered} \tag{6.31}$$

where

$$P_n = \frac{(n!)^2}{(2n+1)!}, \quad Q_n = 2P_n\frac{(2n)!n!}{(3n+1)!}.$$

Notice that

$$2P_{n+1} = P_n\frac{1}{1+\frac{1}{n+1}}$$

and

$$\lim_{n\to\infty} P_n = 0, \quad \lim_{n\to\infty} Q_n = 0, \quad \lim_{n\to\infty} \frac{Q_n}{P_n} = 0.$$

Since the value b is arbitrary, equation (6.31) can be split into a series of equations by equating to zero the coefficients of b^k, $(k = 0, 1, \ldots)$ in the left-hand side of equation (6.31).

For $k = 0$ the corresponding coefficient in the left–hand side of equation (6.31) vanishes because of the first equation of (6.29).

For $k = 1$, equation (6.31) yields:

$$x\left(-p_0(x,t) + 2\int_0^1 (1-(1-s)^n)p_0(xs,t)\,ds\right) - n\frac{\partial q_0(x,t)}{\partial t} = 0.$$

Since n is arbitrary, then

$$p_0(x,t) = 0, \quad \frac{\partial q_0(x,t)}{\partial t} = 0.$$

Hence, $\hat{\zeta}(0,t) = 0$.

Similarly, for $k = 2$ one obtains the equation

$$\begin{gathered}x\left(-p_1(x,t) - p_1(0,t) + 2\int_0^1 (1-(1-s)^n s^n)p_1(xs,t)\,ds\right.\\ \left.+P_n\left(p_1(x,t) - \tfrac{\partial r_0(x,t)}{\partial t}\right)\right) - n\tfrac{\partial q_1(x,t)}{\partial t} - 2nP_n q_0(x,t)\\ +2n\int_0^1 (1-s)^n s^{n-1})q_0(xs,t)\,ds = 0.\end{gathered}$$

Consecutively dividing by n, P_n and letting $n \to \infty$, one obtains

$$p_1(x,t) = c_0 + c_1 x, \quad \frac{\partial q_1(x,t)}{\partial t} = 0, \quad q_0(x,t) = c_2 x, \quad \frac{\partial r_0(x,t)}{\partial t} = -c_0,$$

where c_0, c_1, c_2 are arbitrary constants.

For $k = 3$, one has

$$\begin{gathered}x^{n+1}\left(-p_2(x,t) - p_2(0,t) + 2\int_0^1 (1-(1-s)^n s^{2n})p_2(xs,t)\,ds-\right.\\ P_n\tfrac{\partial r_1(x,t)}{\partial t} + 2P_n p_2(x,t) + 2P_n\int_0^1 (1-s)^n s^{2n})r_0(xs,t)\,ds\\ \left.-Q_n r_0(x,t)\,\right) + x^n\left(-n\tfrac{\partial q_2(x,t)}{\partial t} - 2nP_n q_1(x,t) + 2n\int_0^1 (1-s)^n s^{2n-1})\right.\\ \left.q_1(xs,t)\,ds\right) - nP_n q_1(x,t) = 0.\end{gathered}$$

Similar to the previous case ($k = 2$) one finds

$$q_1(x,t) = 0, \quad \frac{\partial q_2(x,t)}{\partial t} = 0, \quad p_2(x,t) = 0, \quad \frac{\partial r_1(x,t)}{\partial t} = 0, \quad r_0(x,t) = -c_0 t + c_3,$$

where c_3 is an arbitrary constant.

For $k = 4 + \alpha$, $(\alpha = 0, 1, \ldots)$, equation (6.31) yields

$$\begin{gathered} x^{n+1}\Big(\tfrac{\partial p_{\alpha+4}(x,t)}{\partial t} - 2\int_0^1 (1-s)^n s^{(3\alpha)n} p_{3+\alpha}(xs,t)\,ds + \\ (3+\alpha)P_n p_{3+\alpha}(x,t) - P_n \tfrac{\partial r_{2+\alpha}(x,t)}{\partial t} - (\alpha+1)P_n^2 r_{\alpha+1}(x,t) + \\ 2P_n \int_0^1 (1-s)^n s^{(3+\alpha)n} r_{\alpha+1}(xs,t)\,ds - Q_n r_{\alpha+1}(x,t)\Big) + \\ nx^n\Big(-\tfrac{\partial q_{\alpha+3}(x,t)}{\partial t} - 2P_n q_{\alpha+2}(x,t) + 2\int_0^1 (1-s)^n s^{(\alpha+3)n-1} q_{\alpha+2}(xs,t)\,ds\Big) - \\ n(\alpha+2)P_n q_{\alpha+2}(x,t) = 0. \end{gathered}$$

From this equation one obtains

$$p_{\alpha+3}(x,t) = 0, \quad q_{\alpha+2}(x,t) = 0, \quad r_{\alpha+1}(x,t) = 0, \quad (\alpha = 0, 1, \ldots).$$

Thus, from the above equations, one finds

$$\xi = c_2 x, \quad \eta = c_3 - c_0 t, \quad \zeta = (c_1 x + c_0)\varphi$$

with the arbitrary constants c_0, c_1, c_2, c_3.

Therefore equation (6.27) admits the four–dimensional Lie algebra L^4 spanned by the generators

$$X_1 = \partial_t, \; X_2 = x\varphi\partial_\varphi, \; X_3 = x\partial_x, \; X_4 = \varphi\partial_\varphi - t\partial_t.$$

3.2 Invariant solutions

For constructing an invariant solution one has to choose a subalgebra. Since any subalgebra is equivalent to one of the representatives of an optimal system of admitted subalgebras, it is enough to study invariant solutions of the optimal system. Choosing a subalgebra from the optimal system of subalgebras, finding invariants of the subalgebra, and assuming dependence between these invariants, one obtains the representation of an invariant solution. Substituting this representation into equation (6.27) one gets the reduced equations: for the invariant solutions the original equation is reduced to the equation for a function with a single independent variable.

The optimal system of one–dimensional subalgebras of L^4 consists of the subalgebras

$$X_1, \; X_4 + cX_3, \; X_2 - X_1, \; X_4 \pm X_2, \; X_1 + X_3, \tag{6.32}$$

where c is an arbitrary constant. The corresponding representations of the invariant solutions are the following.

The invariants of the subalgebra $\{X_1\}$ are φ and x. Hence, an invariant solution has the representation $\varphi = g(x)$, where the function g has to satisfy the equation

$$g(x)g(0) - \int_0^1 g(xs)g(x(1-s))\,ds = 0. \tag{6.33}$$

The Maxwell solution $\varphi = ke^{\lambda x}$ is an invariant solution with respect to this subalgebra. Let a solution of equation (6.33) be represented through the formal series $g(x) = \sum_{j\geq 0} a_j x^j$. For the coefficients of the formal series one obtains

$$a_0\left(1 - \frac{2}{(k+1)!}\right)a_k = \sum_{j=1}^{k-1} \frac{j!(k-j)!}{(k+1)!} a_j a_{k-j}, \quad (k = 2, 3, ...).$$

Noticing that the value $a_0 = 0$ leads to the trivial case $g = 0$, so that one has to assume $a_0 \neq 0$. Because equation (6.33) admits scaling of the function g, one can set $a_0 = 1$. Since the multiplication by the function $e^{\lambda x}$ transforms any solution of equation (6.33) into another solution, one also can set $a_1 = 0$. All other coefficients become $a_j = 0$, $(j = 2, 3, ...)$. Thus the general solution of equation (6.33) (up to multiplication by the function $ke^{\lambda x}$) is $g = 1$.

In the case of the subalgebra $\{X_4 + cX_3\}$ the representation of an invariant solution is $\varphi = t^{-1}g(y)$, where $y = xt^c$, and the function g has to satisfy the equation

$$cyg'(y) - g(y) + g(y)g(0) - \int_0^1 g(ys)g(y(1-s))\,ds = 0. \tag{6.34}$$

Assuming that a solution is represented through the formal series $g(y) = \sum_{j\geq 0} a_j y^j$, one obtains the equations for the coefficients

$$a_0 = 0, \ (c-1)a_1 = 0, \ (ck-1)a_k = \sum_{j=0}^{k} \frac{j!(k-j)!}{(k+1)!} a_j a_{k-j}, \quad (k = 2, 3, ...).$$

The case where $ck \neq 1$ for all k, $(k = 1, 2, ...)$ leads to the trivial solution $g = 0$ of equation (6.34). If $c = \alpha^{-1}$ where α is integer, then $a_k = 0$, $(k = 1, 2, ..., \alpha-1)$, the coefficient a_α is arbitrary, and for other coefficients a_k, $(k = \alpha+1, \alpha+2, ...)$ one obtains the recurrence formula

$$(\alpha^{-1}k - 1)a_k = \sum_{j=1}^{k-1} \frac{j!(k-j)!}{(k+1)!} a_j a_{k-j}.$$

The representation of an invariant solution of the subalgebra $\{X_2 - X_1\}$ is $\varphi = e^{-xt}g(x)$, where the function g satisfies the equation

$$-xg(x) + g(x)g(0) - \int_0^1 g(xs)g(x(1-s))\,ds = 0.$$

If one assumes that a solution can be represented through the formal series $g(x) = \sum_{j\geq 0} a_j x^j$, the first two terms of the series, obtained after substitution, are

$$a_0 = 0,\ a_1(6 + a_1) = 0.$$

The case $a_1 = 0$ leads to the trivial solution $g = 0$. If $a_1 \neq 0$, then the other coefficients are defined by the recurrent formula

$$(1 - \frac{6}{k(k+1)})a_{k-1} = -\sum_{j=1}^{k-2} \frac{j!(k-j)!}{(k+1)!} a_j a_{k-j},\ (k = 3, 4, ...).$$

An invariant solution of the subalgebra $\{X_1 + X_3\}$ has the form $\varphi = g(y)$, where $y = xe^{-t}$. The function g has to satisfy the equation

$$-yg'(y) + g(y)g(0) - \int_0^1 g(ys)g(y(1-s))\,ds = 0. \tag{6.35}$$

The solution of this equation $g = 6e^y(1-y)$ is known as the BKW-solution [9, 90][18]. This solution was obtained by assuming that the series $g(y) = e^y \sum_{j\geq 0} a_j y^j$ can be terminated. In fact, substituting the function $g(y) = e^y \sum_{j\geq 0} a_j y^j$ into equation (6.35) for the coefficients a_k one obtains the equations

$$\begin{gathered} a_0 + a_1 = 0,\ 2\,(a_0 - 6)\,a_2 = a_1(6 + a_1),\ 6\,(a_0 - 6)\,a_3 = a_2(12 + a_1), \\ \left(a_0(1 - \tfrac{2}{(k+1)}) - k\right)a_k = a_{k-1}\left(1 + \tfrac{1}{k(k+1)}a_1\right) + \tfrac{2}{k(k^2-1)}a_{k-2}a_2 + \\ \textstyle\sum_{j=2}^{k-2} \frac{j!(k-j)!}{(k+1)!} a_j a_{k-j},\ (k = 4, 5, ...). \end{gathered} \tag{6.36}$$

One can check that the choice $a_0 = 6$, $a_1 = -6$, and $a_k = 0, -(k = 2, 3, ...)$ satisfies equations (6.36).

A representation of an invariant solution of the subalgebra $\{X_4 \pm X_2\}$ is $\varphi = t^{-(1\pm x)}g(x)$, where the function g has to satisfy the equation

$$-(1 \pm x)g(x) + g(x)g(0) - \int_0^1 g(xs)g(x(1-s))\,ds = 0.$$

[18] This solution is usually considered as invariant solution with respect to the similar subalgebra $X_2 - X_3 + c^{-1}X_1$.

Remark 6.3. *A similar approach was applied to the Fourier–image of the Boltzmann equations, describing the homogeneous relaxation of the N–component gas mixture with the Maxwell molecule interactions and the Smolukhovsky kinetic equation of coagulation. The results are as follows.*

The admitted Lie group of the Fourier–image of the Boltzmann equations

$$\frac{\partial\varphi_i}{\partial t}(x_i,t) = \sum_{j=1}^{N} \nu_{ij}\int_0^1 ds[\varphi_i(x_i(1-\varepsilon_{ij}s),t)\varphi_j(x_i\varepsilon_{ij}s),t) \\ -\varphi_i(x_i,t)\varphi_j(0,t)],\ (i=1,2,\ldots,N).$$

corresponds to the Lie algebra with the generators [64]

$$X_1=\partial_t,\ X_2=\sum_\alpha x_\alpha\varphi_\alpha\partial_{\varphi_\alpha},\ X_3=\sum_\alpha x_\alpha\partial_{x_\alpha},\ X_4=\sum_\alpha \varphi_\alpha\partial_{\varphi_\alpha}-t\partial_t,$$

where ε_{ij}, ν_{ij} are constant. Invariant solutions of these subalgebras were studied in [64].

Equation (6.27) is structurally similar (for some kernels of coagulation) to the Smolukhovsky kinetic equation of coagulation

$$\varphi_\tau(x,\tau) = x^\gamma\int_0^1 s^{k_1}(1-s)^{k_2}\varphi(x(1-s),\tau)\varphi(xs),\tau)\,ds,$$

which is studied in kinetic theory of disperse systems such as atmospheric aerosols, colloid solutions, suspensions. The admitted Lie group of this equation consists of the transformations, corresponding to the generators [62]

$$X_1=\partial_\tau,\ X_2=x\varphi\partial\varphi,\ X_3=x\partial_x-\gamma\varphi\partial_\varphi,\ X_4=\varphi\partial_\varphi-\tau\partial_\tau.$$

4. One-dimensional motion of a viscoelastic medium

The one-dimensional motion of a viscoelastic medium is described by the equations [144]

$$v_t=\sigma_x,\ e_t=v_x,\ \varphi(e)=\sigma+\int_z^t K(t,\tau)\sigma(x,\tau)\,d\tau,\ \varphi'(e)\neq 0, \quad (6.37)$$

where the time t and the distance x are the independent variables, the stress σ, the velocity v, and the strain e are the dependent variables, $z=0$ or $z=-\infty$. For $K=0$ equations (6.37) are differential equations[19]. Hence, it is assumed that $K\neq 0$.

An infinitesimal generator of a Lie group admitted by equations (6.37) is assumed in the form

$$X=\zeta^e\partial_e+\zeta^v\partial_v+\zeta^\sigma\partial_\sigma+\xi^x\partial_x+\xi^t\partial_t$$

[19]The case of differential equations ($K=0$) was studied in [28].

with the coefficients $\zeta^e, \zeta^v, \zeta^\sigma, \xi^x, \xi^t$, which are functions of the independent and dependent variables. The determining equations are

$$\left(D_t\widehat{\zeta^v} - D_x\widehat{\zeta^\sigma}\right)_{|(S)} = 0, \ \left(D_t\widehat{\zeta^e} - D_x\widehat{\zeta^v}\right)_{|(S)} = 0, \tag{6.38}$$

$$\left(\varphi'\widehat{\zeta^e} - \widehat{\zeta^\sigma} - \int_0^t K(t,\tau)\widehat{\zeta^\sigma}(x,\tau)\,d\tau\right)_{|(S)} = 0, \tag{6.39}$$

where

$$\widehat{\zeta^e} = \zeta^e - \xi^x e_x - \xi^t e_t, \ \widehat{\zeta^v} = \zeta^v - \xi^x v_x - \xi^t v_t, \ \widehat{\zeta^\sigma} = \zeta^\sigma - \xi^x \sigma_x - \xi^t \sigma_t$$

with the functions $e(x,t), \ v(x,t), \ \sigma(x,t)$ which satisfy equations (6.37) substituted. The complete set of solutions of the determining equations is sought under the assumption that there exists a solution of the Cauchy problem[20]

$$e(x_o,t) = e_0(t), \ v(x_o,t) = v_o(t), \ \sigma(x_o,t) = \sigma_o(t)$$

with arbitrary functions $e_0(t), v_o(t), \sigma_o(t)$.

The derivatives of the functions $e(x,t), v(x,t), \sigma(x,t)$ at the point $x = x_o$ are found from equations (6.37)

$$v_t = v_o', \ \sigma_t = \sigma_o', \ \sigma_x = v_o', \ v_x = e_t = \frac{g_1}{\varphi'}, \ \ e_x = \frac{g_2}{\varphi'}, \tag{6.40}$$

where

$$g_1 = \sigma_o' + K(t,t)\sigma_o + \int_z^t K_t(t,\tau)\sigma_o(\tau)\,d\tau, \ g_2 = v_o' + \int_z^t K(t,\tau)v_o'(\tau)d\,\tau.$$

Substituting the derivatives $v_t, \sigma_t, \sigma_x, v_x, e_t, e_x$ into the determining equations (6.38), considered at the point x_o, one obtains

$$\begin{gathered} v_o'\left(\zeta_v^v - \eta_t - \zeta_\sigma^\sigma + \xi_x - \eta_e g_1\right) + (v_o')^2(-\eta_v + \xi_\sigma) + g_2(-\zeta_e^\sigma + \sigma_o'\eta_e) \\ +v_o' g_2\xi_e + \zeta_\sigma^v\sigma_o' + \zeta_t^v - \zeta_x^\sigma + \sigma_o'\eta_x + \\ g_1\left(\zeta_e^v - \zeta_v^\sigma + \sigma_o'\eta_v - \xi_\sigma\sigma_o' - \xi_t - \xi_e g_1\right) = 0, \end{gathered}$$

$$\begin{gathered} v_o'(\zeta_v^e - \zeta_\sigma^v + \eta_x + \xi_\sigma g_1) + (v_o')^2\eta_\sigma + g_2(-\xi_\sigma\sigma_o' - \zeta_e^v - \xi_t) \\ +v_o' g_2\left(-\xi_v + \eta_e\right) + \zeta_\sigma^e\sigma_o' + \zeta_t^e - \zeta_x^v + g_1(\zeta_e^e - \eta_\sigma\sigma_o' - \eta_t - \zeta_v^v \\ +\xi_x + g_1(-\eta_e + \xi_v)) = 0 \end{gathered}$$

These equations can be split with respect to $v_o, v_o', v_o' + \int_z^t K(t,\tau)v_o'(\tau)d\tau$. In fact, let $z = 0$. Setting

$$v_o(t) = a_1 + a_2(t - t_o) + a_3(t_o - t)^{n+1}/(n+1), \ (n \geq 1),$$

[20]These conditions are boundary conditions, rather than initial conditions.

one finds at the time $t = t_o$

$$v_o(t_o) = a_1, \ v_o'(t_o) = a_2, \ v_o'(t_o) + \int_0^{t_o} K(t_o, \tau)v_o'(\tau)d\tau = \\ a_2\left(1 + \int_0^{t_o} K(t_o, \tau)\,d\tau\right) - a_3\int_0^{t_o} K(t_o, \tau)(t_o - \tau)^n\,d\tau. \tag{6.41}$$

Since the set of the functions $(t_o - t)^n$ is complete in the space $L_2[0, t_o]$, and t_o is such that $K(t_o, \tau) \neq 0$, there exists n for which $\int_0^{t_o} K(t_o, \tau)(t_o - \tau)^n d\tau \neq 0$. Hence, for the given values $v_o(t_o), v_o'(t_o), \int_0^{t_o} K(t_o, \tau)v_o'(\tau)d\tau$ one can solve equations (6.41) with respect to the coefficients a_1, a_2, a_3. This means that one can split the determining equations with respect to v_o, v_o', $v_o' + \int_z^t K(t, \tau)v_o'(\tau)d\tau$. Similarly, when $z = -\infty$, the function $v_o(t)$ is chosen in the form

$$v_o(t) = e^{(t-t_o)/2}(a_1 + a_2(t - t_o) + a_3(t_o - t)^{n+1}/(n + 1), \quad (n \geq 1),$$

and in this case the set of the functions $(t_o - t)^n e^{(t-t_o)/2}$, $(n \geq 0)$ is complete in the space $L_2(-\infty, t_o]$.

Splitting the determining equations with respect to v_o, v_o', $v_o' + \int_z^t K(t, \tau)$ $v_o(\tau)d\tau$, one finds

$$\xi_v = \xi_e = \xi_\sigma = 0, \quad \eta_v = \eta_e = \eta_\sigma = 0, \ \zeta_e^v = -\xi_t, \\ \xi_x - \eta_t = \zeta_\sigma^\sigma - \zeta_v^v, \quad \zeta_e^\sigma = 0, \quad \zeta_v^e - \zeta_\sigma^v = -\eta_x, \tag{6.42}$$

$$(\zeta_\sigma^v + \eta_x)\sigma_o' + \zeta_t^v - \zeta_x^\sigma = g_1(2\xi_t + \zeta_v^\sigma)), \\ \zeta_\sigma^e\sigma_o' + \zeta_t^e - \zeta_x^v = g_1(\eta_t + \zeta_v^v - \xi_x - \zeta_e^e). \tag{6.43}$$

Equations (6.43) also can be split with respect to $\sigma_o(t_o)$, $\sigma_o'(t_o)$, $e(t_o)$, $\sigma_o'(t_o) + K(t_o, t_o)\sigma_o(t_o) + \int_z^{t_o} K_t(t_o, \tau)\sigma_o(\tau)d\tau$. In fact, let $z = 0$ and $\sigma_o(\tau) = a_1 + a_2(\tau - t_o) + (t_o - \tau)^2(a_3\psi_1(\tau) + a_4\psi_2(\tau))$. If the determinant

$$\Delta = \left(\int_0^{t_o} K(t_o, \tau)(t_o - \tau)^2\psi_1(\tau)d\tau\right)\left(\int_0^{t_o} K_t(t_o, \tau)(t_o - \tau)^2\psi_2(\tau)d\tau\right) - \\ \left(\int_0^{t_o} K(t_o, \tau)(t_o - \tau)^2\psi_2(\tau)d\tau\right)\left(\int_0^{t_o} K_t(t_o, \tau)(t_o - \tau)^2\psi_1(\tau)d\tau\right)$$

is equal to zero for all functions $\psi_1, \psi_2 \in L_2[0, t_o]$, then by virtue of $K(t, \tau) \neq 0$ one obtains that there exists a function $f(t)$ such that

$$K_t(t, \tau) = f(t)K(t, \tau). \tag{6.44}$$

This means, that in some neighborhood of the point $t = t_o$ the kernel

$$K(t, \tau) = h(t)g(\tau), \tag{6.45}$$

where $f(t) = h'(t)/h(t)$. The kernels of the type[21] (6.45) are excluded from the study, because for these kernels the last equation of system (6.37) is reduced

[21] They are called degenerate kernels.

to the differential equation

$$\sigma_t = \varphi'(e)e_t + \sigma(f - hg) - f\varphi(e). \tag{6.46}$$

Thus, for nondegenerate kernels, equations (6.43) can be split with respect to the considered values. Splitting them, one has

$$\begin{gathered}\zeta^v_\sigma + \eta_x = 0,\ \ \zeta^v_t - \zeta^\sigma_x = 0, 2\xi_t + \zeta^\sigma_v = 0,\ \ \zeta^e_\sigma = 0,\\ \zeta^e_t - \zeta^v_x = 0,\ \ \eta_t + \zeta^v_v - \xi_x - \zeta^e_e = 0.\end{gathered} \tag{6.47}$$

For the case $z = -\infty$ one also obtains equations (6.47).

Integrating equations (6.42), (6.47), one finds

$$\begin{gathered}\xi = t(c_1x + c_2) + c_3x^2 + c_5x + c_6,\ \ \eta = x(c_3t + c_4) + c_1t^2 + c_7t + c_8,\\ \zeta^v = -e(c_1x + c_2) - \sigma(c_3t + c_4) - v(2c_1t + 2c_3x + c_5 - c_9) + \lambda_{xt},\\ \zeta^\sigma = -\sigma(3c_1t + c_3x + c_7 - c_9) - 2v(c_1x + c_2) + \lambda_{tt},\\ \zeta^e = -e(c_1t + 3c_3x + 2c_5 - c_7 - c_9) - 2v(c_3t + c_4) + \lambda_{xx}.\end{gathered} \tag{6.48}$$

Here c_i, $(i = 1, 2, \ldots, 9)$ are arbitrary constants, and $\lambda(x, t)$ is an arbitrary function of two arguments.

For studying the remaining determining equations (6.39) it is convenient to write

$$\begin{gathered}z_o = \zeta^\sigma + 2v\xi_t = -\sigma(3c_1t + c_3x + c_7 - c_9) + \lambda_{tt},\\ z_1 = \zeta^e + 2v\eta_x = -e(c_1t + 3c_3x + 2c_5 - c_7 - c_9) + \lambda_{xx}.\end{gathered} \tag{6.49}$$

As before it is necessary to consider two cases: $z = 0$ and $z = -\infty$.

4.1 The case $z = 0$

Let $z = 0$. Substituting (6.40) into (6.39) and evaluating some integrals by parts, one obtains

$$\begin{gathered}\varphi'z_1 - z_o - \int_0^t K(t,\tau)z_o(\tau)d\tau + 2v_o(\xi_t - \varphi'\eta_x) + v_o(0)K(t,0)(\xi(t)-\\ \xi(0)) + \int_0^t v_o(\tau)\left((\xi(t) - \xi(\tau))K_\tau(t,\tau) + \xi_t(\tau)K(t,\tau)\right)d\tau - K(t,0)\eta(0)\\ \sigma_o(0) - \int_0^t \sigma_o(\tau)\left(K_\tau(t,\tau)\eta(\tau) + K_t(t,\tau)\eta(t) + K(t,\tau)\eta_t(\tau)\right)d\tau = 0.\end{gathered} \tag{6.50}$$

Because of the arbitrariness of the function $v_o(t)$, from the last equation one finds

$$K(t,0)(\xi(t) - \xi(0)) = 0, \tag{6.51}$$

$$\xi_t - \varphi'\eta_x = 0, \tag{6.52}$$

$$(\xi(t) - \xi(\tau))K_\tau(t,\tau) + \xi_t(\tau)K(t,\tau) = 0, \tag{6.53}$$

$$\varphi' z_1 - z_o - \int_0^t K(t,\tau) z_o(\tau)\, d\tau - K(t,0)\eta(0)\sigma_o(0) - \\ \int_0^t \sigma_o(\tau)\left(K_\tau(t,\tau)\eta(\tau) + K_t(t,\tau)\eta(t) + K(t,\tau)\eta_t(\tau)\right) d\tau = 0. \tag{6.54}$$

Substituting (6.48) into (6.52) and splitting them with respect to x, one obtains

$$c_1 = 0, \quad c_3 = 0, \tag{6.55}$$

$$c_2 = \varphi' c_4. \tag{6.56}$$

Equations (6.51)–(6.54) become

$$c_4 K(t,0) = 0, \tag{6.57}$$

$$c_4\left((t-\tau)K_\tau(t,\tau) + K(t,\tau)\right) = 0, \tag{6.58}$$

$$\varphi'\left(\lambda_{xx} + e_o(c_7 + c_9 - 2c_5)\right) + c_7\sigma_o - \lambda_{tt} - \int_0^t K(t,\tau)\lambda_{tt}(\tau)\, d\tau - c_9\varphi(e_o) - \\ c_8 K(t,0)\sigma_o(0) - \int_0^t \sigma_o(\tau)(c_4 x + c_7\tau + c_8)K_\tau(t,\tau) + \\ (c_4 x + c_7 \tau + c_8) K_t(t,\tau)\, d\tau = 0. \tag{6.59}$$

If there exist functions $\psi_i(\tau) = (t-\tau)^{n_i}$, $(i = 1, 2)$ such that the determinant

$$\Delta_1 = \left(\int_0^t z_3(t,\tau,x)\psi_1(\tau)\tau(t-\tau)\, d\tau\right)\left(\int_0^t K(t,\tau)\tau(t-\tau)\psi_2(\tau)\, d\tau\right) - \\ \left(\int_0^t z_3(t,\tau,x)\psi_2(\tau)\tau(t-\tau)\, d\tau\right)\left(\int_0^t K(t,\tau)\tau(t-\tau)\psi_1(\tau)\, d\tau\right)$$

is not equal to zero, then choosing the function $\sigma_o(\tau)$ one can obtain contradictory relations. Hence, $\Delta_1 = 0$ for all functions $\psi_i(\tau)$. Here $z_3(t,\tau,x) = (c_4 x + c_7\tau + c_8)K_\tau + (c_4 x + c_7 t + c_8)K_t)$. Because $K(t,\tau) \neq 0$ and the system of the functions $(t-\tau)^n$ is complete in $L_2[0,t]$, there exists a function $f_1(t,x)$ such that

$$z_3(t,\tau,x) = f_1(t,x)K(t,\tau). \tag{6.60}$$

Substituting (6.60) into (6.59), using (6.38), and splitting with respect to $\sigma_o(0)$, $\sigma_o(t)$ and $e_o(t)$, one obtains

$$c_7 + f_1 = 0, \tag{6.61}$$

$$c_8 K(t,0) = 0, \tag{6.62}$$

$$\varphi'\left(\lambda_{xx} + e_o(c_7 + c_9 - 2c_5)\right) + \varphi(c_7 - c_9) - \lambda_{tt} - \int_0^t K(t,\tau)\lambda_{tt}(\tau)\, d\tau = 0. \tag{6.63}$$

Splitting (6.60) with respect to x, and because of (6.61), one finds

$$c_4(K_\tau + K_t) = 0, \tag{6.64}$$

$$(c_7 t + c_8)K_t + (c_7\tau + c_8)K_\tau = -c_7 K. \tag{6.65}$$

Regarding (6.57), (6.58) and (6.64), one obtains

$$c_4 = 0. \tag{6.66}$$

If $c_7^2 + c_8^2 \neq 0$, then from equations (6.62), (6.65), one finds that $c_8 = 0$ and $K = (1/(c_7 t))R(\tau/t)$. The kernels of this type are excluded from the study, because they have a singularity at the time $t = 0$. Hence,

$$c_7 = 0, \quad c_8 = 0, \tag{6.67}$$

and the group classification of equations (6.37), (6.38) in the case $z = 0$ is reduced to the study of equation (6.63).

From equation (6.63) it follows that the kernel of the admitted Lie groups is generated by the generators

$$X_1 = \partial_x, X_2 = \partial_v. \tag{6.68}$$

Extensions of the kernel (6.68) are obtained for specific functions $\varphi(e)$.

If $\varphi'' \neq 0$, then the classifying equations are

$$\varphi'(c_{10} + e(c_9 - 2c_5)) - c_9\varphi = c_{11}, \tag{6.69}$$

$$\lambda_{tt} + \int_0^t K(t,\tau)\lambda_{tt}(\tau)\,d\tau = c_{11}, \tag{6.70}$$

where c_{10}, c_{11} are arbitrary constants. Hence, the extension of the kernel of the main Lie groups is possible for the following cases. If $\varphi = \alpha + \beta\ln(a + ce)$, then the additional generator is

$$Y_1 = -cx/2\partial_x + cv/2\partial_v + (a + ce)\partial_e + \beta c\mu(t)\partial_\sigma.$$

If $\varphi = \alpha(a + ce)^\beta + \gamma$, $(\beta \neq 1)$, then system of equations (6.37) admits the generator

$$Y_2 = (\beta - 1)cx\partial_x + (\beta + 1)cv\partial_v + 2(ce + a)\partial_e + 2\beta c(\sigma - \gamma\mu(t))\partial_\sigma.$$

If $\varphi = \alpha + \exp(\gamma e)$, $(\gamma \neq 0)$, then there is the additional generator

$$Y_3 = \gamma x\partial_x + \gamma v\partial_v + 2\gamma(\sigma - \alpha\mu(t))\partial_\sigma + 2\partial_e.$$

Here $\alpha, \beta, \gamma, a, c$ are constants, and the function $\mu(t)$ is an arbitrary solution of the equation

$$\mu(t) + \int_z^t K(t,\tau)\mu(\tau)\,d\tau = 1.$$

If the function $\varphi(e)$ is linear $\varphi = Ee + E_1$, then along with the generators X_1, X_2 system (6.37), (6.38) also admits the generators

$$Y_4 = v\partial_v + \sigma\partial_\sigma + e\partial_e, \quad Y_\lambda = \lambda_{xt}\partial_v + \lambda_{tt}\partial_\sigma + \lambda_{xx}\partial_e,$$

where $\lambda(x, t)$ is a solution of the equation

$$E\lambda_{xx} = \lambda_{tt} + \int_z^t K(t, \tau)\lambda_{tt}(\tau)d\tau.$$

4.2 The case $z = -\infty$

In the case $z = -\infty$ the study of the determining equation (6.39) leads to equations (6.55), (6.56), (6.58), (6.61), (6.63)–(6.65).

If $c_4 \neq 0$, then from equations (6.58), (6.64) one finds that the kernel $K(t, \tau)$ has the representation $K(t, \tau) = c(t - \tau)$, (c is constant). Such kernels do not satisfy the principle of memory decay [144], so therefore they are not considered here. Hence,

$$c_4 = 0. \tag{6.71}$$

If $c_7 \neq 0$, then from equation (6.65) one obtains the kernel $K(t, \tau) = (1/(c_7 t + c_8))R((c_7\tau + c_8)/(c_7 t + c_8))$, which has a singularity at the time $t = -c_8/c_7$. Thus

$$c_7 = 0. \tag{6.72}$$

From the last equation it follows that in the case $z = -\infty$ the group classifications with respect to the kernels $K(t, \tau)$ and the function $\varphi(e)$ are independent.

The kernel of the Lie groups admitted by system (6.63), (6.38), as for the case $z = 0$, is generated by the generators X_1 and X_2. The classification relation for the function $\varphi(e)$ is the same as in the case $z = 0$. Thus the extension of the kernel of the admitted Lie groups for arbitrary kernels $K(t, \tau)$ in these two cases coincide. Further extension by the generator $X_3 = \partial_t$ occurs for the kernels $K(t, \tau) = R(t - \tau)$.

5. Delay differential equations

The section is devoted to applications of group analysis to functional differential equations[22]. The simplest type of functional differential equation is a type of delay differential equation, where some derivatives of the unknown functions at time t are expressed through their values at earlier instants. Delay differential equations appear in problems with delaying links where certain information processing is needed, for example, in populations dynamics and

[22]The theory and applications of functional differential equations can be found, for example, in [7, 121, 68, 37, 85].

bioscience problems, in control problems, electrical networks containing lossless transmission lines.

For the sake of simplicity an introduction to functional differential equations is given for a functional differential equations with a single independent variable

$$\Phi(t, x) \equiv x'(t) - F(t, x_t) = 0, \ \ (t \in J). \tag{6.73}$$

Here[23] x_t denotes the function $x(t) \in D \subset R^m$, which is defined in the interval $[t - r, t]$ by

$$x_t(s) = x(t + s), \ \ s \in [-r, 0],$$

D is an open set in R^m, J is an interval in R, F is a functional. For delay differential equations the functional F has the representation

$$F(t, x_t) \equiv f(t, x(g_1(t)), ..., x(g_m(t))),$$

where $f : [t_0, \beta) \times D^m \to R^n$, and $g_j(t) \leq t$ for $t_0 \leq t \leq \beta$ for each $j = 1, ..., m$. The function g_1 is usually chosen to be the identity mapping.

The Cauchy problem for delay equations (6.73) is set as follows. The initial conditions are defined by a function $\psi : [-\gamma, 0] \to D$

$$x(t_0 + s) = \psi(s), \ \ s \in [-\gamma, 0]. \tag{6.74}$$

A continuous function $x(t)$, $t \in [t_0 - \gamma, t_0 + \beta)$ is called a solution of the Cauchy problem (6.73), (6.74) if it is differentiable in the interval (t_0, β), satisfies equations (6.73) in the interval (t_0, β) and conditions (6.74) in the interval $[t_0 - \gamma, t_0]$. The value $x'(t_0)$ is understood as the right-hand derivative. With some requirements[24] for the functional F one can guarantee the existence of the solution of the Cauchy problem (6.73), (6.74).

5.1 Example

Let us consider the integro-differential equation

$$x'(t) = \int_{-r}^{0} x(t + s)\, ds. \tag{6.75}$$

This functional differential equation is an integro-differential equation. According to the definition of the admitted Lie group, the determining equation for (6.75) is

$$\begin{gathered} \eta_t(t, x(t)) + x'(t)\eta_x(t, x(t)) - x''(t)\tau(t, x(t)) - \\ x'(t)\left(\tau_t(t, x(t)) + x'(t)\tau_x(t, x(t))\right) = \\ \int_{-r}^{0} \left(\eta(t + s, x(t + s)) - x'(t + s)\tau(t + s, x(t + s))\right) ds, \end{gathered} \tag{6.76}$$

[23] The notations accepted in literature on functional differential equations are used.

[24] These requirements are similar to the conditions used in ordinary differential equations (see, for example, in [37]).

where $X = \tau(t,x)\partial_t + \eta(t,x)\partial_x$ is the generator of the admitted Lie group. Equation (6.76) has to be satisfied for any solution $x = x(t)$ of equation (6.75). The determining equation is still an integro-differential equation, but it is not easy to split. Notice that differentiating equation (6.75), one obtains the delay differential equation

$$x''(t) = x(t) - x(t-r), \tag{6.77}$$

From some point of view equation (6.77) is easier to study. In the next subsection this will be demonstrated. Notice also that a Cauchy problem for this equation is set up by the initial conditions [121]

$$x'(t_o) = x_1, \quad x_{t_o}(s) = \psi(s), \quad s \in [t_o - r, t_o]. \tag{6.78}$$

with an arbitrary value x_1 and an arbitrary continuous function ψ.

5.2 Admitted Lie group

The example to be studied gives an idea of the application of the definition of an admitted Lie group, developed initially for integro-differential equations, to functional differential equations.

Let there be given a one–parameter Lie group $G^1(X)$ of transformations

$$\bar{t} = f^t(t,x,a), \quad \bar{x} = f^x(t,x,a) \tag{6.79}$$

with the generator

$$X = \tau(t,x)\partial_t + \eta(t,x)\partial_x.$$

Definition 6.4. *A one–parameter Lie group G^1 of transformations (6.79) is a symmetry group admitted by equation (6.73) if G^1 satisfies the determining equations*

$$(\bar{X}\Phi)(t, x(t)) = 0 \tag{6.80}$$

for any solution $x(t)$ of equations (6.73).

Here the operator $\bar{X}$ is the prolongation of the canonical Lie–Bäcklund operator equivalent to the generator X:

$$\bar{X} = \overline{\eta}\partial_x + ...$$

where $\overline{\eta} = \eta - x'\tau$. The actions of the derivatives ∂_x and $\partial_{x^{(j)}}$ are considered in terms of the Frechet derivatives.

Formally[25] the determining equations[26] (6.80) can be constructed similarly to those for integro-differential equations. Assume that the Lie group $G^1(X)$

[25] Why this is a formal construction was discussed in the section devoted to integro-differential equations.
[26] It should be also mentioned that other approaches can be found in [97, 179].

transforms a solution $x_0(t)$ of equation (6.73) into the solution $x_a(t)$ of the same equations. The transformed function $x_a(t)$ is

$$x_a(\bar{t}) = f^x(t, x_0(t), a)$$

with the expression $t = \psi^t(\bar{t}; a)$ substituted, which is found from the relation $\bar{t} = f^t(t, x_0(t), a)$. Differentiating the equations $\Phi(t, x_a)$ with respect to the group parameter a and considering these equations for the value $a = 0$

$$\left(\frac{\partial}{\partial a}\Phi(t, x_a)\right)_{|a=0} = 0,$$

one obtains the determining equations (6.80).

The main features of the determining equations in the given definition is that they must be satisfied for any solution of equations (6.73). This allows splitting the determining equations with respect to arbitrary elements. Since arbitrary elements of delay differential equations are similarly contained in the determining equations as for differential equations, the process of solving the determining equations for delay differential equations is similar to finding the solutions of the determining equations for differential equations. This will be demonstrated in examples.

Notice that the given definition is free from the requirement that the admitted Lie group should transform a solution into a solution, and also it can be applied when finding an equivalence group, contact and Lie-Bäcklund transformations for functional differential equations.

Remark 6.4. *The given definition allows the definition of an admitted Lie group for more general objects. For example, let us consider the functional differential equation studied by Barba [37]*

$$y(x)y'(x) = y\left(y(x)\right). \tag{6.81}$$

According to the given definition of an admitted Lie group, the determining equation for (6.81) is

$$\begin{gathered}\tilde{X}\left(y(x)y'(x) - y\left(y(x)\right)\right) = \left(\eta(x, y(x)) - y'(x)\xi(x, y(x))\right)y'(x) + \\ y(x)(\eta_x(x, y(x)) + y'(x)\eta_y(x, y(x)) - y''(x)\xi(x, y(x)) - y'(x)(\xi_x(x, y(x)) \\ + y'(x)\xi_y(x, y(x)))) - (\eta(x, y(y(x))) - y'(y(x))\xi(x, y(y(x))) + \\ y'(y(x))\left(\eta(x, y(x)) - y'(x)\xi(x, y(x))\right)),\end{gathered}$$

which has to be satisfied for any solution $y = y(x)$ of equation (6.81). Here a generator of the admitted Lie group is $X = \xi(x, y)\partial_x + \eta(x, y)\partial_y$.

5.3 Continuation of the study of equation (6.75)

Let us continue studying equation (6.77). The determining equation for equation (6.77) considered at the point $(t_o + 0)$ (limit from the right handed side), after substitution into it the derivatives

$$x''(t_o) = x(t_o) - x(t_o - r), \quad x'''(t_o) = x_1 - x'(t_o - r),$$

the determining equation becomes

$$\begin{gathered} 2\eta_{tx}(t_o, x_o)x_1 + \eta_{tt}(t_o, x_o) + \eta_{xx}(t_o, x_o)x_1^2 + \eta_x(t_o, x_o)x_o - \eta_x(t_o, x_o)x_2 - \\ 2\tau_{tx}(t_o, x_o)x_1^2 - \tau_{tt}(t_o, x_o)x_1 - 2\tau_t(t_o, x_o)x_o + 2\tau_t(t_o, x_o)x_2 - \tau_{xx}(t_o, x_o)x_1^3 - \\ 3\tau_x(t_o, x_o)x_0x_1 + 3\tau_x(t_o, x_o)x_2x_1 - \eta(t_o, x_o) + \eta(t_o - r, x_2) + \\ (\tau(t_o, x_o) - \tau(t_o - r, x_2))\, x_3 = 0, \end{gathered} \tag{6.82}$$

where $x_o = \psi(t_o)$, $x_2 = \psi(t_o - r)$, $x_3 = \psi'(t_o - r)$. By virtue of the theorem for the existence of a solution of the Cauchy problem (6.77), (6.78) with an arbitrary function $\psi(s)$ and an arbitrary value x_1, the values t_o, x_o, x_1, x_2 and x_3 can be assigned arbitrarily. The arbitrariness of these variables allows splitting the determining equation, and then finding the general solution of the determining equation. Further analysis is similar to the classical analysis of solving determining equations of partial differential equations. Since the values t_o and x_o are arbitrary in the determining equation, they are written as t and x, respectively.

The determining equation (6.82) can be split with respect to x_1 and x_3 into the five equations

$$\begin{gathered} 2\eta_{tx}(t, x) - \tau_{tt}(t, x) - 3\tau_x(t, x)x + 3\tau_x(t, x)x_2 = 0, \\ \eta_{xx}(t, x) - 2\tau_{tx}(t, x) = 0, \ \ \tau_{xx}(t, x) = 0, \ \ \tau(t, x) - \tau(t - r, x_2) = 0, \\ \eta_{tt}(t, x) + \eta_x(t, x)x - \eta_x(t, x)x_2 - 2\tau_t(t, x)x + 2\tau_t(t, x)x_2 - \\ \eta(t, x) + \eta(t - r, x_2) = 0, \end{gathered} \tag{6.83}$$

The first equation can be also split with respect to x_2:

$$\tau_x(t, x) = 0.$$

Hence, the first four equations of (6.82) give

$$2\eta(t, x) = x(c_1 + \tau'(t)) + \varphi(t), \ \ \tau(t) = \tau(t - r),$$

where c_1 is constant. Substituting this representation into the last equation of (6.82), and splitting it with respect to x_o and x_2, one obtains

$$\begin{gathered} \tau'''(t) - 4\tau'(t) = 0, \ \ 3\tau'(t) + \tau'(t - r) = 0, \\ \varphi''(t) - \varphi(t) + \varphi(t - r) = 0. \end{gathered}$$

The general solution of the first two equations is $\tau = c_2$, where c_2 is constant. Then the infinitesimal generator corresponding to the admitted Lie group is

$$X = c_2\partial_t + c_1 x\partial_x + \varphi(t)\partial_x,$$

where the function $\varphi(t)$ is an arbitrary solution of equation (6.77).

6. Group classification of the delay differential equation $u_t + uu_x = g(u, \bar{u})$

Let us study the equation[27]

$$u_t + uu_x = g(u, \bar{u}), \tag{6.84}$$

where x and t are the independent variables, $u = u(x, t)$, $\bar{u} = u(x, t - r)$, $u_t = u_t(x, t)$, $u_x = u_x(x, t)$. Since for $g_u = 0$, equation (6.84) is a partial differential equation, and it is assumed that $g_{\bar{u}} \neq 0$. A unique solution of (6.84) is defined [15] by the initial values

$$u(x, s) = \psi(x, s), \;\; s \in [t_0 - r, t_0],$$

where t_0 and the function $\psi(x, s)$, $s \in [t_0 - r, t_0]$ are arbitrary. Because of this arbitrariness one can assume that the functions $\bar{u}_t = u_t(x, t - r)$, $\bar{u}_x = u_x(x, t - r)$, $u_x(x, t)$, $u(x, t)$, $\bar{u}(x, t)$ can be assigned arbitrary values at a fixed point (x, t).

According to the definition of an admitted Lie group, the determining equation of the admitted Lie group of equation (6.85) is

$$\begin{gathered}\bar{u}_t g_{\bar{u}}(\bar{\tau} - \tau) + u_x(-\eta_t - \eta_u g - \eta_x u + \tau_t u + \tau_u gu + \tau_x u^2 + \zeta) + \bar{u}_x g_{\bar{u}} \\ (\bar{\eta} - \eta) - g_u\zeta - g_{\bar{u}}\bar{\zeta} - \tau_t g - \tau_u g^2 - \tau_x gu + \zeta_t + \zeta_u g + \zeta_x u = 0,\end{gathered} \tag{6.85}$$

where $\bar{\tau} = \tau(x, t - r, \bar{u})$, $\bar{\eta} = \eta(x, t - r, \bar{u})$, $\bar{\zeta} = \zeta(x, t - r, \bar{u})$, $\tau = \tau(x, t, u)$, $\eta = \eta(x, t, u)$, $\zeta = \zeta(x, t, u)$ are values of the coefficients of the generator of this group

$$X = \tau(x, t, u)\partial_t + \eta(x, t, u)\partial_x + \zeta(x, t, u)\partial_u.$$

Splitting the determining equation with respect to $\bar{u}_t$, $\bar{u}_x$ and u_x gives

$$\begin{gathered}g_{\bar{u}}(\bar{\tau} - \tau) = 0, \quad g_{\bar{u}}(\bar{\eta} - \eta) = 0, \\ \zeta - \eta_t - \eta_u g - \eta_x u + \tau_t u + \tau_u gu + \tau_x u^2 = 0, \\ -g_u\zeta - g_{\bar{u}}\bar{\zeta} - \tau_t g - \tau_u g^2 - \tau_x gu + \zeta_t + \zeta_u g + \zeta_x u = 0.\end{gathered} \tag{6.86}$$

Since $g_{\bar{u}} \neq 0$, then $\tau(x, t - r, \bar{u}) = \tau(x, t, u)$, $\eta(x, t - r, \bar{u}) = \eta(x, t, u)$. By virtue of the arbitrariness of u and $\bar{u}$ in the last equalities, the coefficients $\tau(x, t, u)$ and $\eta(x, t, u)$ do not depend on the variable u:

$$\tau(x, t - r) = \tau(x, t), \quad \eta(x, t - r) = \eta(x, t). \tag{6.87}$$

The third equation of (6.86) defines the function $\zeta(x, t, u)$:

$$\zeta = \eta_t + u(\eta_x - \tau_t) - \tau_x u^2. \tag{6.88}$$

[27] Some particular results for this equation are presented in [165].

The last equation of (6.86) becomes

$$-(g_u + g_{\bar{u}})\eta_t - (g_u u + g_{\bar{u}}\bar{u} - g)(\eta_x - \tau_t) + (g_u u^2 + g_{\bar{u}}\bar{u}^2 - 3gu)\tau_x - g\tau_t + \eta_{tt} - \tau_{xx}u^3 + (\eta_{xx} - 2\tau_{tx})u^2 + (2\eta_{tx} - \tau_{tt})u = 0. \tag{6.89}$$

This shows that the kernel of admitted Lie groups consists of the transformations corresponding to the generators is

$$X_1 = \partial_x, \quad X_2 = \partial_t.$$

Extensions of the kernel are for a specific function $g(u, \bar{u})$.

Differentiating (6.89) with respect to $\bar{u}$, and dividing it by g, one finds

$$\tau_t = -a\eta_t - b(\eta_x - \tau_t) + c\tau_x, \tag{6.90}$$

where

$$a = \frac{g_u + g_{\bar{u}}}{g_{\bar{u}}}, \quad b = \frac{g_u u + g_{\bar{u}}\bar{u} - g}{g_{\bar{u}}}, \quad c = \frac{g_u u^2 + g_{\bar{u}}\bar{u}^2 - 3gu}{g_{\bar{u}}}.$$

Differentiating equation (6.90) with respect to $\bar{u}$, one obtains

$$-a_{\bar{u}}\eta_t - b_{\bar{u}}(\eta_x - \tau_t) + c_{\bar{u}}\tau_x = 0. \tag{6.91}$$

Further analysis of equation (6.91) is similar to the group classification of the gas dynamics equations [130]. Notice that if the function $g(u, \bar{u})$ is such that one can find there noncomplanar vectors $(a_{\bar{u}}, b_{\bar{u}}, c_{\bar{u}})$ (with u and $\bar{u}$ changing), then it corresponds to the trivial case

$$\eta_t = 0, \ \eta_x - \tau_t = 0, \ \tau_x = 0.$$

Hence, an extension of the kernel can occur if the vector $(a_{\bar{u}}, b_{\bar{u}}, c_{\bar{u}})$ lies in either two, one or zero dimensional spaces with u and $\bar{u}$ changing. For the sake of simplicity only the two-dimensional case is studied here.

6.1 Two dimensional case

In this case there exists a constant vector $(k_1, k_2, k_3) \neq 0$ such that

$$a_{\bar{u}}k_1 + b_{\bar{u}}k_2 + c_{\bar{u}}k_3 = 0. \tag{6.92}$$

Because the vector $(-\eta_t, -(\eta_x - \tau_t), \tau_x)$ is also orthogonal to the vector $(a_{\bar{u}}, b_{\bar{u}}, c_{\bar{u}})$ (6.91), the vectors $(-\eta_t, -(\eta_x - \tau_t), \tau_x)$ and (k_1, k_2, k_3) are parallel, and, hence,

$$\eta_t = -\lambda k_1, \ \eta_x - \tau_t = -\lambda k_2, \ \tau_x = \lambda k_3 \tag{6.93}$$

for some function $\lambda(x, t) \neq 0$. Substituting these relations into (6.90), one obtains

$$\tau_t = \lambda(k_1 a + k_2 b + k_3 c).$$

Differentiating it with respect to u, and by virtue of (6.92), one finds

$$ak_1 + bk_2 + ck_3 = k, \tag{6.94}$$

where k is constant. Hence,

$$\tau_t = \lambda k.$$

Integrating equation (6.94) with respect to $\bar{u}$, one arrives at

$$g_u(k_1+k_2u+k_3u^2)+g_{\bar{u}}(k_1+k_2\bar{u}+k_3\bar{u}^2) = g(k_2+3uk_3+k)+h(u), \tag{6.95}$$

where the function $h(u)$ is an arbitrary function. Equation (6.89) becomes

$$\lambda_t k_1 + u^3\lambda_x k_3 + u^2(\lambda_x k_2 + \lambda_t k_3) + u(\lambda_x k_1 + \lambda_t k_2) = \lambda h. \tag{6.96}$$

Since $\lambda \neq 0$, the function $h(u)$ has to be a polynomial of third degree with constant coefficients

$$h(u) = a_3u^3 + a_2u^2 + a_1u + a_0.$$

Splitting equation (6.96) with respect to u, one finds that

$$\lambda_t k_1 = \lambda a_0,\ \lambda_x k_1 + \lambda_t k_2 = \lambda a_1,\ \lambda_x k_2 + \lambda_t k_3 = \lambda a_2,\ \lambda_x k_3 = \lambda a_3. \tag{6.97}$$

Let $k_1 \neq 0$. From the first equation of (6.97) one obtains $\lambda = \mu(x)e^{ta_o/k_1}$. Since of $\lambda = -\eta_t/k_1$, and the periodicity (6.87) of the function $\eta(x,t)$, the function $\lambda(x,t)$ is also periodic. It leads to $a_0 = 0$, and then to the contradiction $\lambda k_1 = 0$. Hence, $k_1 = 0$, $a_0 = 0$, and $\eta_t = 0$.

Let $k_2 \neq 0$. Since of $\lambda = (\tau_t - \eta_x)/k_2$, and periodicity of the functions $\eta(x,t)$ and $\tau(x,t)$, the function $\lambda(x,t)$ is also periodic. It leads to $a_1 = 0$, and then to the contradiction $\lambda_t = 0$. Because $\tau_{tt} = k_2\lambda_t$ and periodicity of the function $\tau(x,t)$, one finds $\tau_t = 0$. Hence, $k = 0$, $\lambda = -\eta_x/k_2$ and $\eta_x \neq 0$. The remaining equations (6.97) and (6.93) become

$$\eta_{xx} = \frac{a_2}{k_2}\eta_x,\ a_3 - \frac{k_3}{k_2}a_2 = 0,\ \tau_x + \frac{k_3}{k_2}\eta_x = 0. \tag{6.98}$$

Thus the function $h = u^2\dfrac{a_2}{k_2}(k_2 + uk_3)$ and the general solution of equation (6.95) is

$$g = u(k_2 + k_3u)^2\left(\psi + \frac{a_2}{k_2^2}\frac{u}{(k_2 + k_3u)}\right),$$

where the function $\psi = \psi\left(\dfrac{u(k_2 + k_3\bar{u})}{\bar{u}(k_2 + k_3u)}\right)$ is an arbitrary function of a single argument. Further analysis depends on the value of a_2. If $a_2 \neq 0$, then the extension of the kernel is given by the generator

$$X = e^{xa_2/k_2}\left(k_3\partial_t - k_2\partial_x - u(k_2 + k_3u)\partial_u\right).$$

If $a_2 = 0$, then the extension of the kernel is given by the generator

$$X = k_3 x \partial_t - k_2 x \partial_x - u(k_2 + k_3 u)\partial_u.$$

Let $k_1 = 0,\ k_2 = 0$. In this case the analysis, similar to the previous case, gives

$$a_0 = 0,\ a_1 = 0,\ a_2 = 0,\ \eta_t = 0,\ \eta_x = 0,\ \tau_t = 0,\ \tau_{xx} = \tau_x a_3/k_3,$$
$$k = 0,\ \ g = u^2\left(u\psi - \tfrac{a_3}{k_3}\right),$$

where the function $\psi = \psi\left(\dfrac{\bar{u} - u}{u\bar{u}}\right)$ is an arbitrary function of a single variable. If $a_3 \neq 0$, then the extension of the kernel is given by the generator

$$X = e^{xa_3/k_3}\left(k_3\partial_x - a_3 u^2 \partial_u\right).$$

If $a_3 = 0$, then the extension of the kernel is given by the generator

$$X = x\partial_x - u^2\partial_u.$$

6.2 An equivalence group

The method for finding an admitted Lie group can be simplified by introducing new variables denoting nonlocal terms. This is demonstrated in this section, where the equivalence Lie group of equation (6.84) is found.

Let us denote the function $u(t - r, x)$ by the new dependent variable $v(x, t)$. Equation (6.84) becomes the partial differential equation

$$u_t + uu_x = g(u, v) \tag{6.99}$$

with two dependent variables. Assume that the infinitesimal generator of the equivalence group is

$$X = \tau(t, x, u, v, g)\partial_t + \eta(t, x, u, v, g)\partial_x + \zeta(t, x, u, v, g)\partial_u +$$
$$\chi(t, x, u, v, g)\partial_v + \theta(t, x, u, v, g)\partial_g.$$

Applying to equation (6.99) the methods developed for partial differential equations, one finds

$$\begin{gathered}\zeta_t = 0,\ \zeta_x = 0,\ \chi_t = 0,\ \chi_x = 0,\ \theta_t = 0,\ \theta_x = 0,\\ \eta_v = \tau_v u,\ \eta_g = \tau_g u,\ \zeta_v = \tau_v g,\ \zeta_g = \tau_g g,\\ \zeta = \eta_u g + \eta_x u + \eta_t - \tau_u ug - \tau_x u^2 - \tau_t u,\\ \theta = -\tau_u g^2 - \tau_x ug - \tau_t g + \zeta_u g.\end{gathered} \tag{6.100}$$

Equating the mixed derivatives $(\zeta_v)_g = (\zeta_g)_v$, one finds

$$\tau_v = 0,$$

and, hence,

$$\eta_v = 0, \ \zeta_v = 0.$$

These equations have to be supplemented by the determining equation corresponding to the invariance of the equation[28]

$$v(x,t) = u(t-r,x). \tag{6.101}$$

The determining equation corresponding to (6.101) is obtained according to the developed algorithm, and it is

$$\begin{aligned} \chi(u,v,g(u,v)) - \zeta(\bar{u},g(\bar{u},\bar{v})) + (\tau(t,x,u,g(u,v)) \\ -\tau(t-r,x,\bar{u},g(\bar{u},\bar{v})))\, v_t - (\eta(t,x,u,g(u,v)) \\ -\eta(t-r,x,\bar{u},g(\bar{u},\bar{v})))\, v_x = 0, \end{aligned} \tag{6.102}$$

where $\bar{v} = v(t-r,x)$. Notice also that $\bar{u} = v$.

Splitting equation (6.102) with respect to v_t and v_x, one obtains

$$\begin{gathered} \chi(u,v,g(u,v)) = \zeta(\bar{u},g(\bar{u},\bar{v})), \ \tau(t,x,u,g(u,v)) = \tau(t-r,x,\bar{u},g(\bar{u},\bar{v})), \\ \eta(t,x,u,g(u,v)) = \eta(t-r,x,\bar{u},g(\bar{u},\bar{v})) \end{gathered} \tag{6.103}$$

Differentiating the first equation of (6.102)[29] with respect to $\bar{v}$, $\chi(u,v,g(u,v)) = \zeta(v,g(v,\bar{v}))$, and one obtains

$$\zeta_g = 0. \tag{6.104}$$

Hence, $\chi = \chi(v) = \zeta(u)_{|u=v}$. Taking into account equations 6.100), equation (6.104) leads to

$$\eta_g = 0, \ \tau_g = 0.$$

The second and third equations of (6.102) become

$$\tau(t,x,u) = \tau(t-r,x,v), \ \eta(t,x,u) = \eta(t-r,x,v).$$

Differentiating them with respect to u, one finds that

$$\tau = \tau(t,x) = \tau(t-r,x), \ \eta = \eta(t,x) = \eta(t-r,x). \tag{6.105}$$

Because of the last equations of (6.100)

$$\zeta = \eta_x u + \eta_t - \tau_x u^2 - \tau_t u, \ \theta = -\tau_x u g - \tau_t g + \zeta_u g,$$

---

[28] Appending this equation when finding an admitted Lie group of equation (6.84) is very important. In the approach developed in [97] such a requirement is absent.

[29] In the $\zeta(v,g(v,\bar{v}))$ the first argument of the function $\zeta = \zeta(u,g)$ is v: $\zeta(v,g(v,\bar{v})) = \zeta(u,g(u,v)_{|u=v,v=\bar{v}}$.

and the remaining equations of (6.100) $\zeta_t = 0,\ \zeta_x = 0,\ \theta_t = 0,\ \theta_x = 0$ give

$$\tau_{tx} = 0,\ \tau_{tt} = 0,\ \tau_{xx} = 0,\ \eta_{tx} = 0,\ \eta_{tt} = 0,\ \eta_{xx} = 0,$$

or

$$\tau = c_1 t + c_2 x + c_3,\ \eta = c_4 t + c_5 x + c_6,$$

since of equations (6.105), $c_1 = 0,\ c_4 = 0$. Therefore, the equivalence group consists of the transformations corresponding to the generators

$$Y_1 = \partial_t,\ Y_2 = \partial_x,\ Y_3 = x\partial_x + u\partial_u + v\partial_v + g\partial_g,$$
$$Y_4 = -x\partial_t + u^2\partial_u + v^2\partial_v + 3ug\partial_g.$$

The operators Y_1 and Y_2 define the shifts with respect to t and x, respectively. The operator Y_3 defines the dilations of the variables x, u, v_x and g. The operator Y_4 corresponds to the transformation

$$t' = t - ax,\ x' = x,\ u' = \frac{u}{1 - au},\ g' = \frac{g}{(1 - av)^3}.$$

7. Stochastic differential equations

Stochastic differential equations are in general used to describe random phenomena. There are a lot of different classes of stochastic processes, and Wiener processes in particular are often applied to defining a wide class of stochastic differential equations. In contrast to deterministic differential equations, only few attempts to apply group analysis to stochastic differential equations can be found in the literature [120, 1, 50, 102, 173, 174, 49].

Before defining an admitted symmetry for stochastic differential equations, we give a short introduction to stochastic differential equations.

Let Ω be a set of elementary events ω, $\mathcal{F}$ a σ-algebra of subsets of Ω and $\mathcal{P}$ be a probability (or a probability measure) on $\mathcal{F}$. The triple $(\Omega, \mathcal{F}, \mathcal{P})$ is called a probability space. Let $\mathcal{B}$ be the Borel σ-algebra on R^N. A mapping $X : \Omega \to R^N$ is called an N-dimensional random variable if for each $B \subset \mathcal{B}$ the set $X^{-1}(B) \in \mathcal{F}$. A collection $\{X(t)\}_{t\geq 0}$ of random variables is called a stochastic process.

A scalar (standard) Brownian motion or Wiener-Lévy process is a stochastic process $B(t)$ satisfying the following properties:

(i) $\mathcal{P}(\{B(0) = 0\}) = 1$;

(ii) for any finite partition $\{t_i\}_{i=0}^n$ $t_i < t_{i+1}$ of $I = [0, T],\ T > 0$, the random variables $B(t_{i+1}) - B(t_i)$ are independent;

(iii) for all $t, s \in I$, the probability distribution $B(t) - B(s)$ is Gaussian with $E(B(t) - B(s)) = 0$ and $E([B(t) - B(s)]^2) = \mu^2|t - s|$, where μ is a nonzero constant.

An N-dimensional Brownian motion is a stochastic process $B(t) = (B_1(t), B_2(t), ..., B_N(t))$ where $B_i(t)$, $(i = 1, 2, ..., N)$ are independent scalar Brownian motions.

Let $\{X(t)\}_{t\geq 0}$ be a stochastic process such that $||X(t)|| < \infty$, $\forall t \in I$, where

$$||X(t)|| = \left(\int_\Omega |X(t)|^2 \, d\mathcal{P}\right)^{1/2}.$$

For a partition $t_0 < t_1 < ... < t_n = T$ with $\Delta_n = \max(t_{k+1} - t_k)$ one defines the random variable

$$Y_n = \sum_{i=1}^{n-1} X(t_k) \left(B(t_{k+1}) - B(t_k)\right).$$

If there exists a random variable Y such that

$$\lim_{n\to\infty \Delta_n \to 0} ||Y_n - Y|| = 0,$$

then Y is called the Ito integral of $X(t)$ and is denoted by $\int_{t_0}^T X(t)\, dB_t$.

The stochastic process

$$Y(t) = X(t_0) + \int_{t_0}^t f(s, X(s))dt + \int_{t_0}^t g(s, X(s))dB_s \tag{6.106}$$

is called an Ito process, and formally it is defined by

$$dY_t = f(t, X_t)dt + g(t, X_t)dB_t. \tag{6.107}$$

The vector function $f(t, x) = (f_1(t, x), f_2(t, x), ..., f_N(t, x))$ is called a drift vector, the matrix $g(t, x) = (g_{ij}(t, x))$ is called a diffusion matrix. The equation

$$dX_t = f(t, X_t)dt + g(t, X_t)dB_t. \tag{6.108}$$

is called a stochastic differential equation.

Let us consider a Lie group of transformations with the infinitesimal generator

$$X = \tau(t, x)\partial_t + \xi_i(t, x)\partial_{x_i}. \tag{6.109}$$

Following [173, 174] a Lie group corresponding to (6.109) is called admitted by stochastic differential equation (6.108) if the generator of the Lie group satisfies the determining equations

$$\frac{\partial \xi_i}{\partial t} + \frac{\partial \xi_i}{\partial x_j} f_j - \frac{\partial f_i}{\partial x_j}\xi_j - \tau\frac{\partial f_i}{\partial t} - f_i(\frac{\partial \tau}{\partial t} + \frac{\partial \tau}{\partial x_j} f_j)$$
$$+\frac{1}{2}\left(\frac{\partial^2 \xi_i}{\partial x_j \partial x_k} - f_i \frac{\partial^2 \tau}{\partial x_j \partial x_k}\right) g_{j\alpha} g_{k\alpha} = 0,$$

$$\frac{\partial \xi_i}{\partial x_j} g_{j\alpha} - \frac{\partial g_{i\alpha}}{\partial x_j} \xi_j - \tau \frac{\partial g_{i\alpha}}{\partial t} - \frac{g_{i\alpha}}{2} \left(\frac{\partial \tau}{\partial t} + \frac{\partial \tau}{\partial x_j} f_j + \frac{1}{2} \frac{\partial^2 \tau}{\partial x_j \partial x_k} g_{j\beta} g_{k\beta} \right) = 0,$$

$$g_{j\alpha} \frac{\partial \tau}{\partial x_j} = 0.$$

The justification of the determining equations is given on the basis of the Ito formula: the evolution of a scalar function $F(t, x)$ satisfies the formula

$$dF = \left(\frac{\partial F}{\partial t} + f_j \frac{\partial F}{\partial x_j} + \frac{1}{2} g_{j\alpha} g_{k\alpha} \frac{\partial^2 F}{\partial x_j \partial x_k} \right) dt + \left(\frac{\partial F}{\partial x_j} g_{j\alpha} \right) dB_\alpha.$$

Chapter 7

SYMBOLIC COMPUTER CALCULATIONS

One of the features of a compatibility analysis of differential equations is the extensive analytical manipulations involved in the calculations. These manipulations consist of sequentially executing such operations as prolongations of a system, substitution of complicated expressions, and matrix calculations. However, the cumbersome part of these calculations (or certainly part of it) can be entrusted to a computer. With the advent of sophisticated programming languages, applications of computer symbolic calculations became a reality[1]. Symbolic manipulation programs are capable of doing infinite precision rational arithmetic, algebraic simplification, expanding and factoring, differentiating and integrating, finding greatest common denominators and other operations. Computer algebra systems[2] have become an important computational tool. It is worth to mentioning that improvements in software and hardware encourage more extensive use of symbolic manipulation technology. The goal of this chapter is to show the necessity and usefulness of using computer symbolic calculations when studying the analysis of compatibility[3]. This is demonstrated by solving the problem of linearization of a third order ordinary differential equation.

Computer algebra systems can be conveniently divided into two categories, special purpose and general purpose. The systems Axiom, Derive, Macsyma, Maple, Mathematica, MuPAD and Reduce are the general purpose symbol systems. Some of these packages integrate a numeric and symbolic computational engine, a graphics system, a programming language, a documentation system, and advanced connectivity to other applications.

---

[1]An excellent discussion about computer algebra (past, present and future) is given in [33]

[2]Symbol manipulation programs are also called computer algebra systems.

[3]This chapter has to be considered as giving only introductory remarks on this subject. The author thinks that symbolic calculations tremendously simplify the task of obtaining new results in the area of analytical research. This is the main reason for including this chapter in the book.

A large body of literature exists on the topic of application of computer algebra systems. An excellent survey of the different packages presently available, a discussion of their strengths and applications to group analysis is given by W.Hereman in [72]. Since the study of compatibility is involved in group analysis, part of this survey is devoted to applications of computer algebra systems to study of the compatibility of overdetermined systems of partial differential equations. A review of the implementations of algorithms designed to compute the integrability conditions of an overdetermined system can also be found, for example, in [104, 105]. In this section we only mention some programs of compatibility study which are not included in these reviews.

One of the first programs of compatibility analysis on a computer was the code [154]. This program was realized on a computer Strela: it made some analytical manipulations of the equality of mixed derivatives.

As noticed in section 4.1 there are two approaches for studying compatibility. These approaches are related to the works of E.Cartan and C.H.Riquier.

The Cartan approach [24][4] is based on the calculus of exterior differential forms. The problem of the compatibility of a system of partial differential equations is reduced to the problem of the compatibility of a system of exterior differential forms. An application of the Cartan method requires the transition of an original system of partial differential equations into a system of exterior forms. It is related with the requirement of more memory of a computer that performs the calculations. In general, the running out of computer memory is the weakest point of symbolic calculations.

Another approach for the analysis of compatibility started from Riquier's works [146]. The Riquier methods applied by M.Janet and improved by J.M. Thomas and J.F.Ritt proceeds in a different way. The modern state of this approach is presented in [93, 138]. From the symbolic calculations point of view, the main operations of the study of compatibility in the Riquier approach are: prolongations of a system of partial differential equations (differentiations of complex functions), reducing similar terms, different groupings of selected terms, the calculations of ranks of matrices, and solving linear systems of equations.

The first realization on a computer of the Cartan algorithm [5] was programmed in the computer system named Auto-Analytic [6]. The program [168, 169], realized in the computer language Refal [171], has better characteristics. These two realizations were not capable for solving continuum mechanics problems because of memory lack of a computer. In this sense the second method of compatibility analysis is more preferable. Even the first version of the program [52] realized in Refal was more powerful.

As was noted, the main problem of using a computer for compatibility analysis is the lack of computer memory. For overcoming this obstacle other

[4]See also [44, 16].

approaches were developed. For example, in [53] the step which requires a lot of memory was excluded from the total program [55]: it has to be made by the user before running the program. Another example is the program realized in [54] which was developed for quasilinear systems of equations.

1. Introduction to Reduce

In this section Reduce [69][5] is taken as an example of a computer algebra system[6].

Reduce is a universal system of analytic calculations for solving of engineering and scientific problems. It has the following capabilities:

expansion and ordering of polynomials and rational functions;
substitutions and pattern matching in a wide variety of forms;
automatic and user controlled simplification of expressions;
calculations with symbolic matrices;
arbitrary precision integer and real arithmetic;
facilities for defining new functions and extending program syntax;
analytic differentiation and integration;
factorization of polynomials;
facilities for the solution of a variety of algebraic equations;
facilities for the output of expressions in a variety of formats;
facilities for generating optimized numerical programs from symbolic input;
calculations with a wide variety of special functions;
Dirac matrix calculations of interest to high energy physicists.

Reduce is based on a dialect of Lisp called Standard Lisp. The code of a user can be written in Reduce algebraic mode, Reduce symbolic mode, or in Standard Lisp. The simplest is the Reduce algebraic mode (usually called Reduce). The Reduce symbolic mode (which is also called RLisp) is convenient for solving problems which require an extension of the system. Codes written in Reduce or RLisp are interpreted and evaluated in a program of Standard Lisp, then it is executed and the result is displayed in mathematical form.

1.1 Reduce commands

Reduce is an interactive system. Every Reduce statement is terminated by semicolon ";" or "$", where ";" is used in order to display a result of calculations while "$" is opposite. Let us consider some Reduce commands[7] (Reduce algebraic mode commands) which are sufficient for understanding codes presented in Appendix[8].

---

[5]Reduce is chosen because of the author's preferences. Analysis of the most overdetermined systems presented in the book were made by using Reduce. At present the latest version of Reduce is 3.8.
[6]Maple and Mathematica have similar commands.
[7]Details can be found in [69]
[8]Examples are given from the codes presented in Appendix.

Declarations.

The command **OPERATOR** declares prefix operators. This declaration allows the use of operators as functions.

Example. **OPERATOR coeffin,invar,ua,sk;**

In calculations the operators **coeffin,yy,invar** can be used as **coeffin(j), invar(3), ua(1,3,0,0)**.

The command **DEPEND** declares dependence for the purpose of differentiation.

Example. **DEPEND ff,uk(1),ua(1,2,0,0);**

It declares that the first variable **ff** depends on others: **uk(1)** and **ua(1,2,0,0)**.

The command **NODEPEND** is opposite to the **DEPEND**.

Example. **NODEPEND ff,uk(1);**

Assignment and substitutions.

The command <lhs>**:=**<rhs> assigns the left hand side the value of a the right hand side. Here the left hand side can be a simple variable or an array element, the right hand side is an expression which is assigned to the left hand side. The assignment can be cancelled by the command **CLEAR**.

Example. **ms := ms+1;**

The value **ms** is cancelled by the command: **CLEAR ms**.

The command **SUB** is a local substitution: it replaces some variables only in this command.

Example. **SUB(ff=sskk(kj),sexpr3);**

It replaces the variables **ff** by **sskk(kj)**, and then evaluates the value **sexpr3**.

The substitution **LET** is a global substitution, which is cancelled by the command **CLEAR**.

Example. **LET ua(1,0,2,0)**3=uk(1)+coeffin(3)**2+1;**

Conditional statements.

The conditional statements are similar to many computer languages.

Example. **IF sk(1)=0 THEN ua(1,0,2,0) := uk(1);**

The **FOR** statements are used for iterating over number or lists. The various forms of loops can be used. Some of them are presented in the examples.

Example. **FOR j:=1:nua DO DEPEND ff,ua(j);**

It declares that the variable **ff** depends on the variables **ua(1),ua(2),. . . ,ua(nua)**.

Example. **WHILE NOT(ssexpr2={}) DO BEGIN**

< body of the loop >

END;

It is used when the number of repetitions is not known in advance.

Example. **FOR ALL k,l,m LET ua(k,l,m,1)=0, ua(k,l,m,2)=0;**

It assigns 0 for all variables **ua(k,l,m,1)** and ua(k,l,m,2).

Differentiation.

Example. **DF(ff,ua(2,1,1,0));**

The command **DF** performs differentiation of **ff** with respect to **ua(2,1,1,0)**.

Parts of expressions.

These commands are useful for working with parts of expressions. They are, for example, **FIRST, SECOND, PART, DEN, NUM**. The commands **FIRST, SECOND, PART** take part of a list. The command **DEN, NUM** take, respectively, denominator and numerator of expressions.

Output of expressions.

There are many switches which control the output format of expressions. They also assist when presenting result in a more convenient form. Let us mention two of them: **FACTOR** and **NAT**. The switch **FACTOR** displays a polynomial in a factorized form. The switch **NAT** displays expressions in the form which can be used for the next run of the program. Switches are turned on by the statement **ON**, and turned off by the statement **OFF**. The format of output also can be changed by the command **FACTOR**, which differs from the switch **FACTOR**. It displays a polynomial in the normal polynomial form with respect to selected variables. Cancelling the command **FACTOR** is performed by the command **REMFAC**.

Example. **ON FACTOR;**

It turns the switch **FACTOR** on. The command **FACTOR ua(1);** displays the polynomial expression $x * ua(1) * *2 + x * *2 * ua(1) * *2 + 10 * x * y$ in the form: $ua(1) * *2 * x * (x + 1) + 10 * x * y$.

1.2 Some remarks

After deciding to solve a problem by a computer algebra system, the original problem becomes a new problem. This problem includes writing a code and obtaining the final result. Sometimes after unsuccessful struggling with a computer, user returns to the original problem and finds an analytical solution of the problem without using a computer. There are also problems that humans can do easily and the programs are incapable of doing. Although using computer algebra system allows experimenting in analytical calculations: this essentially accelerates the solution of a problem.

Another remark is related with the task of writing a complete program which solves all possible problems. It is easier to separate the original problem into steps, and to use a computer algebra system for some of the steps. Sometimes there are parts of the original problem that cannot be programmed. For example, in compatibility theory this is the problem of finding a quasiregular coordinate system. Sometimes, only using computer algebra system for one step is enough to solve the problem. For example, in finding an admitted Lie group the most cumbersome step is obtaining of a system of determining equations. Analysis of compatibility of these overdetermined systems can be made in a few runs of the program which constructs determining equations: after each run of the program a user analyzes the obtained results, chooses the simplest relations, includes them into the program, and runs the program one more time. This strategy gives more results in the analysis of overdetermined systems of determining equations than using a program which makes a complete analysis of obtaining an admitted Lie group.

One feature of the application of computer algebra systems is intermediate swelling of the calculation. This leads to the requirement of large amounts of memory and time for running a code. Thus intermediate analysis for simplifying results of calculations can assists. There is an opinion that if the final result is too cumbersome, then it is necessary to recheck the accuracy of a code.

The last remark is concerned with the testing of a code. The more testing steps involved in a code, the more reliable it is. Ideally a code is tested of each step.

1.3 Example of a code

In the Appendix the code of procedures for solving a linear partial differential equation of a special type is presented. Let us integrate a linear partial differential equation

$$f_x + a_1 f_{y_1} + a_2 f_{y_2} + ... + a_n f_{y_n} = 0, \tag{7.1}$$

where $f = f(x, y_1, y_2, ..., y_n)$, the coefficients a_i, $(i = 1, 2, ..., n)$ are linear functions of the independent variables $y_1, y_2, ..., y_{i-1}$:

$$a_i = \beta_{i,1}(x)y_1 + \beta_{i,2}(x)y_2 + ... + \beta_{i,i-1}(x)y_{i-1} + \gamma_i(x).$$

The characteristic system for equation (7.1) is

$$\frac{dx}{1} = \frac{dy_1}{\gamma_1(x)} = \frac{dy_2}{\beta_{2,1}(x)y_1 + \gamma_2(x)} = \frac{dy_3}{\beta_{3,1}(x)y_1 + \beta_{3,2}(x)y_2 + \gamma_3(x)} = ...$$

Integration of the characteristic system can be made by sequentially calculating the invariants J_i. To obtain the invariants one can use the following algorithm. This algorithm is like the well-known sweep method of solving a three diagonal system of linear equations. At first one finds the variables

$$y_1 = J_1 + F_1(x), \ y_2 = J_2 + F_2(x, J_1), \ldots, y_n = J_n + F_n(x, J_1, J_2, \ldots, J_{n-1}),$$

where $F_i(x, J_1, J_2, \ldots, J_{i-1}) = \int a_i(x, y_1(x), y_2(x) \ldots, y_{i-1}(x))\, dx$ with the variables $y_1(x), y_2(x), ..., y_{i-1}(x)$ found in the previous step of calculations. Here the variables J_i, $(i = 1, 2, ..., n-1)$ are considered as constant variables. The invariants J_i, $(i = 1, 2, ..., n)$ are defined by the inverse substitutions

$$J_n = y_n - F_n, \ J_{n-1} = y_{n-1} - F_{n-1}, ..., \ J_1 = y_1 - F_1.$$

The procedures for solving equation (7.1) are used in the program for simplifying (solving) a system of linear homogeneous equations. After solving one equation of the type (7.1) the other equations of the system are changed according to the invariants J_i, $(i = 1, 2, ..., n)$ that were found which become the new independent variables. The program was used for finding invariants of equivalence groups which are considered in the next section.

For example, let the equation to be solved be fu(17) = 0, where

```
fu(17):= - df(ff,ua(6,1,1,0))- 3*df(ff,ua(5,0,2,0))$
```

with the function ff depending on the variables ua(6,1,1,0),ua (5,0,2,0). All independent variables of the function ff are collected in the list full_list_ind, which

Before applying the procedure one has to define the following variables:

```
full_list_ind, nequat, ms, excludings, nuk
```

Here the variable full_list_ind defines the list of the independent variables of the function ff, the number nequat is the number of the equations fu(j) = 0, (j = 1, 2, ..., nequat) for solving, ms is the number of parametric variables sk(j), (j = 1, 2, ..., ms): the function ff does not depend on these variables), the list excludings defines the set of variables which are excluded from the calculations, nuk is the number of the invariants J_i, (i = 1, 2, ..., nuk) after solving the previous equations fu(j)= 0. The number ms serves for splitting remaining equations with respect to the variables sk(j), (j=1,2,...,ms). These variables play the role of the variable x after integrating equation (7.1). The list excludings and the number nuk one needs for reconstituting the invariants J_k, (k = 1, 2, ..., nuk), which are denoted in the program by uk(j) through the original independent variables. For starting, these numbers and the list excludings can be assigned:

```
ms:=0;
nuk:=1;
excludings:={};
```

Application of the procedure is given by the following commands

```
%-------------------------------------------------------
1:   sexpr3:=fu(17);
2:   choose_first_ind:=ua(6,1,1,0);
3:   ssresult:=the_main(choose_first_ind, sexpr3,
     nuk, full_list_ind)$
4:   nuk:=first ssresult$
5:   full_list_ind:=second ssresult$
6:   nfull:=length full_list_ind;
7:   fu(17);
%-------------------------------------------------------
```

The first command assigns the expression `fu(17)` for solving. The second command chooses the leading variable for solving: this variable corresponds to x in equation (7.1), the other variables are ordered by the program according to the form of equation[9] (7.1). The third command applies the procedure `the_main` with the defined parameters:

`choose_first_ind, sexpr3, nuk, full_list_ind.`

The result of the calculations is collected in the list `ssresult`. This list is used for the next run of the procedure `the_main`. The last command of the example is for checking: the expression `fu(17)` has to be equal to zero at this step.

2. Application of a computer algebra system to the problem of linearization a third order ordinary differential equation

2.1 Introduction to the problem

Many methods of solving differential equations use a change of variables that transforms a given differential equation into another equation with known properties. Since the class of linear equations is considered to be the simplest class of equations, there is the problem of transforming a given differential equation into a linear equation. This problem, which is called a linearization problem, is a particular case of the equivalence problem. The equivalence problem can be formulated as follows. Let a set of invertible transformations be given. One can introduce the equivalence property according to these transformations: two differential equations are equivalent if there is a transformation of the given set which transforms one equation into another. The equivalence property separates all differential equations into classes of equivalent equations. Assume that there are two equations. The equivalence problem is: do these two equations belong to the same class. This problem involves a number of related problems such as defining a class of transformations, finding invariants of these transformations, obtaining the equivalence criteria, and constructing the transformation.

For the linearization problem one studies the classes of equations equivalent to linear equations. The first linearization problem for ordinary differential equations was solved by S.Lie [94]. He found the general form of all ordinary differential equations of second order that can be reduced to a linear equation by changing the independent and dependent variables. He showed that any linearizable second-order equation should be at most cubic in the first-order derivative and provided a linearization test in terms of its coefficients. The linearization criterion is written through relative invariants of the equivalence group. A.M.Tresse [170] treated the equivalence problem for second order

[9] If it is not possible, the program will be cycled.

ordinary differential equations in terms of relative invariants of the equivalence group of point transformations. In [74] an infinitesimal technique for obtaining relative invariants were applied to the linearization problem.

S.Lie also noted that all second order equations can be transformed to each other by means of contact transformations, and that this is not so for third order equations.

A different approach for tackling the equivalence problem of second order ordinary differential equations was developed by E.Cartan [23]. The idea of his approach was to associate with every differential equation a uniquely defined geometric structure of a certain form. The Cartan approach was further applied by S.-S.Chern [25] to third order differential equations. Since none of the conditions given in [25] are implicit expressions that could be used as tests for deciding about the type of the studied equation, in a series of articles [59, 66, 36, 35, 122] the linearization problem was also considered. Linearization with respect to point transformations is studied in [59], with respect to contact transformations in [13, 66, 36, 35, 122]. The linearization problem were also investigated with respect to generalized Sundman transformation [42, 41].

2.1.1 Second order equation: the Lie linearization test

For simplicity of understanding the problem let us start from a second order ordinary differential equation.

Lemma 7.1. *(S.Lie [94]). Any second order ordinary differential equation obtained from a linear equation by a change of the independent and dependent variables is cubic in the first derivative.*

Proof[10].

Notice that the Laguerre-Forsyth canonical form of a second order linear equation with the independent variable t and the dependent variable u is

$$u'' = 0. \tag{7.2}$$

Assume that the equation $y'' = F(x, y, y')$ is obtained from the linear ordinary differential equation (7.2) by the change of the variables

$$t = \varphi(x, y), \quad u = \psi(x, y). \tag{7.3}$$

The derivatives are changed by the formulae

$$\begin{array}{c} u' = g(x, y, y') = \frac{D_x\psi}{D_x\varphi} = \frac{\psi_x + y'\psi_y}{\varphi_x + y'\varphi_y}, \\ u'' = P(x, y, y', y'') = \frac{D_x g}{D_x\varphi} = \frac{g_x + y' g_y + y'' g_{y'}}{\varphi_x + y'\varphi_y} = \\ = (\varphi_x + y'\varphi_y)^{-3}(y''(\varphi_x\psi_y - \varphi_y\psi_x) + y'^3(\varphi_y\psi_{yy} - \varphi_{yy}\psi_y) + \\ y'^2(\varphi_x\psi_{yy} - \varphi_{yy}\psi_x + 2(\varphi_y\psi_{xy} - \varphi_{xy}\psi_y)) + \\ y'(\varphi_y\psi_{xx} - \varphi_{xx}\psi_y + 2(\varphi_x\psi_{xy} - \varphi_{xy}\psi_x)) + \varphi_x\psi_{xx} - \varphi_{xx}\psi_x). \end{array} \tag{7.4}$$

[10]See also the proof in [74].

Here

$$\Delta = \varphi_x \psi_y - \varphi_y \psi_x \neq 0,$$

subscript means a derivative, for example, $\varphi_x = \partial\varphi/\partial y$, $\varphi_y = \partial\varphi/\partial y$. Since the Jacobian of the change of variables $\Delta \neq 0$, equation (7.2) becomes

$$y'' + a(x, y)y'^3 + b(x, y)y'^2 + c(x, y)y' + d(x, y) = 0, \qquad (7.5)$$

where

$$\begin{aligned} a &= \Delta^{-1}(\varphi_y \psi_{yy} - \varphi_{yy}\psi_y), \quad b = \Delta^{-1}(\varphi_x \psi_{yy} - \varphi_{yy}\psi_x + 2(\varphi_y \psi_{xy} - \varphi_{xy}\psi_y)), \\ c &= \Delta^{-1}(\varphi_y \psi_{xx} - \varphi_{xx}\psi_y + 2(\varphi_x \psi_{xy} - \varphi_{xy}\psi_x)), \quad d = \Delta^{-1}(\varphi_x \psi_{xx} - \varphi_{xx}\psi_x). \end{aligned} \qquad (7.6)$$

•

If a second order ordinary differential equation is linearizable, then it has the form (7.5). The mapping of this equation into a linear equation is reconstituting by finding the functions $\varphi(x, y)$ and $\psi(x, y)$ that satisfy the relations (7.6). Since for a given equation there are only two unknown functions $\varphi(x, y)$ and $\psi(x, y)$, equations (7.6) form an overdetermined system of partial differential equations. Let us analyze the compatibility of this system.

First assume that[11] $\varphi_y = 0$. From relations (7.6) one defines

$$a = 0, \ \psi_{yy} = \psi_y b, \ 2\psi_{xy} = \varphi_x^{-1}\psi_y\varphi_{xx} + c, \ \psi_{xx} = \varphi_x^{-1}\psi_x\varphi_{xx} + \psi_y d. \qquad (7.7)$$

Comparing the mixed derivatives $(\psi_{xy})_y = (\psi_{yy})_x$ and $(\psi_{xy})_x = (\psi_{xx})_y$, one finds

$$c_y = 2b_x, \quad \varphi_x^{-2}(2\varphi_x\varphi_{xxx} - 3\varphi_{xx}^2) = 4(d_y + bd) - (2c_x + c^2). \qquad (7.8)$$

Because $\varphi_y = 0$, differentiating the last equation with respect to y , one obtains

$$d_{yy} - b_{xx} - b_x c + b_y d + d_y b = 0.$$

Thus, a second order ordinary differential equation of the form (7.5) is linearizable with the function $\varphi = \varphi(x)$ if the coefficients of this equation satisfy the conditions

$$a = 0, \quad c_y = 2b_x, \quad d_{yy} - b_{xx} - b_x c + b_y d + d_y b = 0. \qquad (7.9)$$

The functions $\varphi(x)$ and $\psi(x, y)$ are restituted by solving the involutive overdetermined system of equations (7.7), (7.8).

Relations (7.6) in the case $\varphi_y \neq 0$ are analyzed similarly, but the process is more cumbersome. In fact, from (7.6) one finds

$$\begin{aligned} \varphi_y \psi_{yy} &= (\varphi_{yy}\psi_y + a\Delta), \quad 2\varphi_y^2\psi_{xy} = 2\varphi_{xy}\varphi_y\psi_y - \varphi_{yy}\Delta - (a\varphi_x - b\varphi_y)\Delta, \\ \varphi_y^2\psi_{xx} &= 2\varphi_{xy}\varphi_y\psi_x - \varphi_x\varphi_{yy}\psi_x - \varphi_x^2\psi_x a + \varphi_x\varphi_y\psi_x b + \varphi_y^2(\psi_y d - \psi_x c), \\ \varphi_y^2\varphi_{xx} &= 2\varphi_{xy}\varphi_x\varphi_y - \varphi_x^2\varphi_{yy} - \varphi_x^3 a + \varphi_x^2\varphi_y b - \varphi_x\varphi_y^2 c + \varphi_y^3 d. \end{aligned} \qquad (7.10)$$

[11] A transformation with $\varphi = \varphi(x)$ is called a fiber preserving point transformation.

From the equations $(\psi_{xy})_y = (\psi_{yy})_x$ and $(\psi_{xy})_x = (\psi_{xx})_y$, one gets

$$
\begin{aligned}
2\varphi_y\varphi_{yyy} &= 3(\varphi_{yy}^2 - 2\varphi_{xy}\varphi_y a + 2\varphi_x\varphi_{yy} a + \varphi_x^2 a^2) - 2\varphi_x\varphi_y(a_y + ab) \\
&\quad + \varphi_y^2(2b_y - 4a_x + 4ac - b^2), \\
6\varphi_y^2\varphi_{xyy} &= 3(4\varphi_{xy}\varphi_{yy}\varphi_y - \varphi_x\varphi_{yy}^2 + 2\varphi_x\varphi_{yy}\varphi_y b - 2\varphi_{xy}\varphi_y^2 b) + \\
&\quad 3\varphi_x^3 a^2 + 3\varphi_x\varphi_y^2(-2a_x + 2ac - b^2) + 2\varphi_y^3(-b_x + 2c_y + 3ad).
\end{aligned}
\tag{7.11}
$$

Forming the mixed derivatives $(\varphi_{xyy})_y = (\varphi_{yyy})_z$ and $(\varphi_{xx})_{yy} = (\varphi_{xyy})_x$, one obtains

$$
\begin{aligned}
H &= 3a_{xx} - 2b_{xy} + c_{yy} - 3a_x c + 3a_y d + 2b_x b - 3c_x a - c_y b + 6d_y a = 0, \\
K &= b_{xx} - 2c_{xy} + 3d_{yy} - 6a_x d + b_x c + 3b_y d - 2c_y c - 3d_x a + 3d_y b = 0.
\end{aligned}
\tag{7.12}
$$

Conditions (7.9) form a particular case of the relations (7.12): they are selected by the way of finding a linearizing transformation.

Theorem 7.1. *(S.Lie). A second order ordinary differential equation is linearizable if and only if it has the form (7.5) with the coefficients satisfying the conditions (7.12).*

2.1.2 Invariants of the equivalence group

The form (7.5) is invariant with respect to any change of the independent and dependent variables (7.3). Hence, there is the problem of finding all invariants of the equivalence group of equation (7.5). Using the algorithm for finding an equivalence group with arbitrary elements $a(x,y)$, $b(x,y)$, $c(x,y)$ and $d(x,y)$, one obtains the infinitesimal generator of the equivalence group

$$
\begin{aligned}
X = \xi\partial_x + \eta\partial_y + \left(\xi_{yy} + (\xi_x - 2\eta_y)a - b\xi_y\right)\partial_a + \\
\left(2\xi_{xy} - \eta_{yy} - 3a\eta_x - b\eta_y - 2c\xi_y\right)\partial_b + \\
\left(\xi_{xx} - 2\eta_{xy} - 2b\eta_x - c\xi_x - 3\xi_y d\right)\partial_c + \left((\eta_y - 2\xi_x)d - c\eta_x - \eta_{xx}\right)\partial_d,
\end{aligned}
$$

where the functions $\xi(x,y)$, $\eta(x,y)$ are arbitrary. Because the property for an equation to be linearizable with respect to a change of the independent and dependent variables is invariant with respect to any change of the independent and dependent variables, the equations[12] $H = 0$ and $K = 0$ have to be invariants of the equivalence group. Let us find all invariants of second order of the equivalence group

$$
F = F(x, y, v, v_x, v_y, v_{xx}, v_{xy}, v_{yy}),
$$

where $v = (a, b, c, d)$. For this purpose the generator X is prolonged up to second order derivatives v_{xx}, v_{xy}, v_{yy}. The prolonged generator $\widetilde{X}$ acts on the

[12]The functions H and K are defined by formulae (7.12).

function F:

$$\widetilde{X}F = 0. \tag{7.13}$$

Since the functions $\xi(x, y)$, $\eta(x, y)$ are arbitrary, equations (7.13) are split with respect to them and their derivatives. Thus one obtains 30 equations `fu(i)=0`, ($i = 1, 2, ..., 30$) of first order for the function $F(x, y, v, v_x, v_y, v_{xx}, v_{xy}, v_{yy})$. Solution of these equations is found by a computer program explained in the previous section[13]. First one chooses a simple equation, for example,

```
fu(28)=-df(ff,ua(4,2,0,0))*ua(3)-df(ff,ua(4,1,0,0))-
        2*df(ff,ua(3,2,0,0))*ua(2)-
        3*df(ff,ua(2,2,0,0))*ua(1);
```

Here the notations are used `ua(1)`$= a$, `ua(2)`$= b$, `ua(3)`$= c$, `ua(4)`$= d$, `ua(j,i,k,0)` are derivatives of `ua(j)` with respect to x (i times) and with respect to y (k times). For solving the equation `fu(28)`$= 0$ one writes the code:

```
sexpr3:=fu(28);
choose_first_ind:=ua(4,1,0,0);
ssresult:=the_main(choose_first_ind,sexpr3,nuk,
full_list_ind)$
nuk:=first ssresult;
full_list_ind:=second ssresult$
nfull:=length full_list_ind;
fu(28);
```

Analyzing the remaining equations, one chooses the next equation to be solved. After some runs of the program one obtains:

```
fu(1):= df(ff,uk(17))*uk(15)$
fu(2):= df(ff,uk(17))*uk(17) +
        2*df(ff,uk(15))*uk(15)$
fu(3):= 2*df(ff,uk(17))*uk(17) +
        df(ff,uk(15))*uk(15)$
fu(4):= df(ff,uk(15))*uk(17)$
```

where

$$\texttt{uk(15)} = K, \quad \texttt{uk(17)} = H.$$

Thus, for finding invariants one has to solve the equations

$$K\frac{\partial F}{\partial H} = 0, \;\; H\frac{\partial F}{\partial H} + 2K\frac{\partial F}{\partial K} = 0, \;\; 2H\frac{\partial F}{\partial H} + K\frac{\partial F}{\partial K} = 0, \;\; H\frac{\partial F}{\partial K} = 0.$$

These equations were obtained in [74]. The functions H, K are relative invariants of the equivalence group.

---

[13] See also Appendix 1.

2.2 Third order equation: linearizing point transformations

In this section the necessary and sufficient conditions for linearization of third-order equations by means of point transformations are presented [14].

Using the Laguerre-Forsyth canonical form, any linear third-order equation with the independent variable t and the dependent variable u can be written in the form

$$u''' + \alpha(t)u = 0. \tag{7.14}$$

The change of the independent and dependent variables (7.3) leads to the change of the derivatives (7.4) and

$$\begin{aligned} u''' &= Q(x, y, y', y'', y''') = \frac{D_x P}{D_x \varphi} \\ &= \frac{\Delta}{(\varphi_x + y'\varphi_y)^5} \left((\varphi_x + y'\varphi_y)\, y''' - 3\varphi_y\, (y'')^2 \right) + \cdots, \end{aligned}$$

here the other terms being at most linear in y''. Hence, the transformations (7.3) with $\varphi_y = 0$ and $\varphi_y \neq 0$, respectively, provide two distinctly different candidates for linearization.

If $\varphi_y = 0$, equation (7.14) becomes

$$y''' + (A_1\, y' + A_0)y'' + B_3\, y'^3 + B_2\, y'^2 + B_1\, y' + B_0 = 0, \tag{7.15}$$

where $A_i = A_i(x, y)$ and $B_i = B_i(x, y)$ are expressed through the functions $\varphi(x, y)$, $\psi(x, y)$, and their derivatives:

$$A_1 = 3\, \psi_y^{-1}\, \psi_{yy}\,, \quad A_0 = 3\, (\varphi_x \psi_y)^{-1} (\varphi_x\, \psi_{xy} - \psi_y\, \varphi_{xx}), \tag{7.16}$$

$$B_3 = \psi_y^{-1}\, \psi_{yyy}\,, \qquad B_2 = 3\, (\varphi_x \psi_y)^{-1} (\varphi_x\, \psi_{xyy} - \psi_{yy}\, \varphi_{xx}), \tag{7.17}$$

$$B_1 = (\varphi_x^2 \psi_y)^{-1} (3\varphi_{xx}^2 \psi_y - \varphi_{xxx}\varphi_x \psi_y - 6\varphi_{xx}\varphi_x \psi_{xy} + 3\varphi_x^2 \psi_{xxy}), \tag{7.18}$$

$$B_0 = (\varphi_x^2 \psi_y)^{-1} (3\varphi_{xx}^2 \psi_x - \varphi_{xxx}\varphi_x \psi_x - 3\varphi_{xx}\varphi_x \psi_{xx} + \varphi_x^2 \psi_{xxx} + \alpha \psi \varphi_x^5). \tag{7.19}$$

Notice that the point transformation (7.3) with $\varphi_y = 0$ leaves any equation of the form (7.15) in the same class.

If $\varphi_y \neq 0$, setting $r(x, y) = \varphi_x / \varphi_y$, equation (7.14) arrives at the equation:

$$\begin{aligned} y''' + \frac{1}{y' + r} \Big(&-3(y'')^2 + (C_2\, y'^2 + C_1\, y' + C_0)y'' \\ &+ D_5\, y'^5 + D_4\, y'^4 + D_3\, y'^3 + D_2\, y'^2 + D_1\, y' + D_0 \Big) = 0. \end{aligned} \tag{7.20}$$

---

[14] Results of this section were obtained in close collaboration with N.H.Ibragimov [77].

where $r = r(x, y)$, $C_i = C_i(x, y)$ and $D_i = D_i(x, y)$ are functions of x, y. The coefficients of equation (7.20) and the functions $\varphi(x, y)$ and $\psi(x, y)$ in the transformation (7.3) with $\varphi_y \neq 0$ are related by the following equations:

$$C_2 = 3(\varphi_y \Delta)^{-1}(\varphi_y \Delta_y - 2\varphi_{yy}\Delta), \tag{7.21}$$

$$C_1 = 3(\varphi_y \Delta)^{-1}\left((r\,\Delta_y + \Delta_x)\varphi_y - 4\Delta(r\,\varphi_y)_y\right), \tag{7.22}$$

$$C_0 = 3(\varphi_y \Delta)^{-1}\left(r\,\varphi_y \Delta_x - 2\varphi_{yy}r^2\Delta - 2\varphi_y r_x \Delta - 2r\,\varphi_y r_y \Delta\right), \tag{7.23}$$

$$D_5 = (\varphi_y \Delta)^{-1}\left(\varphi_y(\varphi_y \psi_{yyy} - \varphi_{yyy}\psi_y) + 3\varphi_{yy}(\varphi_{yy}\psi_y - \varphi_y \psi_{yy}) + \alpha\psi(\varphi_y)^5\right), \tag{7.24}$$

$$\begin{aligned} D_4 &= (\varphi_y\,\Delta)^{-1}\{(4\Delta - 5r\,\varphi_y\psi_y)\varphi_{yyy} + 3[5r\,\varphi_{yy}\psi_y - 5\Delta\,\varphi_{yy}\varphi_y^{-1} \\ &\quad -5r\,\varphi_y\psi_{yy} + 4\Delta_y]\varphi_{yy} + 5r\,\varphi_y^2\psi_{yyy} - 3\varphi_y\Delta_{yy} + 5\alpha\,\psi\,r\,\varphi_y^5\}, \end{aligned} \tag{7.25}$$

$$\begin{aligned} D_3 &= (\varphi_y\,\Delta)^{-1}\{2r(8\Delta - 5r\,\varphi_y\psi_y)\varphi_{yyy} + [6(\Delta_x + 7r\,\Delta_y - 3\Delta r_y) \\ &\quad +30r\,(r\varphi_{yy}\psi_y - 2\Delta\varphi_{yy}\varphi_y^{-1} - r\,\varphi_y\psi_{yy})]\varphi_{yy} + 10r^2{\varphi_y}^2\psi_{yyy} \\ &\quad +3(2\Delta r_{yy} + 4r_y\Delta_y - \Delta_{xy} - 3r\,\Delta_{yy})\varphi_y + 10\alpha\psi r^2\varphi_y^5\}, \end{aligned} \tag{7.26}$$

$$\begin{aligned} D_2 &= (\varphi_y\,\Delta)^{-1}\{2r^2(12\Delta - 5r\,\varphi_y\psi_y)\varphi_{yyy} + 10r^3{\varphi_y}^2\psi_{yyy} \\ &\quad + 6[3r\,\Delta_x + 9r^2\Delta_y - 5r^3\varphi_y\psi_{yy} - \Delta r_x - 8r\,\Delta r_y]\varphi_{yy} \\ &\quad + 30r^2(r\,\psi_y - 3\Delta{\varphi_y}^{-1}){\varphi_{yy}}^2 + 10\alpha\,\psi\,r^3{\varphi_y}^5 \\ &\quad + [4(\Delta r_x)_y + 14r\,\Delta r_{yy} - 16\Delta{r_y}^2 + 8r_y\Delta_x \\ &\quad + 24r r_y\Delta_y - 7r\,\Delta_{xy} - \Delta_{xx} - 10r^2\Delta_{yy}]\varphi_y\}, \end{aligned} \tag{7.27}$$

$$\begin{aligned} D_1 &= (\varphi_y\,\Delta)^{-1}\{r^3(16\Delta - 5r\,\varphi_y\psi_y)\varphi_{yyy} + 5r^4\,{\varphi_y}^2(\psi_{yyy} + \alpha\,\psi\,{\varphi_y}^3) \\ &\quad +3r\,[6r\,\Delta_x + 10r^2\Delta_y - 5r^3\varphi_y\psi_{yy} - 4\Delta r_x - 14r\,\Delta r_y]\varphi_{yy} \\ &\quad +15r^3(r\,\psi_y - 4\Delta{\varphi_y}^{-1}){\varphi_{yy}}^2 + [(r_{xx} + 6r\,r_{xy} + 11r^2 r_{yy} \\ &\quad -19r\,{r_y}^2 - 13r_x r_y)\Delta + 3r_x\Delta_x + 5r\,r_x\Delta_y + 13r\,r_y\Delta_x \\ &\quad +15r^2 r_y\Delta_y - 5r^2\Delta_{xy} - 2r\,\Delta_{xx} - 5r^3\Delta_{yy}]\varphi_y\}, \end{aligned} \tag{7.28}$$

$$\begin{aligned} D_0 &= (\varphi_y\,\Delta)^{-1}\{r^4(4\Delta - r\,\varphi_y\psi_y)\varphi_{yyy} + r^5\,{\varphi_y}^2(\psi_{yyy} + \alpha\,\psi\,{\varphi_y}^3) \\ &\quad +3r^2[2r\,\Delta_x + 2r^2\Delta_y - r^3\psi_{yy} - 2\Delta r_x - 4r\,\Delta r_y]\varphi_{yy} \\ &\quad +3r^4(r\,\psi_y - 5\Delta{\varphi_y}^{-1}){\varphi_{yy}}^2 + [(3r^3 r_{yy} + 2r^2 r_{xy} + r\,r_{xx} \\ &\quad -6r^2{r_y}^2 - 3{r_x}^2 - 7r\,r_x r_y)\Delta + r(3r_x\Delta_x + r\,r_x\Delta_y \\ &\quad +5r\,r_y\Delta_x - r\,\Delta_{xx} + 3r^2 r_y\Delta_y - r^2\Delta_{xy} - r^3\Delta_{yy})]\varphi_y\}. \end{aligned} \tag{7.29}$$

Therefore, it has been shown that every linearizable third-order equation by point transformations belongs either to the class of equations (7.15) with the linear dependence on the second-order derivative or to the class of equations (7.20) that are at most quadratic in y'' with a specific dependence on y'.

2.2.1 The linearization test for equation (7.15)

Consider equation (7.15). The linearizing transformations have the form

$$t = \varphi(x), \quad u = \psi(x, y). \tag{7.30}$$

Theorem 7.2. *Equation (7.15)*

$$y''' + (A_1\, y' + A_0)y'' + B_3\, y'^3 + B_2\, y'^2 + B_1\, y' + B_0 = 0$$

is linearizable if and only if its coefficients obey the following five equations:

$$A_{0y} - A_{1x} = 0, \quad (3B_1 - A_0^2 - 3A_{0x})_y = 0, \tag{7.31}$$

$$3B_2 = 3A_{1x} + A_0 A_1, \quad 9B_3 = 3A_{1y} + A_1^2, \tag{7.32}$$

$$27B_{0yy} = (9B_1 - 6A_{0x} - 2A_0^2)A_{1x} + 9(B_{1x} - A_1\, B_0)_y + 3B_{1y}A_0. \tag{7.33}$$

Provided that the conditions (7.31)-(7.33) are satisfied, the linearizing transformation (7.30) is defined by the third order ordinary differential equation for the function $\varphi(x)$:

$$3(2\chi_x - \chi^2) = 3B_1 - A_0^2 - 3A_{0x} \tag{7.34}$$

and by the following involutive system of partial differential equations for $\psi(x, y)$:

$$3\psi_{yy} = A_1\, \psi_y, \qquad 3\psi_{xy} = (3\chi + A_0)\psi_y, \tag{7.35}$$

$$\psi_{xxx} = 3\chi\, \psi_{xx} + B_0\, \psi_y - \frac{1}{6}(3A_{0x} + A_0^2 - 3B_1 + 9\chi^2)\, \psi_x - \Omega\psi, \tag{7.36}$$

where

$$\chi = \varphi_x^{-1}\varphi_{xx} \tag{7.37}$$

and

$$\Omega = \frac{1}{54}\left(9A_{0xx} + 18A_{0x}A_0 + 54B_{0y} - 27B_{1x} + 4A_0^3 - 18A_0B_1 + 18A_1B_0\right). \tag{7.38}$$

Finally, the coefficient α of the resulting linear equation (7.14) is given by

$$\alpha = \Omega\, \varphi_x^{-3}. \tag{7.39}$$

Remark 7.1. *Another representation of equation (7.33) is $\Omega_y = 0$. This equation follows from equation (7.39) since $\varphi_y = 0$ and hence $\alpha(\varphi(x))_y = 0$.*

Remark 7.2. *Assume that $\Omega \neq 0$. Substituting into equation (7.34) the derivatives φ_{xx} and φ_{xxx}, found from (7.39), equation (7.34) becomes*[15]

$$\alpha^{-8/3}\Omega^{8/3}\left(6\alpha\alpha'' - 7(\alpha')^2\right) = 6\Omega_{xx}\Omega - 7\Omega_x^2 - 9\beta\Omega^2 \qquad (7.40)$$

where $\beta = (3B_1 - A_0^2 - 3A_{0x})/3$, and α' is the derivative of the function $\alpha(t)$ with respect to t.

In [59] it is stated that the conditions (7.32), (7.33) and (7.40) (for $\Omega \neq 0$) provide sufficient conditions. However, for sufficiency one needs to append the condition (7.33). The equation $y''' + y^2 = 0$ provides a counterexample[16] *of sufficiency the conditions (7.32), (7.33) and (7.40).*

Proof.

The proof of the linearization theorem requires the study of integrability conditions for the unknown functions $\varphi(x)$ and $\psi(x, y)$. The functions $\varphi(x)$ and $\psi(x, y)$ satisfy equations (7.16)- (7.19) with given coefficients $A_i = A_i(x, y)$ and $B_i = B_i(x, y)$.

The expressions (7.16) for A_1 and A_0 can be rewritten in the following form:

$$A_0 = 3\,\frac{W_x}{W}, \qquad A_1 = 3\,\frac{H_y}{H},$$

where $W = \psi_y/\varphi_x$ and

$$H = \psi_y. \qquad (7.41)$$

Equation (7.41) and the definition of W yield:

$$\varphi_x = HW^{-1}. \qquad (7.42)$$

Differentiating equation (7.42) with respect to y, one finds

$$W_y = H_y W H^{-1}. \qquad (7.43)$$

Equations (7.17) and (7.18) are rewritten:

$$B_3 = H^{-1}H_{yy}, \quad B_2 = 3W^{-1}H^{-2}(H_{xy}HW - H_xH_yW + H_yW_xH)$$

and

$$B_1 = (HW)^{-2}(2H_{xx}HW^2 - 3H_x^2W^2 + 2H_xW_xHW + W_{xx}H^2W + W_x^2H^2),$$

[15] Equation (7.40) was obtained in [59] with a misprint, namely, the factor 7 was missing.

[16] A detailed study of this example is given later in the subsection devoted to examples.

respectively. Equation (7.19) for B_0 becomes

$$S \equiv (H^2WW_{xx} - HW^2H_{xx} + 3W^2H_x^2 - 4HWH_xW_x + H^2W_x^2)W\psi_x \\ +3(HW_x - WH_x)HW^2\psi_{xx} + H^2W^3\psi_{xxx} - H^3W^3B_0 + \alpha H^5\psi = 0. \tag{7.44}$$

Noticing that the equation

$$HS_y - 5H_yS = 0$$

does not contain the functions $\varphi(x)$, $\psi(x, y)$ and their derivatives, one finds the coefficient α from this equation:

$$\alpha = H^{-6}W\Big(H^3W^2B_{0y} - H_{xxx}H^2W^2 + 4H_{xx}H_xHW^2 - 3H_{xx}W_xH^2W - \\ 3H_x^3W^2 + 4H_x^2W_xHW - H_xW_{xx}H^2W - H_xW_x^2H^2 + H_yB_0H^2W^2\Big). \tag{7.45}$$

Since $\varphi = \varphi(x)$, the derivative $\alpha_y = 0$, and equation (7.45) yields:

$$W^2\Big(H^3B_{0yy} + H^2H_yB_{0y} + H^3(H^{-1}H_y)_yB_0 - 3H_{xy}H_x^2 + HH_{xy}H_{xx} - \\ H^2H_{xxxy} + HH_{xxx}H_y + 3HH_{xxy}H_x - 4H_{xx}H_xH_y + 3H^{-1}H_x^3H_y\Big) - \\ W\Big(4H_x^2H_yW_x - HH_xH_yW_{xx} - 4HH_{xy}H_xW_x - 3HH_{xx}H_yW_x + \\ H^2H_{xy}W_{xx} + 3H^2H_{xxy}W_x\Big) - HW_x^2(HH_{xy} - H_yH_x) = 0. \tag{7.46}$$

Rewriting equation (7.42) in the form

$$H = W\varphi_x \tag{7.47}$$

and using $\varphi = \varphi(x)$, the representations for B_2 and B_3 become

$$B_2 = 3W^{-1}W_{xy}, \quad B_3 = W^{-1}W_{yy}.$$

The representation for B_1, after denoting $\chi = \varphi_x^{-1}\varphi_{xx}$, leads to equation (7.34):

$$3(2\chi' - \chi^2) = 3B_1 - 3A_{0x} - A_0^2. \tag{7.48}$$

Since φ, and hence χ does not depend on y, differentiation of equation (7.48) with respect to y yields the second equation (7.31):

$$(3B_1 - A_0^2 - 3A_{0x})_y = 0. \tag{7.49}$$

Using equation (7.43) and the expressions for A_0 and A_1, one determines the first-order derivatives of W:

$$W_x = \frac{1}{3}WA_0, \quad W_y = \frac{1}{3}WA_1. \tag{7.50}$$

Equations (7.32) become:

$$3B_2 = 3A_{1x} + A_0A_1, \quad 9B_3 = 3A_{1y} + A_1^2.$$

Equating the mixed derivatives $(W_x)_y$ and $(W_y)_x$ obtained from equations (7.50), one arrives at the first equation (7.31):

$$A_{0y} = A_{1x}. \tag{7.51}$$

Using equations (7.41), (7.50), the functions H and W, together with their derivatives, are eliminated from equation (7.46) and this equation becomes (7.33). Equations (7.35) are provided by (7.50) whereas equation (7.36) is obtained from (7.44). Thus, all third order derivatives of the function $\psi(x, y)$ can be determined: equation (7.36) gives ψ_{xxx}, the remaining derivatives ψ_{xxy}, ψ_{xyy} and ψ_{yyy} are obtained by differentiating equations (7.35). Calculations show that all mixed fourth order derivatives found from these expressions by different way are equal. In fact, it is necessary to check only the equation $(\psi_{xxx})_y = (\psi_{xy})_{xx}$. This means that equations (7.35)-(7.36) for $\psi(x, y)$ are completely integrable. Finally, equations (7.38)-(7.39) are obtained from (7.45). •

The search for second order invariants of the equivalence group (after running the program) leads to the following equations

$$\begin{gathered}S_1 = F_{J_1}J_1 + F_{J_2}J_2 + F_{J_5}J_5 + F_{J_6}J_6 + F_{J_7}J_7 + F_{J_8}J_8 + 2F_{J_9}J_9 + 2F_{J_{10}}J_{10} \\ +2F_{J_{11}}J_{11} + 2F_{J_{12}}J_{12} + 3F_{J_{13}}J_{13} + 2F_{J_{14}}J_{14} + 3F_{J_{15}}J_{15} + 3F_{J_{16}}J_{16} = 0,\end{gathered}$$

$$\begin{gathered}S_2 = F_{J_1}J_1 + F_{J_2}J_2 + 2F_{J_3}J_3 + 3F_{J_4}J_4 + 2F_{J_5}J_5 + 2F_{J_6}J_6 + 3F_{J_7}J_7+ \\ 2F_{J_8}J_8 + 2F_{J_9}J_9 + F_{J_{10}}J_{10} + F_{J_{11}}J_{11} + F_{J_{12}}J_{12} + 2F_{J_{14}}J_{14} + F_{J_{15}}J_{15}+ \\ F_{J_{16}}J_{16} = 0,\end{gathered}$$

$$\begin{gathered}S_3 = 6F_{J_5}J_3 + 6F_{J_6}J_3 + 3F_{J_7}J_4 + 3F_{J_8}J_3 + 3F_{J_9}(4J_8 + J_5)+ \\ 3F_{J_{10}}(J_1 - J_2) + 6F_{J_{11}}J_2 + 6F_{J_{12}}J_1 + 3F_{J_{14}}(3J_8 + J_6) + 3F_{J_{15}} \\ (J_{12} - J_{11} + 3J_{10}) + F_{J_{16}}(6J_{10} - J_{11}) = 0,\end{gathered}$$

$$\begin{gathered}S_4 = 3F_{J_9}J_4 + 3J_3(F_{J_{10}} + F_{J_{11}} + F_{J_{12}}) + 3F_{J_{13}}(7J_1 - J_2)+ \\ 3F_{J_{15}}(J_8 + J_6) + F_{J_{16}}(3J_5 - J_8) = 0\end{gathered}$$

$$\begin{gathered}S_5 = 18J_3(F_{J_1} + F_{J_2}) + 18J_4(F_{J_5} + F_{J_6})+ \\ 6F_{J_9}(6J_7 + J_3J_2 + J_3J_1) + 18F_{J_{10}}J_8 + 18F_{J_{11}}(2J_6 - J_8)+ \\ 18F_{J_{12}}(J_6 + J_5 - J_8) + 18F_{J_{13}}(2J_{12} + J_{11} + 3J_{10})+ \\ 3F_{J_{14}}(6J_7 - J_3J_2 - J_3J_1) + 3F_{J_{15}}(6J_9 + 3J_2^2 + 2J_2J_1 - J_1^2)+ \\ 2F_{J_{16}}(-6J_{14} + 9J_9 + J_2^2 + 4J_2J_1 + 3J_1^2) = 0\end{gathered}$$

where $F = F(J_1, J_2, ..., J_{16})$ is the function which defines the invariants. Here

$$\begin{aligned}J_1 &= A_{0y} - B_2 + A_0A_1/3, & J_2 &= A_{1x} - B_2 + A_0A_1/3, \\ J_3 &= A_{1y} - 3B_3 + A_1^2/3, & J_4 &= A_{1yy} - 3B_{3y} + 2A_1B_3 - 2A_1^3/9, \\ J_5 &= A_{0yy} + A_1A_{1x} - 3B_{3x} - A_{0y}A_1/3, & J_6 &= A_{1xy} - 3B_{3x} + 2A_1A_{1x}/3,\end{aligned}$$

$$J_7 = B_{2yy} - 3B_{3xy} + A_1A_{1xy} - A_1^2A_{1x} - A_1B_{2y} + 2A_1B_{3x} + 3A_{1x}B_3 - 2A_{0y}B_3 - (A_0A_{1yy} + A_{0yy}A_1 + A_{1y}B_2)/3 + A_1(4A_0A_{1y} + 2A_1B_2 + 5A_{0y}A_1)/9 - 2A_0A_1^3/27,$$

$$J_8 = A_1A_{1x} + B_{2y} - 3B_{3x} - (A_0A_{1y} + A_{0y}A_1 + A_1B_2)/3 + A_0A_1^2/9,$$

$$J_9 = A_1A_{1xx} + B_{1yy} - 3B_{3xx} + 2A_{1x}B_2 - A_{1y}B_1 + 3B_1B_3 - A_0B_{3x} - A_{0x}B_3 - (A_0A_{0yy} + 2A_0^2B_3 + A_0A_1A_{1x} - A_{0x}A_{1y} + 4A_{0y}B_2 + A_1^2B_1 + A_1B_{1y} + 2B_2^2)/3 + (2A_0^2A_{1y} + 5A_0A_{0y}A_1 + 4A_0A_1B_2 + A_{0x}A_1^2)/9,$$

$$J_{10} = B_{1y} - B_{2x} + (A_0A_{1x} - A_0A_{0y} - A_0B_2 + A_{0x}A_1)/3 + A_0^2A_1/9,$$

$$J_{11} = A_{1xx} - B_{2x} + (2A_0A_{1x} - A_0B_2 + A_{0x}A_1)/3 + A_0^2A_1/9,$$

$$J_{12} = A_{0xy} - B_{2x} + (A_0A_{0y} + A_0A_{1x} - A_0B_2 + A_{0x}A_1)/3 + A_0^2A_1/9,$$

$$J_{13} = 3\Omega = A_{0xx} + 2A_0A_{0x} - 2A_0B_1 + 2A_1B_0 + 6B_{0y} - 3B_{1x} + 4A_0^3/9,$$

$$J_{14} = A_1A_{1xx} + B_{2xy} - 3B_{3xx} - A_0B_{3x} - (A_0^2B_3 + A_0A_1B_2 + A_0A_{1xy} - A_0B_{2y} + A_{0xy}A_1 + A_{0y}B_2 + A_1B_{2x} - 4A_{1x}B_2 + 2B_2^2)/3 - (A_{0x}B_3 + A_{1y}B_1 - 3B_1B_3)/2 + (A_{0x}A_1^2 - A_{0x}A_{1y} - A_1^2B_1)/6,$$

$$J_{15} = B_{1xy} - B_{2xx} - A_0B_{2x} - A_{1y}B_0 + 3B_0B_3 + +(A_{0xx}A_1 - A_0A_{0xy} + A_0A_{1xx} + 2A_0B_{1y} - A_{0x}B_2 - A_1^2B_0)/3 + (2A_0^2A_{1x} - A_0^2A_{0y} - 2A_0^2B_2 + 3A_0A_{0x}A_1)/9 + (A_{0x}A_{1x} - A_{0y}B_1 + A_{1x}B_1)/2 - A_{0x}A_{0y}/6 + 2A_0^3A_1/27,$$

$$J_{16} = B_{0yy} + B_0B_3 - A_{0y}B_1 + (A_{0x}A_{0y} - A_0B_{2x} + A_1B_{0y} + A_{1x}B_1 + B_1B_2 - B_{2xx})/3 + (2A_0^2A_{0y} - A_0A_1B_1 + A_0A_{1xx} + A_{0x}A_{1x} - 2A_{0x}B_2 + A_{0xx}A_1 - A_1^2B_0)/9 + A_0(A_0A_{1x} - 4A_0B_2 + 5A_{0x}A_1)/27 + 4A_0^3A_1/81.$$

The system of equations $S_i = 0$, $(i = 1, 2, 3, 4, 5)$ is a complete system. Even if one appends to this system the equation

$$S_6 = F_{J_{13}} = 0,$$

it will still be a complete system.

From the equations $S_i = 0$, $(i = 1, 2, 3, 4, 5)$ one can find that $J_k = 0$, $(k = 1, 2, ..., 16)$ are relative invariants of the equivalence group. Further analysis[17] of the equations $S_i = 0$, $(i = 1, 2, 3, 4, 5)$ requires studying different cases that depend on equating to zero of some of the variables J_k.

[17] The complete analysis is very cumbersome and it is beyond the goals of our study. The way of analyzing is presented in the section related with contact transformations.

2.2.2 The linearization test for equation (7.20)

Since the way of studying this case is similar to the previous one the results are presented without details.

Theorem 7.3. *Equation (7.20)*

$$
\begin{aligned}
&y''' + \frac{1}{y'+r}\big(-3(y'')^2 + (C_2\,y'^2 + C_1\,y' + C_0)y''\\
&+D_5\,y'^5 + D_4\,y'^4 + D_3\,y'^3 + D_2\,y'^2 + D_1\,y' + D_0\big) = 0.
\end{aligned}
$$

is linearizable if and only if its coefficients obey the following equations:

$$C_0 = 6r\,r_y - 6r_x + r\,C_1 - r^2 C_2, \tag{7.52}$$

$$6r_{yy} = \frac{\partial C_2}{\partial x} - \frac{\partial C_1}{\partial y} + r\frac{\partial C_2}{\partial y} + C_2 r_y, \tag{7.53}$$

$$
\begin{aligned}
18D_0 &= 3r^3\frac{\partial C_1}{\partial y} - 6r^2\frac{\partial C_1}{\partial x} - 3r^3\frac{\partial C_2}{\partial x} + 9r^4\frac{\partial C_2}{\partial y} - 36r^2 r_{xy} + 18r\,r_{xx}\\
&+90r\,r_x r_y - 36r^2{r_y}^2 + 6r(3C_1 - rC_2)r_x + 9r^2(rC_2 - 2C_1)r_y\\
&-54{r_x}^2 - 2r^2C_1^2 + 2r^3C_1C_2 + 4r^4C_2^2 + 18r^4D_4 - 72r^5D_5,
\end{aligned}
\tag{7.54}
$$

$$
\begin{aligned}
18D_1 &= 9r^2\frac{\partial C_1}{\partial y} - 12r\frac{\partial C_1}{\partial x} - 27r^2\frac{\partial C_2}{\partial x} + 33r^3\frac{\partial C_2}{\partial y} - 36r\,r_{xy}\\
&+18r_{xx} + 6(3C_1 + 4rC_2)r_x - 3r(6C_1 + 7rC_2)r_y + 18r{r_y}^2\\
&-18r_x r_y - 4rC_1^2 - 2r^2C_1C_2 + 20r^3C_2^2 + 72r^3D_4 - 270r^4D_5,
\end{aligned}
\tag{7.55}
$$

$$
\begin{aligned}
9D_2 &= 3r\frac{\partial C_1}{\partial y} - 3\frac{\partial C_1}{\partial x} - 21r\frac{\partial C_2}{\partial x} + 21r^2\frac{\partial C_2}{\partial y} + 15C_2 r_x\\
&-15r\,C_2 r_y - C_1^2 - 5rC_1C_2 + 14r^2C_2^2 + 54r^2D_4 - 180r^3D_5,
\end{aligned}
\tag{7.56}
$$

$$3D_3 = 3r\frac{\partial C_2}{\partial y} - 3\frac{\partial C_2}{\partial x} - C_1C_2 + 2rC_2^2 + 12rD_4 - 30r^2D_5, \tag{7.57}$$

$$
\begin{aligned}
54\frac{\partial D_4}{\partial x} &= 18\frac{\partial^2 C_1}{\partial y^2} + 3C_2\frac{\partial C_1}{\partial y} - 72\frac{\partial^2 C_2}{\partial x\partial y} - 39C_2\frac{\partial C_2}{\partial x} + 18r\frac{\partial^2 C_2}{\partial y^2}\\
&-3rC_2\frac{\partial C_2}{\partial y} + 72\frac{\partial C_2}{\partial y}r_y + 378\,r\frac{\partial D_5}{\partial x} - 108r^2\frac{\partial D_5}{\partial y} + 270D_5 r_x\\
&+33C_2^2 r_y + 108D_4 r_y - 540rD_5 r_y + 36\,r\,C_1D_5 - 8\,r\,C_2^3\\
&-36\,r\,C_2D_4 + 108\,r^2\,C_2D_5 + 54\,r\,H,
\end{aligned}
\tag{7.58}
$$

and

$$H_x = 3Hr_y + rH_y, \tag{7.59}$$

where the function H is defined by the equation

$$\begin{aligned} H &= \frac{\partial D_4}{\partial y} - 2\frac{\partial D_5}{\partial x} - 3r\frac{\partial D_5}{\partial y} - 5D_5 r_y - 2rC_2D_5 \\ &+\frac{1}{3}\Big[\frac{\partial^2 C_2}{\partial y^2} + 2C_2\frac{\partial C_2}{\partial y} - 2C_1D_5 + 2C_2D_4\Big] + \frac{4}{27}C_2^3. \end{aligned} \tag{7.60}$$

Provided that the conditions (7.52)-(7.59) are satisfied, the transformation (7.3)

$$t = \varphi(x, y), \quad u = \psi(x, y)$$

with $\varphi_y(x, y) \neq 0$ mapping equation (7.20) into a linear equation (7.14) is obtained by solving the following compatible system of equations for the functions $\varphi(x, y)$ and $\psi(x, y)$:

$$\varphi_x = r\varphi_y, \qquad \psi_x = -\varphi_y W + r\psi_y, \tag{7.61}$$

$$6\varphi_y\varphi_{yyy} = 9\varphi_{yy}{}^2 + \left(15rD_5 - 3D_4 - C_2^2 - 3\frac{\partial C_2}{\partial y}\right)\varphi_y{}^2, \tag{7.62}$$

$$\begin{aligned} \psi_{yyy} &= W D_5 \varphi_y + \frac{1}{6}\Big[15r D_5 - C_2^2 - 3D_4 - 3\frac{\partial C_2}{\partial y}\Big]\psi_y \\ &-\frac{1}{2}H\psi + 3\varphi_{yy}\psi_{yy}\varphi_y{}^{-1} - \frac{3}{2}\varphi_{yy}{}^2\psi_y\varphi_y{}^{-2}, \end{aligned} \tag{7.63}$$

where the function W is defined by the equations

$$3W_x = \left(C_1 - rC_2 + 6r_y\right)W, \qquad 3W_y = C_2W. \tag{7.64}$$

The coefficient α of the resulting linear equation (7.14) is given by

$$2\alpha = \varphi_y^{-3}H, \tag{7.65}$$

where H is the function defined in (7.60).

Remark 7.3. *The necessary and sufficient conditions comprise eight differential equations (7.52)-(7.59) for ten coefficients of equation (7.20). The linearizing change of variables (7.3) is determined by equations (7.61)-(7.64) for the functions $\varphi(x, y)$ and $\psi(x, y)$.*

Proof.

The problem is: for the given coefficients $C_i(x, y)$, $D_i(x, y)$ of equation (7.20) find the integrability conditions for the functions $\varphi(x, y)$ and $\psi(x, y)$.

According to the introduced notations and the condition $\alpha(x, y) = \alpha(\varphi(x, y))$ one has

$$\varphi_x = r\varphi_y, \quad \psi_x = \varphi_y^{-1}(\psi_y\varphi_z - \Delta). \tag{7.66}$$

$$\alpha_x = \varphi_y^{-1}\varphi_x\,\alpha_y. \tag{7.67}$$

Since,

$$(\Delta\varphi_y)^{-1}(\Delta_y\varphi_y - 2\Delta\varphi_{yy}) = (\varphi_y^{-2}\Delta)^{-1}(\varphi_y^{-2}\Delta)_y,$$

setting

$$\Delta = W\,\varphi_y^2, \tag{7.68}$$

equations (7.21)-(7.23) for the coefficients C_i are simplified:

$$C_2 = 3\,\frac{W_y}{W}, \quad C_1 = 3\,\frac{W_x + rW_y}{W} - 6r_y, \quad C_0 = 3\,\frac{rW_x - 2Wr_x}{W}. \tag{7.69}$$

Substituting the expression (7.68) for Δ into equations (7.24) and (7.25), one arrives at the following equations:

$$\varphi_y^2\psi_{yyy} = (\varphi_{yy}\varphi_{yyy}\psi_y - 3\varphi_{yy}^2\psi_y + 3\varphi_y\varphi_{yy}\psi_{yy} - \alpha\varphi_y^5\psi + W\varphi_y^3 D_5), \tag{7.70}$$

$$2W\varphi_y\varphi_{yyy} = (3W\varphi_{yy}^2 - 3\varphi_y^2 W_{yy} - W\varphi_y^2 D_4 + 5\varphi_y^2 rWD_5). \tag{7.71}$$

The expressions (7.26)-(7.29) for D_0, D_1, D_2, D_3 become:

$$\begin{aligned}
D_3 &= W^{-1}\Big(6r_yW_y - 3W_{xy} + 3W_{yy}r + 4D_4rW - 10D_5r^2W\Big),\\
D_2 &= W^{-1}\Big(2r_{xy}W + 4r_xW_y - 2r_{yy}rW - 4r_y^2W + 4r_yW_x\\
&\quad +10r_yW_yr - 7W_{xy}r - W_{xx} + 8W_{yy}r^2 + 6D_4r^2W - 20D_5r^3W\Big),\\
D_1 &= W^{-1}\Big(2r_{xy}rW + r_{xx}W - 7r_xr_yW + 3r_xW_x\\
&\quad +5r_xW_yr - 3r_{yy}r^2W - r_y^2r\,W + 5r_yW_xr + 5r_yW_yr^2\\
&\quad -5W_{xy}r^2 - 2W_{xx}r + 7W_{yy}r^3 + 4D_4r^3W - 15D_5r^4W\Big),\\
D_0 &= W^{-1}\Big(r_{xx}rW - 3r_x^2W - r_xr_yrW + 3r_xW_xr\\
&\quad +r_xW_yr^2 - r_{yy}r^3W + r_yW_xr^2 + r_yW_yr^3\\
&\quad -W_{xy}r^3 - W_{xx}r^2 + 2W_{yy}r^4 + D_4r^4W - 4D_5r^5W\Big).
\end{aligned} \tag{7.72}$$

Using equations (7.66), (7.70) and (7.71) one can find all third order derivatives of the functions φ and ψ. Equating all mixed derivatives, one obtains: from the equation $(\varphi_x)_{yyy} = (\varphi_{yyy})_x$

$$\begin{aligned}
\frac{\partial D_4}{\partial x} &= r\,\frac{\partial D_4}{\partial y} + 5r\,\frac{\partial D_5}{\partial x} - 5r^2\frac{\partial D_5}{\partial y} + 5D_5(r_x - 3r\,r_y) - 2r_{yyy} + 2D_4r_y\\
&\quad +3W^{-1}(2r_yW_{yy} - W_{xyy} + r\,W_{yyy}) + 3W^{-2}W_{yy}(W_x - r\,W_y),
\end{aligned}$$

and from the equation $(\psi_x)_{yyy} = (\psi_{yyy})_x$

$$S \equiv 2\alpha\,\varphi_y^3 - H = 0, \tag{7.73}$$

where

$$H = \frac{\partial D_4}{\partial y} - 2\frac{\partial D_5}{\partial x} - 3r\frac{\partial D_5}{\partial y} - D_5 r_y + W^{-1}\left(W_{yyy} - 2D_5 W_x + (2D_4 - 8r D_5 + 3W^{-1}W_{yy})W_y\right).$$

Since $\varphi_y \neq 0$, equation (7.73) yields (7.65):

$$2\alpha = \varphi_y^{-3} H.$$

The equation $\alpha_x - r\alpha_y = 0$ leads to equation (7.59)

$$H_x - r H_y - 3H r_y = 0.$$

Using the expressions for C_1 and C_2 given in (7.69) one finds the first derivatives of W

$$W_y = \frac{1}{3} W C_2, \quad W_x = \frac{1}{3} W(C_1 - rC_2 + 6r_y). \tag{7.74}$$

Equating the mixed derivatives W_{xy} and W_{yx}, one obtains (7.54)

$$C_{2x} - C_{1y} + C_2 r_y + rC_{2y} - 6r_{yy} = 0.$$

Eliminating the derivatives W_x and W_y from the obtained expressions for the functions D_0, D_1, D_2, D_3, and D_{4x}, one arrives at the linearization conditions summarized in the theorem. This completes the proof of the theorem. •

Remark 7.4. *The study of second order invariants of the equivalence group is very cumbersome. Notice only that the result is similar to the study of the invariants of the equivalence group of equation (7.15): there are five linear homogeneous equations which form a complete system; the linearization conditions (7.52)-(7.59) are relative invariants of this group.*

2.2.3 Applications of the linearization theorems

Example 1. The equation

$$y''' - \left(\frac{6}{y} y' + \frac{3}{x}\right) y'' + \frac{6}{y^2} y'^3 + \frac{6}{xy} y'^2 + \frac{6}{x^2} y' + 6\frac{y}{x^3} = 0 \tag{7.75}$$

is an equation of the form (7.15) with the coefficients

$$A_1 = -\frac{6}{y}, \ A_0 = -\frac{3}{x}, \ B_3 = \frac{6}{y^2}, \ B_2 = \frac{6}{xy}, \ B_1 = \frac{6}{x^2}, \ B_0 = \frac{6y}{x^3}. \tag{7.76}$$

One can verify that the coefficients (7.76) obey the conditions (7.31)-(7.33), and $\Omega = 0$. Hence, equation (7.76) is linearizable with $\alpha = 0$. The linearizing transformation is obtained as follows.

Since

$$3B_1 - A_0^2 - 3A_{0x} = 0, \tag{7.77}$$

and $\chi = \varphi_x^{-1}\varphi_{xx}$, equation (7.34) is

$$2\varphi_x\varphi_{xxx} - 3\varphi_{xx}^2 = 0.$$

Let us take its simplest solution $\varphi = x$. Equations (7.35), (7.36) become

$$\psi_y^{-1}\psi_{yy} = -\frac{2}{y}, \quad \psi_y^{-1}\psi_{xy} = -\frac{1}{x}, \quad \psi_{xxx} = \frac{6y}{x^3}\psi_x.$$

A particular solution of these equations is

$$\psi = \frac{1}{xy}.$$

Therefore, the transformation

$$t = x, \quad u = \frac{1}{xy} \tag{7.78}$$

maps equation (7.75) to the linear equation

$$u''' = 0.$$

Example 2. Consider the following equation of the form (7.15):

$$y''' + \frac{3}{y}y'y'' - 3y'' - \frac{3}{y}y'^2 + 2\,y' - y = 0. \tag{7.79}$$

The coefficients of this equation are

$$A_1 = \frac{3}{y}, \; A_0 = -3, \; B_3 = 0, \; B_2 = -\frac{3}{y}, \; B_1 = 2, \; B_0 = -y.$$

These coefficients obey the linearization conditions (7.31)-(7.33),

$$3B_1 - A_0^2 - 3A_{0x} = -3,$$

and $\Omega = -2$. Hence, equation (7.34) is

$$2\varphi_x\varphi_{xxx} - 3\varphi_{xx}^2 + \varphi_x^2 = 0.$$

A particular solution of this equation is $\varphi = \mathrm{e}^x$. Equations (7.35), (7.35) have the forms

$$\psi_{yy} - \frac{1}{y}\psi_y = 0, \quad \psi_{xy} = 0, \quad \psi_{xxx} - 3\psi_{xx} + y\psi_y + \frac{3}{2}\psi_x - 2\psi = 0.$$

The simplest solution of these equations is $\psi = y^2$. Therefore, the change of variables:

$$t = \mathrm{e}^x, \quad u = y^2. \tag{7.80}$$

maps equation (7.79) into the linear equation

$$u''' - \frac{2}{t^3}u = 0. \tag{7.81}$$

Remark 7.5. *In the previous examples, when determining the linearizing transformations, the calculations were confined to particular solutions of equations (7.34)-(7.36). The general solution of equations (7.34)-(7.36) involves the symmetry and equivalence transformations for the original and linearized equations.*

Example 3. Consider the non-linear equation

$$y''' + \frac{1}{y'}\left(-3y''^2 - xy'^5\right) = 0. \tag{7.82}$$

It has the form (7.20) with the following coefficients:

$$\begin{gathered} r = 0, \quad C_0 = C_1 = C_2 = 0, \\ D_0 = D_1 = D_2 = D_3 = D_4 = 0, \quad D_5 = -x. \end{gathered} \tag{7.83}$$

Because of the coefficients (7.83), $H = 2$, and the linearizing conditions (7.52)-(7.59) are satisfied. Hence, equation (7.82) is linearizable, and it can be developed further. Equations (7.64) are written

$$\frac{\partial W}{\partial x} = 0, \quad \frac{\partial W}{\partial y} = 0$$

and yield $W =$ const, for example, let us set $W = -1$. Therefore, equations (7.61) have the form

$$\varphi_x = 0, \quad \psi_x = \varphi_y,$$

and hence:

$$\varphi = \varphi(y), \quad \psi = x\,\varphi'(y) + \omega(y). \tag{7.84}$$

The third order equations (7.62) and (7.63) are reduced to the equations

$$2\varphi'\varphi''' = 3\,\varphi''^2, \;\; 2\varphi'^2(\omega + \omega''') - 6\varphi'\varphi''\,\omega'' + 3\,\varphi''^2\,\omega' = 0. \tag{7.85}$$

These equations are satisfied by letting $\varphi = y$ and $\omega = 0$. Since relation (7.65) yields $\alpha = 1$, the change of variables

$$t = y, \quad u = x$$

reduces (7.82) into the linear equation

$$u''' + u = 0.$$

Example 4. Let us study the equation

$$y''' + y^2 = 0. \tag{7.86}$$

The coefficients for this equation are

$$A_0 = 0,\ A_1 = 0,\ B_0 = y^2,\ B_1 = 0,\ B_2 = 0,\ B_3 = 0. \tag{7.87}$$

In this case

$$\Omega = 2y,\ \beta = (3B_1 - A_0^2 - 3A_{0x})/3 = 0,$$

the conditions (7.31)-(7.32) are satisfied, and equation (7.40) is reduced to the equation

$$6\alpha\alpha'' - 7(\alpha')^2 = 0,$$

where α' is the derivative of the function $\alpha(t)$ with respect to t. Setting, for example, $\alpha = 1$ the sufficient conditions of [59] are satisfied. Thus, according to [59], equation (7.86) has to be linearizable by the point transformation

$$t = \varphi(x),\ u = \psi(x, y).$$

However, since $\Omega_y \neq 0$ condition (7.33) is not satisfied, and hence, equation (7.86) is not linearizable. Let us prove it directly. Assuming that (7.86) is linearizable, the functions $\varphi(x)$ and $\psi(x, y)$ and the coefficients (7.87) are related by formulae (7.16)-(7.19) which accept the following form (for proving enough to study the representations for A_0, A_1, B_0):

$$\psi_{yy} = 0, \quad \varphi_x\psi_{xy} - \psi_y\varphi_{xx} = 0$$
$$3\varphi_{xx}^2\psi_x - \varphi_{xxx}\varphi_x\psi_x - 3\varphi_{xx}\varphi_x\psi_{xx} + \varphi_x^2\psi_{xxx} + \alpha\psi\varphi_x^5 - y^2\varphi_x^2\psi_y = 0.$$

The general solution of the first two equations is

$$\psi(x, y) = k(y\varphi_x(x) + h(x)),$$

where $k \neq 0$ is constant. From the last equation one finds

$$\alpha = \frac{y^2 - y\gamma + \lambda}{\varphi_x^2(y\varphi_x + h)}, \tag{7.88}$$

where

$$\gamma = (\varphi_x^{-1}\varphi_{xxxx} + 3\varphi_x^{-3}\varphi_{xx}^3 - 4\varphi_{xxx}\varphi_{xx}\varphi_x^{-2}),$$
$$\lambda = \varphi_x^{-3}(\varphi_{xxx}\varphi_x - 3\varphi_{xx}^2)h' + 3\varphi_x^{-2}\varphi_{xx}h'' - \varphi_x^{-1}h'''.$$

Since the dependence of the coefficient α with respect to the variables x, y is defined by the composition $\alpha \circ \varphi$, this coefficient does not depend on y. This contradicts the representation in (7.88).

2.3 Third order equation: linearizing contact transformations

S.Lie showed [95] that, contrary to second order ordinary differential equations, not every third order ordinary differential equation is related to the simplest equation

$$u''' = 0. \tag{7.89}$$

He also proved that any third order ordinary differential equation related to equation (7.89) via general contact transformations

$$t = \varphi(x, y, p),\; u = \psi(x, y, p),\; q = g(x, y, p), \tag{7.90}$$

are at most cubic in the second order derivative, i.e., it has the form

$$y''' + a(x, y, y')y''^{3} + b(x, y, y')y''^{2} + c(x, y, y')y'' + d(x, y, y') = 0. \tag{7.91}$$

S.Lie did not investigate further the problem of linearization of third order equations either by contact or by point transformations.

S.-S.Chern [25] applied the Cartan approach for third order ordinary differential equations. Since his conditions are given in implicit form, linearization by contact transformations were also studied in [13, 66, 36, 35, 122].

The contact transformation (7.90) preserves the tangent conditions

$$du - q\,dt = 0,\; dq - u''\,dt,\; du'' - u'''\,dt = 0.$$

Here the following notations are used

$$p(x) = y'(x), \qquad q(t) = u'(t).$$

The first tangent condition $du - q\,dt = 0$ gives

$$(g\varphi_p - \psi_p)dp + (\psi_x + p\psi_y - g(\varphi_x + p\varphi_y))dx = 0.$$

Because of arbitrariness of dx and dp, one obtains

$$g\varphi_p = \psi_p, \qquad \psi_x + p\psi_y = g(\varphi_x + p\varphi_y). \tag{7.92}$$

By virtue of (7.92), the Jacobian of the contact transformation (7.90) is

$$(\varphi_y g - \psi_y)\left(\varphi_p(g_x + g_y p) - g_p(\varphi_x + \varphi_y p)\right) \neq 0.$$

The tangent conditions $du' = u''\,dt$ and $du'' = u'''\,dt$ give the representation of the transformed second and third derivatives, respectively,

$$u'' = \frac{D_x g}{D_x\varphi} = \frac{g_x + pg_y + y''g_p}{\varphi_x + p\varphi_y + y''\varphi_p} = P(x, y, p, y''),$$
$$u''' = \frac{D_x P}{D_x\varphi} = \frac{P_x + pP_y + y''P_p + P_{y''}y'''}{\varphi_x + p\varphi_y + y''\varphi_p},$$

Since the Jacobian of the contact transformation is not equal to zero, one has

$$P_{y''} = \frac{g_p D_x\varphi - \varphi_p D_x g}{(D_x\varphi)^2} = \frac{g_p(\varphi_x + \varphi_y p) - \varphi_p(g_x + g_y p)}{(\varphi_x + p\varphi_y + y''\varphi_p)^2} \neq 0.$$

The equation

$$u'''(t) + \alpha(t)u(t) = 0 \tag{7.93}$$

is transformed by the contact transformation (7.90) into the equation

$$y''' + \frac{P_p}{P_{y''}}y'' + \frac{P_y}{P_{y''}}p + \frac{P_x}{P_{y''}} + \frac{\alpha\psi}{P_{y''}}D_x\varphi = 0,$$

which has the form

$$y''' + a(x, y, p)y''^3 + b(x, y, p)y''^2 + c(x, y, p)y'' + d(x, y, p) = 0. \tag{7.94}$$

Representations for the coefficients a, b, c, d through the functions φ, ψ, g are

$$(\varphi_p(g_x + g_y p) - g_p(\varphi_x + \varphi_y p))a = \varphi_{pp}g_p - \varphi_p g_{pp} - \alpha\psi\varphi_p^3, \tag{7.95}$$

$$\begin{aligned}(\varphi_p(g_x + g_y p) - g_p(\varphi_x + \varphi_y p))b = 2g_p(\varphi_{px} + \varphi_{py}p) + \varphi_{pp}(g_x + g_y p) - \\ 2\varphi_p(g_{px} + g_{py}p) - \varphi_p g_y + \varphi_y g_p - g_{pp}(\varphi_x + \varphi_y p) - 3\alpha\psi\varphi_p^2(\varphi_x + \varphi_y p)\end{aligned} \tag{7.96}$$

$$\begin{aligned}(\varphi_p(g_x + g_y p) - g_p(\varphi_x + \varphi_y p))c = 2(\varphi_{px} + \varphi_{py}p)(g_x + g_y p) - \\ \varphi_p(g_{xx} + 2g_{xy}p + g_{yy}p^2) + g_p(\varphi_{xx} + 2\varphi_{xy}p + \varphi_{yy}p^2) - \\ 2(g_{px} + g_{py}p)(\varphi_x + \varphi_y p) + \varphi_y g_x - \varphi_x g_y - 3\alpha\psi\varphi_p(\varphi_x + \varphi_y p)^2\end{aligned} \tag{7.97}$$

$$\begin{aligned}(\varphi_p(g_x + g_y p) - g_p(\varphi_x + \varphi_y p))d = (\varphi_{xx} + 2\varphi_{xy}p + \varphi_{yy}p^2)(g_x + g_y p) - \\ (g_{xx} + 2g_{xy}p + g_{yy}p^2)(\varphi_x + \varphi_y p) - \alpha\psi(\varphi_x + \varphi_y p)^3\end{aligned} \tag{7.98}$$

2.3.1 Second order invariants of the equivalence group

As was obtained in the previous section the second order invariants play a key role in the problem of linearization by point transformations of third order ordinary differential equation. Therefore, before the study of the linearization problem, let us find invariants of second order of the equivalence group. The equivalence group $\mathcal{E}$ of equation (7.94) is the general group of contact transformations with the generator

$$X = \xi\frac{\partial}{\partial x} + \eta\frac{\partial}{\partial y} + \zeta\frac{\partial}{\partial p} + \zeta^a\frac{\partial}{\partial a} + \zeta^b\frac{\partial}{\partial b} + \zeta^c\frac{\partial}{\partial c} + \zeta^d\frac{\partial}{\partial d},$$

where

$$\xi = -W_p, \eta = (W - pW_p), \zeta = (W_x + pW_y),$$
$$\zeta^a = -3W_{py}pa - 3W_{px}a - W_{ppp} - W_{pp}b - 2W_y a,$$
$$\zeta^b = -W_{py}pb - 3W_{py} - W_{px}b - 3W_{ppy}p - 3W_{ppx} + 2W_{pp}c - 6W_{yx}pa - 3W_{yy}p^2 a - W_y b - 3W_{xx}a,$$
$$\zeta^c = -6W_{pyx}p - 3W_{pyy}p^2 + W_{py}pc - 3W_{pxx} + W_{px}c + 3W_{pp}d - 4W_{yx}pb - 3W_{yx} - 2W_{yy}p^2 b - 3W_{yy}p - 2W_{xx}b,$$
$$\zeta^d = 3W_{py}pd + 3W_{px}d - 3W_{yxx}p - 2W_{yx}pc - W_{yyy}p^3 - 3W_{yyx}p^2 - W_{yy}p^2 c + W_y d - W_{xxx} - W_{xx}c.$$

The characteristic function $W(x, y, p)$ is an arbitrary function.

The search for second order invariants of the equivalence group (after running the program) leads to the following equations

$$S_1 = -3\frac{\partial F}{\partial J_1}J_2 + \frac{\partial F}{\partial J_3}J_4 + 2\frac{\partial F}{\partial J_2}J_3 = 0,$$
$$S_2 = 3\frac{\partial F}{\partial J_1}J_1 + \frac{\partial F}{\partial J_3}J_3 + 2\frac{\partial F}{\partial J_2}J_2 = 0,$$
$$S_3 = 3\frac{\partial F}{\partial J_4}J_4 + 2\frac{\partial F}{\partial J_3}J_3 + \frac{\partial F}{\partial J_2}J_2 = 0,$$
$$S_4 = 3\frac{\partial F}{\partial J_4}J_3 + 2\frac{\partial F}{\partial J_3}J_2 - \frac{\partial F}{\partial J_2}J_1 = 0,$$

where

$$J_1 = 27Da_p - 18a_p c - 18bDa + 81a_y + 18b_p b - 9b_{pp} + 18aDb - 36c_p a - 54a^2 d + 18abc - 4b^3,$$

$$J_2 = -18a_p d - 18cDa + 9Da_x + 9pDa_y + 6b_p c + 3Db_p + 6bDb + 24b_y - 6c_{pp} - 36d_p a - 18abd + 12ac^2 - 2b^2 c,$$

$$J_3 = 36dDa - 6Db_x - 6pDb_y - 6c_p c + 3Dc_p - 6bDc - 21c_y + 18d_p b + 9d_{pp} + 18aDd - 18acd + 12b^2 d - 2bc^2,$$

$$J_4 = -36dDb + 18c_p d + 18cDc + 9Dc_x + 9pDc_y - 18d_p c - 27Dd_p - 18bDd + 54d_y + 54ad^2 - 18bcd + 4c^3,$$

and the operator D is

$$D = \frac{\partial}{\partial x} + p\frac{\partial}{\partial y}.$$

To find the invariants let us analyze the invariant manifolds.

For any invariant manifold (Φ) defined by the equations:

$$F_i(J_1, J_2, J_3, J_4) = 0, \quad (i = 1, \ldots, k)$$

the equations

$$S_j(F_i)_{|(\Phi)} = 0, \quad (j = 1, 2, 3, 4; \quad i = 1, \ldots, k).$$

have to be satisfied.

Assume that on this manifold $J_4 = 0$. Setting $F_1 = J_4$, one obtains $J_3 = 0$ from the equation $S_4(F_1)_{|(\Phi)} = 0$. This leads to $S_4(J_3)_{|(\Phi)} = 2J_{2|(\Phi)} = 0$, and the last leads to $S_4(J_2)_{|(\Phi)} = -J_{1|(\Phi)} = 0$. Hence[18], if $J_4 = 0$ for an invariant manifold, then it implies $J_1 = 0,\ J_2 = 0,\ J_3 = 0$, i.e., $J_1 = 0,\ J_2 = 0,\ J_3 = 0$ and $J_4 = 0$ are relative invariants of the contact transformations.

If $J_4 \neq 0$, then the general solution of the equation $S_1 = 0$ is

$$F = F(J_4, J_5, J_6),$$

where $3J_5 = J_2 - \frac{J_3^2}{J_4}$, $J_6 = -\frac{1}{9}\left(J_1 + \frac{J_3^3}{J_4^2}\right) - \frac{J_3 J_5}{J_4}$. The remaining equations become

$$\begin{aligned} S_2 &= 3\tfrac{\partial F}{\partial J_6} J_6 + 2\tfrac{\partial F}{\partial J_5} J_5 = 0, \\ S_3 &= \tfrac{\partial F}{\partial J_5} J_5 + 3\tfrac{\partial F}{\partial J_4} J_4 = 0, \\ S_4 J_4 + J_3 S_2 - J_3 S_3 &= 3(-2\tfrac{\partial F}{\partial J_6} J_5^2 + \tfrac{\partial F}{\partial J_5} J_4 J_6) = 0. \end{aligned}$$

Integrating the equation $S_3 = 0$, one obtains

$$F = F(J_7, J_6),$$

where $J_7 = 9J_4^{-1}J_5^3$. The function F also has to satisfy the equations

$$\frac{\partial F}{\partial J_6} J_6 + 2\frac{\partial F}{\partial J_7} J_7 = 0, \quad J_5^2\left(-2\frac{\partial F}{\partial J_6} + 27\frac{\partial F}{\partial J_7} J_7\right) = 0.$$

If $J_5 \neq 0$, then $J_7 \neq 0$ and there is no invariants (nonsingular and singular).
If $J_5 = 0$, then there is only one equation

$$\frac{\partial J}{\partial J_6} J_6 = 0,$$

which defines the singular invariant manifold

$$J_5 = 0, \quad J_6 = 0.$$

2.3.2 Conditions for linearization

First of all notice that if $\alpha(t) = 0$, then $J_1 = 0,\ J_2 = 0,\ J_3 = 0,\ J_4 = 0$. Hence, for $\alpha(t) = 0$ they are invariants of the contact transformations. In the case $\alpha(t) \neq 0$, one obtains $J_1 \neq 0$. Therefore the study of the linearization problem has to be divided into two cases: $\alpha(t) = 0$ and $\alpha(t) \neq 0$.

Since the assumption $\varphi_p = 0$ leads to $\psi_p = 0$ which corresponds to the studied class of point transformations, it is assumed that $\varphi_p \neq 0$.

[18] Similarly, one obtains $J_1 = 0,\ J_2 = 0,\ J_4 = 0$ if one starts from $J_3 = 0$.

The linearization test with $\alpha = 0$. The following theorem provides the test for linearization of the first candidate among the third order equations.

Theorem 7.4. *Equation (7.94)*

$$y''' + a(x, y, p)y''^3 + b(x, y, p)y''^2 + c(x, y, p)y'' + d(x, y, p) = 0$$

is linearizable to the equation with $\alpha = 0$ if and only if its coefficients obey the following equations:

$$J_1 = 0,\ J_2 = 0,\ J_3 = 0,\ J_4 = 0. \tag{7.99}$$

Provided that the conditions (7.99) are satisfied, the transformation (7.90) mapping equation (7.94) to a linear equation (7.93) with $\alpha = 0$ is obtained by solving the following compatible system of equations for the functions $\varphi(x, y, p)$, $\psi(x, y, p)$, $g(x, y, p)$, $k(x, y, p)$, and $H(x, y, p)$:

$$D\psi = Hg,\ \psi_y = \varphi_y g + k,\ \psi_p = \varphi_p g$$

$$\varphi_p Dg = Hg_p - k,\ 3g_y\varphi_p^2 = 3\varphi_{pp}k + 3aHk + \varphi_p(3\varphi_y g_p - bk),$$
$$\varphi_p g_{pp} = (\varphi_{pp} g_p + ak)$$

$$3\varphi_p^2 Dk = 3\varphi_{pp}Hk + 3aH^2k - H\varphi_p bk + 3\varphi_p\varphi_y k,$$
$$9\varphi_p^2 k_y = 18\varphi_{py}\varphi_p k - 9\varphi_{pp}\varphi_y k + 3aHk(2\varphi_p c - 3\varphi_y) - 9a\varphi_p^2 dk -$$
$$9DaH\varphi_p k + 3Db\varphi_p^2 k + H\varphi_p k(3b_p - 2b^2) + \varphi_p k(-3c_p\varphi_p + \varphi_p bc + 3\varphi_y b),$$
$$3\varphi_p k_p = 3\varphi_{pp}k + 3aHk - \varphi_p bk,$$

$$54\varphi_p^2\varphi_{yy} = 108\varphi_{py}\varphi_p\varphi_y - 54\varphi_{pp}\varphi_y^2 + 6aH(3\varphi_p^2 bd - 2\varphi_p^2 c^2 + 3\varphi_p^2 Dc +$$
$$6\varphi_p\varphi_y c - 9\varphi_y^2) + 18a\varphi_p^2(2\varphi_p cd - 3\varphi_p Dd - 3\varphi_y d) +$$
$$18DaH\varphi_p(2\varphi_p c - 3\varphi_y) - 108Da\varphi_p^3 d - 18DbH\varphi_p^2 b + 6Db\varphi_p^2(\varphi_p c + 3\varphi_y) +$$
$$H\varphi_p(-6b_p\varphi_p c + 18b_p\varphi_y - 27b_y\varphi_p + 6c_p\varphi_p b + 9(Db)_p\varphi_p + 2\varphi_p b^2 c -$$
$$27\varphi_p D^2 a - 12\varphi_y b^2) + 2\varphi_p(9b_p\varphi_p^2 d - 9c_p\varphi_p\varphi_y + 27c_y\varphi_p^2 - 9(Dc)_p\varphi_p^2 -$$
$$9d_p\varphi_p^2 b - 12\varphi_p^2 b^2 d + 2\varphi_p^2 bc^2 + 6\varphi_p^2 bDc + 9\varphi_p^2 D^2 b + 3\varphi_p\varphi_y bc + 9\varphi_y^2 b)$$

$$54\varphi_p^2\varphi_{ppy} = 108\varphi_{py}\varphi_{pp}\varphi_p + 54\varphi_{py}aH\varphi_p - 18\varphi_{py}\varphi_p^2 b - 27\varphi_{pp}^2\varphi_y -$$
$$54\varphi_{pp}aH\varphi_y + 18\varphi_{pp}\varphi_p\varphi_y b + 9a^2H^2(-2\varphi_p c - 3\varphi_y) + 54a^2H\varphi_p^2 d +$$
$$27aDaH^2\varphi_p - 18aDbH\varphi_p^2 + 3aH^2\varphi_p(-3b_p + 2b^2) +$$
$$6aH\varphi_p(3c_p\varphi_p - \varphi_p bc + 3\varphi_y b) + 9a\varphi_p(-3d_p\varphi_p^2 - 4\varphi_p^2 bd + 2\varphi_p^2 c^2 -$$
$$\varphi_p^2 Dc - 2\varphi_p\varphi_y c - 3\varphi_y^2) + 9Da\varphi_p^2(-4\varphi_p c + 3\varphi_y) + 18Db\varphi_p^3 b -$$
$$54Ha_y\varphi_p^2 + \varphi_p^2(6b_p\varphi_p c - 9b_p\varphi_y + 45b_y\varphi_p - 5c_p\varphi_p b - 9(Db)_p\varphi_p -$$
$$2\varphi_p b^2 c + 27\varphi_p D^2 a + 3\varphi_y b^2)$$

$$6\varphi_p\varphi_{ppp} = 9\varphi_{pp}^2 - 9a^2H^2 + 6aH\varphi_p b - 12a\varphi_p\varphi_y - 3Da\varphi_p^2 -$$
$$6Ha_p\varphi_p + \varphi_p^2(3b_p - b^2)$$

$$3\varphi_p^2 DH = 3\varphi_{pp}H^2 + 3aH^3 - H^2\varphi_p b + H\varphi_p(-\varphi_p c + 3\varphi_y) + 3\varphi_p^3 d,$$
$$18\varphi_p H_y = (18\varphi_{py}H + 6aH^2c - 18aH\varphi_p d - 9DaH^2 +$$
$$6DbH\varphi_p + H^2(3b_p - 2b^2) + 2H\varphi_p(-3c_p + bc) + 9d_p\varphi_p^2 +$$
$$6\varphi_p^2 bd - 2\varphi_p^2 c^2 - 3\varphi_p^2 Dc + 9\varphi_y^2),$$
$$3\varphi_p H_p = 3\varphi_{pp}H + 3aH^2 - 2H\varphi_p b + \varphi_p(\varphi_p c + 3\varphi_y).$$

where for simplicity of calculations the following notations are used

$$H = \varphi_x + p\varphi_y, \; k = \varphi_p g_x + \varphi_p g_y p - \varphi_x g_p - \varphi_y g_p p \neq 0.$$

The linearization test with $\alpha \neq 0$. For this case there is the following theorem.

Theorem 7.5. *Let $J_1 \neq 0$. Equation (7.94), is linearizable if and only if its coefficients obey the following equations:*

$$J_3 = -J_2^2/J_1, \; J_4 = J_2^3/J_1^2 \tag{7.100}$$

$$J_1 DJ_1 = J_{1p}J_2 + 3aJ_2^2 - 2bJ_1J_2 + J_1^2 c \tag{7.101}$$

$$J_1^2 DJ_2 = J_{2p}J_1J_2 + 2aJ_2^3 - bJ_1J_2^2 + J_1^3 d \tag{7.102}$$

$$9J_1^2 J_{2pp} = 9a_pJ_1J_2^2 - 6b_pJ_1^2J_2 + 9J_{1pp}J_1J_2 - 15J_{1p}^2J_2 + 15J_{1p}J_{2p}J_1 +$$
$$J_{1p}(-15aJ_2^2 + 4bJ_1J_2 + J_1^2c) + 3J_{1y}J_1^2 + 6J_{2p}J_1(3aJ_2 - bJ_1) + 3c_pJ_1^3 \tag{7.103}$$

$$18J_1^3J_{2x} = -3b_pJ_1^2J_2^2p - 9J_{1p}^2J_2^2p + 18J_{1p}J_{2p}J_1J_2p +$$
$$6J_{1p}J_2p(-3aJ_2^2 + 2bJ_1J_2 - J_1^2c) - 18J_{1y}J_1^2J_2p - 9J_{2p}^2J_1^2p +$$
$$6J_{2p}J_1(3aJ_2^2p - 2bJ_1J_2p + J_1^2cp + 3J_1J_2) - 9d_pJ_1^4p + 6c_pJ_1^3J_2p -$$
$$9a^2J_2^4p + 12abJ_1J_2^3p + 18aJ_1^3J_2dp - 12aJ_1^2J_2^2cp + 36aJ_1J_2^3 -$$
$$2b^2J_1^2J_2^2p - 6bJ_1^4dp + 2bJ_1^3J_2cp - 18bJ_1^2J_2^2 + J_1^4c^2p + 3J_1^4Dcp +$$
$$18J_1^4d - 6J_1^3J_2Dbp + 9J_1^2J_2^2Dap \tag{7.104}$$

$$18J_1^3J_{2y} = 3b_pJ_1^2J_2^2 + 9J_{1p}^2J_2^2 - 18J_{1p}J_{2p}J_1J_2 +$$
$$6J_{1p}J_2(3aJ_2^2 - 2bJ_1J_2 + J_1^2c) + 18J_{1y}J_1^2J_2 + 9J_{2p}^2J_1^2 +$$
$$6J_{2p}J_1(-3aJ_2^2 + 2bJ_1J_2 - J_1^2c) + 9d_pJ_1^4 - 6c_pJ_1^3J_2 +$$
$$9a^2J_2^4 - 12abJ_1J_2^3 - 18aJ_1^3J_2d + 12aJ_1^2J_2^2c + 2b^2J_1^2J_2^2 + 6bJ_1^4d -$$
$$2bJ_1^3J_2c - J_1^4c^2 - 3J_1^4Dc + 6J_1^3J_2Db - 9J_1^2J_2^2Da \tag{7.105}$$

$$
\begin{gathered}
54a_pJ_{1p}J_1J_2^2 - 54a_pJ_{2p}J_1^2J_2 + 18a_pJ_1J_2(3aJ_2^2 - 2bJ_1J_2 + J_1^2c) - \\
36b_pJ_{1p}J_1^2J_2 + 36b_pJ_{2p}J_1^3 + 3b_pJ_1^2(-9aJ_2^2 + 8bJ_1J_2 - 6J_1^2c) + \\
18J_{1pp}J_{1p}J_1J_2 - 18J_{1pp}J_{2p}J_1^2 + 6J_{1pp}J_1(3aJ_2^2 - 2bJ_1J_2 + J_1^2c) - \\
24J_{1p}^3J_2 + 24J_{1p}^2J_{2p}J_1 + J_{1p}^2(-123aJ_2^2 + 22bJ_1J_2 - 8J_1^2c) - \\
24J_{1p}J_{1y}J_1^2 + 6J_{1p}J_{2p}J_1(30aJ_2 - bJ_1) + 2J_{1p}(-63a^2J_2^3 + 45abJ_1J_2^2 - \\
39aJ_1^2J_2c + 4b^2J_1^2J_2 + bJ_1^3c + 27J_1^2J_2Da) - 6J_{1y}J_1^2(-3aJ_2 + bJ_1) - \\
81J_{2p}^2aJ_1^2 + 6J_{2p}J_1(18a^2J_2^2 - 12abJ_1J_2 + 12aJ_1^2c - 2b^2J_1^2 - 9J_1^2Da) + \\
27d_paJ_1^4 + 54a_yJ_1^3J_2 - 45b_yJ_1^4 + 18J_{1py}J_1^3 - 18c_paJ_1^3J_2 + 6c_pbJ_1^4 + \\
9(Db)_pJ_1^4 - 27a^3J_2^4 + 36a^2bJ_1J_2^3 - 54a^2J_1^3J_2d - 36a^2J_1^2J_2^2c - \\
6ab^2J_1^2J_2^2 + 36abJ_1^4d + 42abJ_1^3J_2c - 33aJ_1^4c^2 + 9aJ_1^4Dc + \\
18aJ_1^3J_2Db + 27aJ_1^2J_2^2Da - 8b^3J_1^3J_2 + 5b^2J_1^4c - 18bJ_1^4Db - \\
36bJ_1^3J_2Da + J_1^4J_2 + 54J_1^4cDa - 27J_1^4D^2a = 0
\end{gathered}
\tag{7.106}
$$

$$
\begin{gathered}
810a_pJ_{1p}J_1J_2^2d - 810a_pJ_{2p}J_1^2J_2d + 270a_pJ_1J_2d(3aJ_2^2 - 2bJ_1J_2 + J_1^2c) - \\
540b_pJ_{1p}J_1^2J_2d + 540b_pJ_{2p}J_1^3d + 9b_pJ_1^2(-45aJ_2^2d + 40bJ_1J_2d - \\
30J_1^2cd - 6J_1^2Dd) + 270J_{1pp}J_{1p}J_1J_2d - 270J_{1pp}J_{2p}J_1^2d + \\
90J_{1pp}J_1d(3aJ_2^2 - 2bJ_1J_2 + J_1^2c) - 360J_{1p}^3J_2d + 360J_{1p}^2J_{2p}J_1d + \\
15J_{1p}^2d(-123aJ_2^2 + 22bJ_1J_2 - 8J_1^2c) - 360J_{1p}J_{1y}J_1^2d + \\
90J_{1p}J_{2p}J_1d(30aJ_2 - bJ_1) + 3J_{1p}(-630a^2J_2^3d + 450abJ_1J_2^2d - \\
390aJ_1^2J_2cd + 40b^2J_1^2J_2d + 10bJ_1^3cd + 270J_1^2J_2Dad - J_1J_2^3) + \\
90J_{1y}J_1^2d(-3aJ_2 + bJ_1) - 1215J_{2p}^2aJ_1^2d + 9J_{2p}J_1(180a^2J_2^2d - \\
120abJ_1J_2d + 120aJ_1^2cd - 20b^2J_1^2d - 90J_1^2Dad + J_1J_2^2) + \\
9d_pJ_1^4(45ad - 2bc + 6Db) + 810a_yJ_1^3J_2d - 486b_yJ_1^4d + 270J_{1py}J_1^3d - \\
270c_paJ_1^3J_2d + 108c_pbJ_1^4d - 90c_yJ_1^4c + 54(Db)_pJ_1^4d + 18(Dc)_pJ_1^4c - \\
243(Dc)_yJ_1^4 + 54(D^2c)_pJ_1^4 + 81d_{py}J_1^4 + 162d_ybJ_1^4 - 405a^3J_2^4d + \\
540a^2bJ_1J_2^3d - 810a^2J_1^3J_2d^2 - 540a^2J_1^2J_2^2cd - 90ab^2J_1^2J_2^2d + \\
594abJ_1^4d^2 + 630abJ_1^3J_2cd - 495aJ_1^4c^2d - \\
54aJ_1^4cDd - 27aJ_1^4Dcd + 162aJ_1^4D^2d + 270aJ_1^3J_2Dbd + \\
405aJ_1^2J_2^2Dad + 9aJ_1J_2^4 - 120b^3J_1^3J_2d + 72b^2J_1^4cd + \\
36b^2J_1^4Dd + 4bJ_1^4c^3 - 144bJ_1^4Dbd - 18bJ_1^4D^2c - 540bJ_1^3J_2Dad - \\
4bJ_1^2J_2^3 + 24J_1^4J_2d - 18J_1^4c^2Db + 702J_1^4cDad - 36J_1^4cD^2b + \\
486J_1^4DaDd - 54J_1^4DbDc - 54J_1^4D^3b - 3J_1^3J_2^2c = 0
\end{gathered}
\tag{7.107}
$$

The coefficient

$$\alpha = J_1/(54\varphi_p^3) \neq 0.$$

The functions $\varphi(x, y, p)$, $\psi(x, y, p)$, $g(x, y, p)$, and $k(x, y, p)$ have to satisfy the equations:

$$J_1D\varphi = \varphi_pJ_2,\ 3J_1^2\varphi_y = \varphi_p(-3J_{1p}J_2 + 3J_{2p}J_1 - 3aJ_2^2 + 2bJ_1J_2 - J_1^2c),$$

$$6J_1^2\varphi_p\varphi_{ppp} = 9\varphi_{pp}^2 J_1^2 + \varphi_p^2(12J_{1p}aJ_2 - 12J_{2p}aJ_1 - 6a_pJ_1J_2 + 3b_pJ_1^2 + 3a^2J_2^2 - 2abJ_1J_2 + 4aJ_1^2c - b^2J_1^2 - 3J_1^2Da)$$

$$\varphi_pJ_1Dg = \varphi_pg_pJ_2 - J_1k,$$
$$3\varphi_p^2J_1^2g_y = 3\varphi_{pp}J_1^2k + \varphi_pJ_1k(3aJ_2 - bJ_1) + \varphi_p^2g_p(-3J_{1p}J_2 + 3J_{2p}J_1 - 3aJ_2^2 + 2bJ_1J_2 - J_1^2c)$$

$$54\varphi_pg_{pp} = 54\varphi_{pp}g_p + 54ak - J_1\psi$$

$$\psi_p = \varphi_pg, \quad J_1D\psi = \varphi_pJ_2g,$$
$$3J_1^2\psi_y = -3J_{1p}\varphi_pJ_2g + 3J_{2p}\varphi_pJ_1g + 3J_1^2k + \varphi_pg(-3aJ_2^2 + 2bJ_1J_2 - J_1^2c),$$

$$3\varphi_pJ_1^2Dk = -3J_{1p}\varphi_pJ_2k + 3J_{2p}\varphi_pJ_1k + 3\varphi_{pp}J_1J_2k + \varphi_pJ_1k(bJ_2 - J_1c)$$

$$108\varphi_pJ_1^4k_y = 54kJ_1(2\varphi_{pp}J_1 - a_p\varphi_pJ_2p)(J_{2p}J_1 - J_{1p}J_2) + 36\varphi_{pp}J_1^2k(-3aJ_2^2 + 2bJ_1J_2 - J_1^2c) +$$
$$\varphi_pk[18a_pJ_1J_2p(3aJ_2^2 - 2bJ_1J_2 + J_1^2c) - 36b_pJ_{1p}J_1^2J_2p + 36b_pJ_{2p}J_1^3p + 3b_pJ_1^2(-9aJ_2^2p + 8bJ_1J_2p - 6J_1^2cp + 12J_1J_2) + 18J_{1pp}J_{1p}J_1J_2p - 18J_{1pp}J_{2p}J_1^2p + 6J_{1pp}J_1p(3aJ_2^2 - 2bJ_1J_2 + J_1^2c) - 24J_{1p}^3J_2p + 24J_{1p}^2J_{2p}J_1p + J_{1p}^2(-123aJ_2^2p + 22bJ_1J_2p - 8J_1^2cp + 72J_1J_2) - 24J_{1p}J_{1y}J_1^2p + 6J_{1p}J_{2p}J_1(30aJ_2p - bJ_1p - 12J_1) + 2J_{1p}(-63a^2J_2^3p + 45abJ_1J_2^2p - 39aJ_1^2J_2cp + 90aJ_1J_2^2 + 4b^2J_1^2J_2p + bJ_1^3cp - 42bJ_1^2J_2 + 12J_1^3c + 27J_1^2J_2Dap) +$$

$$6J_{1y}J_1^2(-3aJ_2p + bJ_1p + 12J_1) - 81J_{2p}^2aJ_1^2p + 6J_{2p}J_1(18a^2J_2^2p - 12abJ_1J_2p + 12aJ_1^2cp - 18aJ_1J_2 - 2b^2J_1^2p + 6bJ_1^2 - 9J_1^2Dap) + 27d_paJ_1^4p + 54a_yJ_1^3J_2p - 45b_yJ_1^4p + 18J_{1py}J_1^3p - 18c_paJ_1^3J_2p + 6c_pbJ_1^4p - 36c_pJ_1^4 + 9(Db)_pJ_1^4p - 27a^3J_2^4p + 36a^2bJ_1J_2^3p - 54a^2J_1^3J_2dp - 36a^2J_1^2J_2^2cp + 108a^2J_1J_2^3 - 6ab^2J_1^2J_2^2p + 36abJ_1^4dp + 42abJ_1^3J_2cp - 108abJ_1^2J_2^2 - 33aJ_1^4c^2p + 9aJ_1^4Dcp - 108aJ_1^4d + 108aJ_1^3J_2c + 18aJ_1^3J_2Dbp + 27aJ_1^2J_2^2Dap - 8b^3J_1^3J_2p + 6b^2J_1^4cp - 18bJ_1^4Dbp - 36bJ_1^3J_2Dap + J_1^4J_2p + 54J_1^4cDap + 36J_1^4Db - 27J_1^4D^2ap - 108J_1^3J_2Da]$$

$$3\varphi_pJ_1k_p = 3\varphi_{pp}J_1k + \varphi_pk(3aJ_2 - bJ_1).$$

Remark 7.6. *The linearization conditions look very massy, but their checking on a computer for a given equation takes only a few seconds.*

Proof of the linearization theorems. Defining the derivatives ψ_p and ψ_x from (7.92), and equating the mixed derivatives $(\psi_p)_x = (\psi_x)_p$, one finds from this equation the derivative ψ_y. Introducing the functions

$$H = \varphi_x + p\varphi_y, \quad k = \varphi_pg_x + \varphi_pg_yp - \varphi_xg_p - \varphi_yg_pp \neq 0$$

one finds

$$\varphi_x = H - p\varphi_y, \; g_x = \varphi_p^{-1}(k - \varphi_p g_y p + \varphi_x g_p + \varphi_y g_p p).$$

Since $\alpha(x, y, p) = \alpha \circ \varphi(x, y, p)$, it gives the relations

$$\alpha_x\varphi_y - \alpha_y\varphi_x = 0, \; \alpha_x\varphi_p - \alpha_p\varphi_x = 0, \; \alpha_p\varphi_y - \alpha_y\varphi_p = 0.$$

The equations $(\psi_x)_y = (\psi_y)_x$ and $(\psi_p)_y = (\psi_y)_p$ can be solved with respect to the derivatives k_y and k_p, respectively. Equations (7.95), (7.96), (7.97) serve for finding the derivatives g_{pp}, g_y and k_y. The equation $(g_x)_y = (g_y)_x$ defines the derivative H_p. The derivative φ_{ppp} is found from the equation $(g_{pp})_x = (g_x)_{pp}$. Equation (7.98) gives the representation for the derivative H_x. The equations $(H_p)_x = (H_x)_p$ and $(H_p)_y = (H_y)_p$ can be solved with respect to the derivatives H_y and φ_{yy}, respectively. Setting $(\varphi_{ppp})_x = (\varphi_x)_{ppp}$, one finds the derivative φ_{ppy}. The equations $(k_p)_y = (k_y)_p$, $(k_x)_y = (k_y)_x$, $(H_x)_y = (H_y)_x$, $(g_{pp})_y = (g_y)_{pp}$, and $(\varphi_{ppy})_p = (\varphi_{ppp})_y$ become, respectively:

$$HJ_1 - \varphi_p J_2 = 0, \; HJ_2 + \varphi_p J_3 = 0, \; HJ_3 + \varphi_p J_4 = 0, \; 54\alpha\varphi_p^3 - J_1 = 0, \tag{7.108}$$

$$J_1 D J_1 = J_{1p} J_2 + 3aJ_2^2 - 2bJ_1J_2 + J_1^2 c$$

Notice that one has found the following derivatives

$$\begin{gathered}\varphi_x, \; \varphi_{yy}, \; \varphi_{ppy}, \; \varphi_{ppp}, \; \psi_x, \; \psi_y, \; \psi_p, \; g_x, \; g_y, \; g_{pp}, \\ H_x, \; H_y, \; H_p, \; k_x, \; k_y, \; k_p\end{gathered} \tag{7.109}$$

through

$$\varphi_y, \; \varphi_p, \; \varphi_{pp}, \; \varphi_{py}, \; g_p, \; \varphi, \; \psi, \; g, \; H, \; k.$$

Thus, one has found all third order derivatives of the function φ, all first order derivatives of the function ψ, and all second order derivatives of the function g. The remaining compatibility conditions are obtained by equating the mixed derivatives (with corresponding orders) of the functions $\varphi(x, y, p)$, $\psi(x, y, p)$, $g(x, y, p)$, $H(x, y, p)$, and $k(x, y, p)$. During this process one can find additional relations.

Next, the calculations are divided into two cases: either $J_1 = 0$ or $J_1 \neq 0$.

Let $J_1 = 0$. Since $\varphi_p \neq 0$, equations (7.108) give

$$J_2 = 0, \; J_3 = 0, \; J_4 = 0, \; \alpha = 0.$$

Equating all mixed derivatives, one obtains only one additional relation

$$\begin{gathered}54a(-bd^2+cDd+3Dcd-3D^2d)+54Da(2cd-9Dd)+\\ 18Db(-3d_p-7bd+c^2+3Dc)+54b_pDd-189b_yd-18c_pbd+\\ 90c_yc+81(Db)_pd-18(Dc)_pc+243(Dc)_y-54(D^2c)_p-81d_{py}+\\ 18d_pbc-162d_yb+18b^2cd-36b^2Dd-4bc^3+18bD^2c+36cD^2b-\\ 405D^2ad+54D^3b=0\end{gathered} \tag{7.110}$$

where D is the operator

$$D=\frac{\partial}{\partial x}+p\frac{\partial}{\partial y}.$$

However, finding d_{pp} and d_{px} from the equations $J_3=0$ and $J_4=0$, the relation $(d_{pp})_x=(d_{px})_p$ coincides with equation (7.110). Thus, the necessary and sufficient conditions are the equations (7.99): $J_i=0$, $(i=1,2,3,4)$.

In the case $J_1\neq 0$ equations (7.108) give

$$H=\varphi_pJ_1^{-1}J_2,\ J_2^2+J_1J_3=0,\ J_2J_3+J_1J_4=0,\ \alpha=J_1\varphi_p^{-3}/54=0.$$

Since the value H have been found, by substituting it into the representations of the derivatives $H_x,\ H_y,\ H_p$, one obtains the additional relations. For example, substituting into the expression for the derivative H_p, one finds

$$3J_1^2\varphi_y+\varphi_p(3J_{1p}J_2-3J_{2p}J_1+3aJ_2^2-2bJ_1J_2+J_1^2c)=0.$$

Since $J_1\neq 0$, the derivative φ_y can be found from this equation. Hence, the list of derivatives found for the functions $\varphi(x,y,p)$, $\psi(x,y,p)$, $g(x,y,p)$, $H(x,y,p)$, and $k(x,y,p)$ consists of the list (7.109) and the derivative φ_y. All other compatibility conditions do not include any of the functions $\varphi(x,y,p)$, $\psi(x,y,p)$, $g(x,y,p)$, $H(x,y,p)$, and $k(x,y,p)$. They are (7.100)-(7.107) which are conditions for the coefficients a,b,c,d of equation (7.94). •

The linearization theorems showed that for contact transformations it is not sufficient to study only second order invariants: in the linearization tests there are also third order invariants[19].

The study of third order invariants of the equivalence group is reduced to the analysis of the complete system of equations

$$\begin{gathered}S_1=4F_{J_{15}}J_7+2F_{J_{14}}(2J_5-J_4)+2F_{J_{13}}(-J_{16}+2J_{15})-2F_{J_{12}}J_3-\\ 6F_{J_{11}}J_1-6F_{J_{10}}J_8-2F_{J_9}J_6-4F_{J_6}J_1-3F_{J_5}J_2=0,\end{gathered}$$

$$\begin{gathered}S_2=4F_{J_{16}}(-J_{11}+J_6)+4F_{J_{15}}J_6-4F_{J_{14}}J_{13}-6F_{J_{13}}J_9-2F_{J_{12}}J_{14}+\\ 4F_{J_{11}}J_{10}-4F_{J_7}J_1+2F_{J_6}J_{10}+6F_{J_5}(J_{16}-J_{15})+2F_{J_4}(2J_{16}-J_{15})+\\ 4F_{J_3}J_5+4F_{J_2}J_7+3F_{J_1}J_8=0,\end{gathered}$$

[19]The linearizing conditions must be invariants, since the property of an equation to be linearized is invariant with respect to the set of contact transformations

$$S_3 = 2F_{J_{16}}J_{16} + 2F_{J_{15}}J_{15} + 3F_{J_{14}}J_{14} + 2F_{J_{13}}J_{13} + 4F_{J_{12}}J_{12} + F_{J_{11}}J_{11} + F_{J_9}J_9 + 2F_{J_7}J_7 + F_{J_6}J_6 + 3F_{J_5}J_5 + 3F_{J_4}J_4 + 4F_{J_3}J_3 + 3F_{J_2}J_2 + F_{J_1}J_1 = 0,$$

$$S_4 = 4F_{J_{16}}J_1 + 2F_{J_{14}}(-2J_{16} + J_{15}) + 2F_{J_{13}}(-2J_{11} + J_6) + 2F_{J_{12}}J_4 - 3F_{J_{11}}J_8 + 2F_{J_9}J_{10} - 6F_{J_5}J_7 - 4F_{J_4}J_7 - 6F_{J_3}J_2 = 0,$$

$$S_5 = F_{J_{14}}J_{14} - F_{J_{13}}J_{13} + 3F_{J_{12}}J_{12} - 2F_{J_{11}}J_{11} - 4F_{J_{10}}J_{10} - 3F_{J_9}J_9 - 3F_{J_8}J_8 + F_{J_7}J_7 - 2F_{J_6}J_6 + 2F_{J_5}J_5 + 2F_{J_4}J_4 + 4F_{J_3}J_3 + 3F_{J_2}J_2 - F_{J_1}J_1 = 0,$$

$$S_6 = 4F_{J_{16}}J_4 + 4F_{J_{15}}(-J_5 + J_4) - 6F_{J_{14}}J_{12} - 4F_{J_{13}}J_{14} + 6F_{J_{11}}(-J_{16} + J_{15}) + 4F_{J_{10}}J_{11} - 2F_{J_9}J_{13} + 4F_{J_8}J_1 + 3F_{J_7}J_2 + 2F_{J_5}(-J_{16} + 2J_{15}) + 4F_{J_5}J_3 + 2F_{J_4}J_3 - 4F_{J_1}J_7 = 0,$$

$$S_7 = -2F_{J_{14}}J_7 - 2F_{J_{13}}J_1 + F_{J_{12}}J_2 + F_{J_9}J_8 = 0.$$

where $F = F(J_1, J_2, ..., J_{16})$. A complete analysis of the solutions of this system is very cumbersome and, even more, the representation of some relative invariants J_i is massive.

2.3.3 Applications of contact transformations to linearization

This section is devoted to examples of applications of the developed criteria. Since the criteria give necessary and sufficient conditions, linearizable and nonlinearizable examples of equations are given.

In this section the c_i are arbitrary constants.

Example 1.

Let us study the linearization problem for equation[20]:

$$y''' - \frac{3y''^2}{2y'} = 0. \tag{7.111}$$

The coefficients of this equation are

$$a = 0, \ b = -\frac{3}{2y'}, \ c = 0, \ d = 0.$$

These coefficients obey the linearization conditions (7.99)-(7.110). Hence, equation (7.111) is transformed to the equation

$$u''' = 0.$$

The linearizing transformation is found as follows. From the equations for the derivatives

$$g_{pp} = \frac{\varphi_{pp}g_p}{\varphi_p}, \quad k_p = k(\frac{\varphi_{pp}}{\varphi_p} + \frac{1}{2p}),$$

[20]Equation (8.32) in vol.3 of [72].

one obtains

$$g = g_1\varphi + g_0, \quad k = k_1\varphi_p\sqrt{p},$$

where $k_1 = k_1(x, y)$, $g_0 = g_0(x, y)$, $g_1 = g_1(x, y)$ are arbitrary functions of the integration. Substituting these representations for g and k into the equations for the derivatives g_x, one finds

$$\varphi_{pp} = \varphi_p(-2g_{0x} - 2g_{1x}\varphi - 3k_1\sqrt{p})/(2k_1p\sqrt{p})$$

Then from the representation for g_y one gets

$$\varphi = -\frac{g_{0x} + g_{0y}p + k_1\sqrt{p}}{g_{1x} + g_{1y}p}.$$

Substituting this into the representation for the derivative φ_{pp}, one has

$$(g_{0x}g_{1y} - g_{1x}g_{0y})^2 - k_1^2 g_{1x}g_{1y} = 0$$

For the sake of simplicity it is assumed $g_{1x} = 0$. Substituting φ into the representation of φ_{ppy}, one finds

$$g_{0yy}g_{1y} - g_{0y}g_{1yy} = 0$$

which has the general solution

$$g_0 = c_1g_1 + c_2.$$

From the equation for the derivative φ_x one finds

$$H = -((k_{1x} + k_{1y}p)g_{1y} - g_{1yy}k_1p)/(\sqrt{p}g_{1y}^2).$$

The equation for k_y gives

$$g_{1yy} = (2g_{1y}k_{1y})/k_1, \tag{7.112}$$

and the equation for φ_{yy} is reduced to

$$k_{1yy} = (2k_{1y}^2)/k_1.$$

Substituting the results so obtained into the equation for H_y, one finds

$$k_{1xy} = (k_{1x}k_{1y})/k_1.$$

Then the equation for H_x becomes

$$k_{1xx} = (2k_{1x}^2)/k_1.$$

Hence,

$$k_1 = (x(c_3y + c_4) + c_5y + c_6)^{-1},$$

where

$$c_3c_6 - c_4c_5 = 0.$$

When determining the linearizing transformations, the symmetry and equivalence transformations for the original and linearized equations are involved. Applying these groups, one can simplify the set of transformations. For example, equation (7.111) admits the transformations corresponding to the generators [72]:

$$\partial_x,\ \partial_y,\ x\partial_x,\ y\partial_y.$$

Assuming $c_3 \neq 0$, and because of the shift with respect to y, one can also assume that $c_4 = 0$, and hence, $c_6 = 0$. Integrating (7.112), one obtains

$$g_1 = c_8 + c_7/(c_3 y).$$

Integrating the equations for ψ_x, ψ_y and ψ_p, one finds

$$\psi = (4p(c_3x + c_5)^2 c_5 c_7^2)^{-1}((2c_3y - c_5p)c_5c_7 + (2c_5c_8y^2 + c_7px^2)c_3^2 + 4(\sqrt{p}c_2c_3y + c_3c_7c_9px + c_5c_7c_9p)(c_3x + c_5)c_5c_7).$$

Here it is also assumed that $c_5 \neq 0$.

Shifting with respect to x, and scaling with respect to x and y

$$x \to (x - c_5)/c_3,\ \ y \to y/c_3,$$

one arrives at

$$\varphi = -c_1 + (c_7x\sqrt{p})^{-1}y, \qquad g = c_2 + (c_7x\sqrt{p})^{-1}(c_7 + c_8y),$$
$$\psi = c_9 + (4c_5c_7)^{-1} + (4pc_7^2x^2)^{-1}(4\sqrt{p}c_2x + 2)yc_7 - 2xc_7p + 2c_8y^2).$$

Since the equations for the functions φ and ψ include only their derivatives, they are defined up to arbitrary constants. Hence, one can assume $c_1 = 0$ and $c_9 + (4c_5c_7)^{-1} = 0$. Notice that the Jacobian of the transformation is

$$(4c_7^2p^2x^4)^{-1}$$

The further dilation

$$x \to \frac{x}{\beta},\ \ y \to \frac{y}{\alpha},\ \ p \to p\frac{\beta}{\alpha}$$

with

$$\beta = c_7, \alpha = c_7^{-1}, c_9 = -\frac{1}{4c_5c_7}$$

gives

$$\varphi = \frac{y}{x\sqrt{p}},\ \psi = (2px^2)^{-1}(-px + y(2c_2x\sqrt{p} + c_8y + 1)),\quad g = c_2 + \frac{c_8y + 1}{x\sqrt{p}}.$$

For example, assuming $c_2 = 0$ and $c_8 = 0$:

$$\varphi = \frac{y}{x\sqrt{p}}, \ \psi = (2px^2)^{-1}(-px + y), \ g = \frac{1}{x\sqrt{p}}.$$

Remark 7.7. *The functions*

$$\varphi = \sqrt{p}, \ \psi = \frac{1}{2}(xp - y), \ g = x\sqrt{p}$$

give a particular solution of the case $g_{1x} \neq 0$. This solution corresponds to the two-step transformation mentioned in [72] (vol.3, p.214-215).

Example 2.

The linearizing conditions (7.99), (7.110) are satisfied for the equation[21]

$$y''' - \frac{3y'y''^{2}}{1 + y'^{2}} = 0. \tag{7.113}$$

Hence, this equation is linearizable to the equation

$$u''' = 0.$$

The linearizing transformation can be obtained as follows.

From the equations for the derivatives

$$g_{pp} = \frac{\varphi_{pp} g_p}{\varphi_p}, \quad k_p = k\left(\frac{\varphi_{pp}}{\varphi_p} + \frac{p}{p^2 + 1}\right),$$

one finds

$$g = g_1\varphi + g_0, \quad k = k_1\varphi_p\sqrt{p^2 + 1},$$

where $k_1 = k_1(x, y), \ g_0 = g_0(x, y), \ g_1 = g_1(x, y)$ are arbitrary functions of the integration. Substituting these representations for g and k into the equations for the derivative g_x, one arrives at

$$\varphi_{pp} = -\varphi_p(\sqrt{p^2 + 1}(g_{0x} + g_{1x}\varphi) + k_1(2p^2 + 1))/(k_1 p(p^2 + 1))$$

From the representation for the derivative g_y one gets

$$\varphi = -\frac{g_{0x} + g_{0y}p + k_1\sqrt{p^2 + 1}}{g_{1x} + g_{1y}p}.$$

Substituting it into the representation for the derivative φ_{pp}, one has

$$(g_{0x}g_{1y} - g_{1x}g_{0y})^2 - k_1^2(g_{1x}^2 + g_{1y}^2) = 0.$$

[21] Equation (8.33) in vol.3 [72].

For the sake of simplicity assume that $g_{1x} = 0$, then the last equation gives

$$g_{0x} \pm k_1 = 0.$$

For example, let us set

$$g_{0x} = k_1.$$

From the equation for k_x one finds

$$g_{0yy} g_{1y} - g_{0y} g_{1yy} = 0, \;\; k_{1x} = 0.$$

The equations for k_y and φ_x are reduced to the equations

$$g_{1y} = c_1 k_1^2, \;\; H = \frac{k_{1y}\sqrt{p^2+1}}{c_1 k_1^2}.$$

Substituting H into the equation for H_y one finds

$$k_{1y} = c_2 k_1^2.$$

Integrating the equations for the functions k_1, g_0 and g_1, one obtains

$$k_1 = -(c_2 y - c_6)^{-1}, \;\; g_0 = c_2^{-1}\left(c_4 + k_1(c_2 x + c_1 c_3)\right), \;\; g_1 = c_2^{-1} c_1 k_1 + c_5.$$

Integrating the equations for ψ_x, ψ_y and ψ_p, one finds the function ψ. The transformation that is found is

$$\varphi = \frac{(1+\sqrt{p^2+1})(c_2 y - c_6) - p(c_1 c_3 + c_2 x)}{c_1 p},$$

$$\begin{aligned}\psi = {} & (2c_1^2 c_2 p^2)^{-1}(2\sqrt{p^2+1}(-c_1 c_2^2 c_3 c_5 p y + c_1 c_2^2 c_4 p y + c_1 c_2 c_3 c_5 c_6 p - \\ & c_1 c_2 c_4 c_6 p - c_1 c_2 y + c_1 c_6 - c_2^3 c_5 p x y + c_2^3 c_5 y^2 + c_2^2 c_5 c_6 p x - \\ & 2c_2^2 c_5 c_6 y + c_2 c_5 c_6^2) + 2c_1^2 c_2 c_7 p^2 + 2c_1 c_2^2 c_3 c_5 p^2 x - 2c_1 c_2^2 c_3 c_5 p y - \\ & 2c_1 c_2^2 c_4 p^2 x + 2c_1 c_2^2 c_4 p y + 2c_1 c_2 c_3 c_5 c_6 p - 2c_1 c_2 c_4 c_6 p - 2c_1 c_2 p^2 y - \\ & 2c_1 c_2 y + 2c_1 c_6 + c_2^3 c_5 p^2 x^2 + c_2^3 c_5 p^2 y^2 - 2c_2^3 c_5 p x y + 2c_2^3 c_5 y^2 - \\ & 2c_2^2 c_5 c_6 p^2 y + 2c_2^2 c_5 c_6 p x - 4c_2^2 c_5 c_6 y + 2c_2 c_5 c_6^2),\end{aligned}$$

$$\begin{aligned}g = {} & (c_1 c_2 p)^{-1}((1+\sqrt{p^2+1})(-c_1 + c_2 c_5(c_2 y - c_6)) \\ & - p c_2 c_5 (c_2 x + c_1 c_3) + c_1 c_2 c_4 p).\end{aligned}$$

The Jacobian of the transformation is

$$\frac{(p^2+2) + 2\sqrt{p^2+1}}{c_1^2 p^4}.$$

Notice that equation (7.113) admits the transformations corresponding to the generators [72]:

$$\partial_x, \ \partial_y, \ x\partial_x + y\partial_y.$$

By applying these transformations one can simplify the set of transformations that is obtained. For example, shifts with respect to x and y, and dilation with respect to them

$$c_2x + c_1c_3 \to c_1x, \ c_2y - c_6 \to c_1y$$

lead to

$$\varphi = \frac{y(1+\sqrt{p^2+1})}{p} - x,$$

$$\psi = p^{-2}(1+\sqrt{p^2+1})y(c_3p - c_2^{-1} - c_5px + c_5y) - c_3x - c_2^{-1}y + c_5(x^2+y^2)/2,$$

$$g = (1+\sqrt{p^2+1})(-c_2^{-1} + c_5y)/p + c_3 - c_5x.$$

Notice also that assuming $g_{1y} = 0$ instead of $g_{1x} = 0$, one can obtain the transformation

$$\varphi = -\left(p + \sqrt{1+p^2}\right), \ \psi = (px - y)\left(p + \sqrt{1+p^2}\right),$$
$$g = y - x\left(p + \sqrt{1+p^2}\right).$$

The Jacobian of this transformation is

$$2p^2 + 1 + 2p\sqrt{1+p^2}.$$

Remark 7.8. *Earlier it was known the only complex transformation*

$$x = z + it, \ y = t + iz$$

which maps the linearizable equation

$$z''' - \frac{3z''^2}{2z'} = 0 \tag{7.114}$$

into equation (7.113). Here we present a set[22] *of contact transformations (over real numbers) linearizing equation (7.113) to the equation*

$$u''' = 0.$$

[22] The transformation presented in [175]:

$$\varphi = -2xg, \ \psi = y - xp/4, \ g^2 = -p$$

relates the equation $u''' = 0$ with the equation:

$$y''' - \frac{3y''^2}{2y'} = 0,$$

but not with equation (7.113).

Example 3.
Assume that the equation

$$y''' + (A_1(x, y)y' + A_0(x, y))y'' + F(x, y, p) \tag{7.115}$$

is linearizable by a contact transformation. Direct calculations show that the invariants $J_1 = 0$ and $J_2 = 0$. Hence, the other invariants J_3 and J_4 have to be zero $J_3 = 0$, $J_4 = 0$, and this equation can be mapped by a contact transformation into the simplest equation $u''' = 0$.

The equation $J_3 = 0$ gives

$$3F_{pp} = 2p(3A_{1y} + A_1^2) + 7A_{0y} - A_{1x} + 2A_0A_1$$

or

$$F(x, y, p) = B_3(x, y)p^3 + B_2(x, y)p^2 + B_1(x, y)p + B_0(x, y)$$

with

$$2(3B_2 - 3A_{1x} - A_0A_1) = 7(A_{0y} - A_{1x}) \quad 9B_3 = 3A_{1y} + A_1^2,$$

where $B_0(x, y)$ and $B_1(x, y)$ are arbitrary functions of the integration. This means that equation (7.115) has the form (7.15).

From the equation $J_4 = 0$ one obtains

$$\begin{gathered} 3(A_{0y} - A_{1x})_y - A_1(A_{0y} - A_{1x}) = 0, \\ 6(A_{0y} - A_{1x})_x + 2A_0(A_{0y} - A_{1x}) - 3(3B_1 - A_0^2 - 3A_{0x})_y = 0, \\ 9A_{0xx} + 18A_{0x}A_0 + 54B_{0y} - 27B_{1x} + 4A_0^3 - 18A_0B_1 + 18A_1B_0 = 0. \end{gathered} \tag{7.116}$$

Thus, equation of the form (7.15) is linearizable by a contact transformation, if and only if, its coefficients satisfy equations (7.116).

Recall that an equation of the form (7.15) is linearizable by a point transformation, if and only if, its coefficients satisfy equations (7.31)-(7.33). Since, of the equations (7.31),

$$A_{0y} - A_{1x} = 0, \quad (3B_1 - A_0^2 - 3A_{0x})_y = 0$$

and (7.38)

$$\Omega = \frac{1}{54}\left(9A_{0xx} + 18A_{0x}A_0 + 54B_{0y} - 27B_{1x} + 4A_0^3 - 18A_0B_1 + 18A_1B_0\right),$$

equation (7.15) is also linearizable by a contact transformation if $\Omega = 0$. Therefore equations of the form (7.15) are linearizable by a contact transformation and by a point transformation, if and only if, its coefficients satisfy the equation $\Omega = 0$ and equations[23] (7.31)-(7.32).

[23] It was noted that equation (7.33) is equivalent to the equation $\Omega_y = 0$.

Equation (7.79) is an example of the equation which can be linearizable by a point transformation, but cannot be linearized by a contact transformation. Let us give an opposite example of the equation of the form (7.15) which can be linearized by a contact transformation, but it cannot be linearized by a point transformation. Direct calculations show that, for example, such equation is

$$y''' + yy'' + \frac{1}{54}(63p^2 + 24y^2p + y^4) = 0.$$

Its coefficients are

$$A_0 = y,\ A_1 = 0,\ B_0 = \frac{1}{54}y^4,\ B_1 = \frac{4}{9}y^2,\ B_2 = \frac{7}{6},\ B_3 = 0.$$

For this equation the necessary and sufficient conditions (7.116) are satisfied, however , since $A_{0y} - A_{1x} \neq 0$, it cannot be linearized by a point transformation.

Let us give example of an equation which is not linearizable either by point or contact transformation. The equation

$$y''' + yy'' + \beta(1 - y'^2) = 0,$$

is extensively exploited in hydrodynamics[24]. For this equation

$$A_0 = y,\ A_1 = 0,\ B_0 = \beta,\ B_1 = 0,\ B_2 = -\beta,\ B_3 = 0.$$

Since $A_{0y} - A_{1x} \neq 0$, this equation is not linearizable by a point transformation, and because of the last equation of (7.116) it is not linearizable by a contact transformation.

[24]The Blasius equation ($\beta = 0$), the Hiemenz flow ($\beta = 1$), the Falkner-Skan equation [148].

Chapter 8

APPENDIX

1. Reduce code for solving systems of linear homogeneous equations

Procedures for solving a linear partial differential equation of the type[1]

$$f_x + a_1 f_{y_1} + a_2 f_{y_2} + ... + a_n f_{y_n} = 0$$

are given here.

The code of the procedures for solving a system of linear partial differential equations is presented here. Let a linear partial differential equation be integrated

$$f_x + a_1 f_{y_1} + a_2 f_{y_2} + ... + a_n f_{y_n} = 0,$$

where $f = f(x, y_1, y_2, ..., y_n)$, the coefficients a_i, $(i = 1, 2, ..., n)$ are linear functions of the independent variables $y_1, y_2, ..., y_{i-1}$:

$$a_i = \beta_{i,1}(x)y_1 + \beta_{i,2}(x)y_2 + ... + \beta_{i,i-1}(x)y_{i-1} + \gamma_i(x).$$

1.1 Procedures for solving linear homogeneous equations

```
operator coeffin;
off echo;
operator ssp,uk,ress,sk;
operator yy,invar;
%------------------------------------------------------
algebraic procedure the_main(choose_first_ind,sexpr3,nuk,full_list_ind);
begin   scalar result,allvar,sexpr1,sexpr2,typey,invariants,ss;
    coe1:=df(sexpr3,df(ff,choose_first_ind));
    ss:=coe1*df(ff,choose_first_ind)-sexpr3;
    if ((ss=0) and not(coe1=0)) then begin
        df(ff,choose_first_ind):=0;
```

[1]Examples of the application of these procedures are given in Section 7.2

```
        for j:=1:nequat do fu(j):=fu(j);
        full_list_ind:=select(not(~w=choose_first_ind),full_list_ind)$
        ms:=ms+1;
        sk(ms):=choose_first_ind;
        go to end_the_main;
    end;
    allvar:=const_all_incl(full_list_ind,choose_first_ind,sexpr3);
    sexpr1:=first allvar;
    sexpr2:=second allvar;
    typey:=third allvar;
    invariants:=basic_constr_inv(sexpr1,typey,sexpr3);
    for j:=1:length(typey) do begin
        ss:=sub(ff=part(invariants,j),sexpr3);
        if not (ss=0) then write ("invariant for ",part(typey,j)," = ",
        part(invariants,j)," after subs = ",ss);
    end;
    ss:=solpde(sexpr1,sexpr2,sexpr3,invariants,typey,nuk,full_list_ind);
    nuk:=first ss;
    full_list_ind:=second ss;
end_the_main:
    result:=append({nuk},{full_list_ind});
return result;
end;
%--------------------------------------------------------

algebraic procedure solpde(sexpr1,sexpr2,sexpr3,invariants,
                           typey,nuk,full_list_ind);
% firpar -- first element
% sexpr1 -- list of elements without first
% sexpr2 -- list of parameters
% invariants
% sexpr1 consists of (first, others)
begin scalar ss,ssinv,coe,coe1,snuk,nnukk,jnuk,sexpr1in,invjnuk;
       if typey={} then go to order_done;
order_done:
    nnukk:=length sexpr1;
    nukk:=nuk+nnukk-2;
    ssp(1):=first sexpr1;
    coe1:=df(sexpr3,df(ff,ssp(1)));
    sexpr1in:=sexpr1:=rest sexpr1;
    snuk:=nuk-1;
    invjnuk:=1;
    if invariants={} then go to 5;
    while not(invariants={}) do begin
        invjnuk:=invjnuk+1;
        snuk:=snuk+1;
        ssp(invjnuk):=first sexpr1;
        sexpr1:=rest sexpr1;
        ssinv:=first invariants;
        invariants:=rest invariants;
        ress(invjnuk):=sskk(snuk):=ssinv-uk(snuk);
    end;
5:  invjnuk:=invjnuk+1;
    for jnuk:=invjnuk:nnukk do begin
        snuk:=snuk+1;
        ssp(jnuk):=first sexpr1;
        sexpr1:=rest sexpr1;
        coe:=-df(sexpr3,df(ff,ssp(jnuk)))/coe1;
```

```
        ress(jnuk):=sskk(snuk):=coe*ssp(1)+ssp(jnuk)-uk(snuk);
    end;
    for kj:=nuk:nukk do depend ff,uk(kj);
    for kj:=nuk:nukk do factor df(ff,uk(kj));
    for kj:=nuk:nukk do full_list_ind:=append(full_list_ind,{uk(kj)});
    for jk:=1:nnukk do begin
        df(ff,ssp(jk)):=diffm(ssp(jk),nuk,nukk);
    end;
    if sexpr2={} then begin
        for j:=1:nequat do fu(j):=fu(j);
        go to 1;
    end;
    while not(sexpr2={}) do begin
        ss:=first sexpr2;
        sexpr2:=rest sexpr2;
        for j:=1:nequat do fu(j):=dfparfun(ss,nuk,nukk,fu(j));
    end;
1:  for jk:=1:nnukk do df(ff,ssp(jk)):=0;
    for jnuk:=2:nnukk do begin
        excludings:=append(excludings,{ssp(jnuk)});
        ssp(jnuk):=linsolv(ssp(jnuk),ress(jnuk),jnuk);
        set(first sexpr1in,ssp(jnuk));
        sexpr1in:=rest sexpr1in;
    end;
    for jnuk:=1:nnukk do full_list_ind:=
        select(not(~w=ssp(jnuk)),full_list_ind);
    ms:=ms+1; sk(ms):=ssp(1);
    write("sk(",ms,") = ",sk(ms));
    nuk:=nukk+1;
return append({nuk},{full_list_ind});
end;
%-------------------------------------------------------

algebraic procedure diffm(indvar,nuk,nukk);
begin
    difexp:= for kj:=nuk:nukk sum df(ff,uk(kj))*df(sskk(kj),indvar);
return difexp;
end;
%-------------------------------------------------------

algebraic procedure dfparfun(spar1,nuk,nukk,ssexp);
begin scalar dfspar,retsub;
    dfspar:=df(ff,spar1)+
        for kj:=nuk:nukk sum df(ff,uk(kj))*df(sskk(kj),spar1);
    retsub:=sub(df(ff,spar1)=dfspar,ssexp);
return retsub;
end;
%-------------------------------------------------------

algebraic procedure linsolv(xx,yy,kj);
begin
    jj:=df(yy,xx);
    xxr:=xx-yy/jj;
return xxr;
end;
%-------------------------------------------------------

algebraic procedure gkk(m,l,q);
```

```
begin scalar h;
    h:=1;
    if m=l and l=q then h:=6 else if l=q or m=l or m=q then h:=2;
return h;
end;

%--------------------------------------------------------

% construction invariants

algebraic procedure basic_constr_inv(sexpr1,typey,sexpr3);
begin scalar x,kk,coeff,s1typey,stypey,ssq;
    x:=first sexpr1;
    jj:=df(sexpr3,df(ff,x));
    sexpr3:=sexpr3/jj$
    kk:=length typey;
    stypey:=typey;
    for j:=1:kk do begin
        s1typey:=part(typey,j);
        coeffin(j):=df(sexpr3,df(ff,s1typey));
    end;
    inv_s:=constr_invariants(typey,x);
    for j:=1:kk do clear coeffin(j);
return inv_s;
end;
%--------------------------------------------------------

algebraic procedure constr_invariants(typey,x);
begin scalar ss,inv_s;
    k:=length typey;
    for j:=1:k do begin
        for l:=1:k do
            coeffin(j):=sub(part(typey,l)=yy(l),coeffin(j));
    end;
    yy(1):=invar(1)+int(coeffin(1),x);
    for j:=2:k do yy(j):=invar(j)+int(coeffin(j),x);
    for j:=k step -1 until 1 do begin
        invar(j):=invar(j)+part(typey,j)-yy(j);
    end;
    inv_s:={};
    for j:=1:k do inv_s:=append(inv_s,{invar(j)});
    for j:=1:k do begin
        ss:=sub(ff=invar(j),sexpr3);
        if not(ss=0) then write("not correct inv(",j,") = ",ss);
    end;
    for j:=1:k do clear yy(j);
    clear yy(1);
    for j:=1:k do clear invar(j);
return inv_s;
end;
%--------------------------------------------------------

% moving typey into the second place after first_element

algebraic procedure moving(sexpr1,typey);
begin scalar ss1,ssr,reform,stypey;
    ss1:=first sexpr1;
    ssr:=rest sexpr1;
```

```
    reform:={};
    stypey:=typey;
    while not(stypey = {}) do begin
        ss:=first stypey; stypey:=rest stypey;
        if (length ssr =1) then begin
             ssr:={};
             reform:=append(reform,{ss});
        end
             else begin
                  ssr:=select(not (~w=ss),ssr);
                  reform:=append(reform,{ss});
             end;
    end;
    sexpr1:=append(append({ss1},reform),ssr);
return sexpr1;
end;
%-------------------------------------------------------

%  construction typey,sexpr1,sexpr2

algebraic procedure const_all_incl
                    (full_list_ind,choose_first_ind,sexpr3);
begin scalar sfull,ss,sss,ssexpr1,ssexpr2,coeff,j,jk,result;
    sexpr1:={};
    sfull:=full_list_ind;
    while not(sfull={}) do begin
        ss:=first sfull;
        sfull:=rest sfull;
        if not (df(ff,ss)=0) then
             if not (df(sexpr3,df(ff,ss)) =0 ) then begin
                  sexpr1:=append(sexpr1,{ss});
             end;
    end;
    sexpr2:={};
    ssexpr1:=sexpr1;
    while not(ssexpr1={}) do begin
        ss:=first ssexpr1;
        ssexpr1:=rest ssexpr1;
        coeff:=df(sexpr3,df(ff,ss));
        sfull:=full_list_ind;
        while not (sfull = {}) do begin
             sss:=first sfull;
             sfull:=rest sfull;
             if not (df(coeff,sss) = 0) then
                  sexpr2:=append(append(sexpr2,{sss}),{ss});
        end;
    end;
%       excluding similar elements from sexpr2
    ssexpr2:=sexpr2;
    while not(ssexpr2={}) do begin
        ss:=first ssexpr2;
        ssexpr2:=rest ssexpr2;
        sexpr2:=append({ss},select(not(~w=ss),sexpr2));
    end;
%       constructing typey
    typey:={};
    ssexpr2:=sexpr2;
    while not(ssexpr2={}) do begin
```

```
        ss:=first ssexpr2;
        ssexpr2:=rest ssexpr2;
        ssexpr1:=sexpr1;
        while not(ssexpr1={}) do begin
             sss:=first ssexpr1;
             ssexpr1:=rest ssexpr1;
             if ss=sss then begin
                  typey:=append(typey,{ss});
                  if sexpr2={} then go to empty_sexpr2;
                  if (length(sexpr2)=1 and  first(sexpr2)=ss)
                      then sexpr2:={}
                           else sexpr2:=select(not(~w=ss),sexpr2);
empty_sexpr2:
             end;
        end;
    end;
    sexpr1:=select(not(~w = choose_first_ind),sexpr1);
    sexpr1:=append({choose_first_ind},sexpr1);

%       excluding choose_first_ind from typey
    stypey:=typey;
    while not(stypey={}) do begin
        ss:=first stypey;
        stypey:=rest stypey;
        if (ss = choose_first_ind)
                then typey:=select(not(~w=ss),typey);
    end;
%       ordering of typey ==> y_1,y_2,...
    if length(typey) > 1 then typey:=ordering(sexpr3,typey);
    sexpr1:=moving(sexpr1,typey);
    result:=append(append({sexpr1},{sexpr2}),{typey});
return result;
end;
%---------------------------------------------------------

% ordering of typey

algebraic procedure ordering(sexpr3,typey);
begin scalar ntypey,stypey,result,sresult,
                    nresult,ss,sss,ssq,spreresult,preresult;
    ntypey:=length typey;
    for jk:=1:ntypey do begin
        ss:=part(typey,jk);
        coeff:=df(sexpr3,df(ff,ss));
        for j:=1:ntypey do begin
             sss:=part(typey,j);
             if not(df(coeff,sss)=0) then ssp(jk,j):=1
                  else ssp(jk,j):=0;
        end;
    end;
%%%%%%%%%%%%%%%%%%%%%%%%% checking
%    for jk:=1:ntypey do for j:=1:ntypey do if ssp(jk,j) = 1 then
%         write("before  ssp(",jk,",",j,") := ",ssp(jk,j),"$",
%               "for ",part(typey,jk)," and ",part(typey,j));
%%%%%%%%%%%%%%%%%%%%%%%%% checking
%       begin of preliminary construction
    result:=typey;
    preresult:={};
```

```
    while(not(result={})) do begin
        if length(result) = 1 then
               <<preresult:=append(preresult,result);
                          go to proc_end1;>>;
        spreresult:={};
        nresult := length result;
        sresult:=result;
        for j:=1:nresult do begin
             sj:=part(result,j);
             numsj:=numposition(typey,sj);
             kk:=0;
             for jk:=1:nresult do begin
                  sjk:=part(result,jk);
                  numsjk:=numposition(typey,sjk);
                  kk:=kk+ssp(numsj,numsjk);
             end;
             if kk = 0 then begin
                  spreresult:=append(spreresult,{sj});
                  if not(length(sresult)=1) then
                       sresult:=select(not(~w=sj),sresult)
                  else sresult:={};
             end;
        end;
        result:=sresult;
        preresult:=append(preresult,spreresult);
    end;
proc_end1:

%       end of preliminary construction
%%%%%%%%%%%%%%%%%%%%%%%% checking  after
    for jk:=1:ntypey do for j:=1:ntypey do begin
        ss:=part(preresult,jk); numss:=numposition(typey,ss);
        sss:=part(preresult,j); numsss:=numposition(typey,sss);
        if ssp(numss,numsss) = 1 then if jk<j then begin
             write("after  ssp(",numss,",",numsss,") := ",
                  ssp(numss,numsss),"$,   for coeff ",ss,"
                       with the variable ",part(typey,j));
        end;
    end;
%%%%%%%%%%%%%%%%%%%%%%%% checking
return preresult;
end;
%-----------------------------------------------------------

%      defining the number of position of the element in the slist

algebraic procedure numposition(slist,element);
begin scalar nlist,position_number;
    nslist:=length slist;
    for j:=1:nslist do if element = part (slist,j) then begin
        position_number:=j;
        go to proc_end;
    end;
proc_end:
return  position_number;
end;
%-----------------------------------------------------------
```

```
algebraic procedure exchangeposition(slist,kk1,kk2);
%       kk1<kk2 neq nslist
begin scalar nslist,result;
    nslist:=length slist;
    result:={};
    for j:=1:(kk1-1) do result:=append(result,{part(slist,j)});
    result:=append(result,{part(slist,kk2)});
    for j:=(kk1+1):(kk2-1) do result:=append(result,{part(slist,j)});
    result:=append(result,{part(slist,kk1)});
    for j:=(kk2+1):nslist do result:=append(result,{part(slist,j)});
return  result;
end;
%----------------------------------------------------------
```

1.2 Reconstitution of the original independent variables

After integration of the equations of the linear system of partial differential equations one has to rewrite all of the remaining independent variables, which are collected in the list `full_list_ind`, in terms of the original independent variables. These variables are reconstituted through the operators `sskk(j)`. Before reconstituting one has to clear all of the variables except for the values of the operators `sskk(j)`. For example, the command
`for j:=1:nuk do clear uk(j)$`
clears the values of the operator `uk`. The variables J_k, $(k = 1, 2, \ldots, \mathrm{nuk})$ (in the program they are denoted by `uk(k)`) are found by the code:

```
nexcludings:=length excludings;
for j:=nexcludings step -1 until 1 do begin
    jj:=df(sskk(j),uk(j));
    write("jj = ",jj,"        sskk(",j,") = ",sskk(j));
    uk(j):=uk(j)-sskk(j)/jj;
    write(" sskk(",j,") :=  ",sskk(j));
end;
for j:=1:nexcludings do write(" uk(",j,") :=   ",uk(j));
```

References

[1] Albeverio, S. and S. Fei: 1995, 'Remark on symmetry of stochastic dynamical systems and their conserved quantities'. *Journal of Physics A: Mathematical and General* **28**, 6363–6371.

[2] Ames, W. F.: 1965, *Nonlinear Partial Differential Equations in Engineering*. New–York: Academic Press.

[3] Ames, W. F.: 1972, 'Ad Hoc Methods for Nonlinear Partial Differential Equations'. *Applied Mechanics Reviews* pp. 1021–1031.

[4] Andreev, V. K., O. Kaptsov, V. Pukhnachov, and A. A. Rodionov: 1998, *Applications of Group-Theoretic Methods in Hydrodynamics*. Dordrecht: Kluwer.

[5] Arais, E. A., V. P. Shapeev, and N. N. Yanenko: 1974, 'Realization of the Cartan exterior forms method'. *Dokl. AS USSR* **214**(4), 737–738.

[6] Arais, E. A. and G. V. Sibiriakov: 1973, *Auto-Analytic*. Novosibirsk: Novosibirsk State University.

[7] Bellman, R. and K. L. Cooke: 1963, *Differential-Difference Equations*. New York: Academic Press.

[8] Bluman, G. W. and J. D. Cole: 1969, 'The general similarity solution of the heat equation'. *J. math. mech.* **18**, 1025–1042.

[9] Bobylev, A. V.: 1975, 'On exact solutions of the Boltzmann equation'. *Dokl. AS USSR* **225**(6), 1296–1299.

[10] Bobylev, A. V.: 1984, 'Exact solutions of the nonlinear Boltzmann equation and the theory of relaxation of a maxwellian gas'. *Teor. and matem. physika* **60**(20), 280–310.

[11] Bobylev, A. V.: 1993, 'The Boltzmann equation and group transformations'. *Math. Models Methods Appl. Sci.* **3**, 443–476.

[12] Bobylev, A. V. and N. H. Ibragimov: 1989, 'Relationships between the symmetry properties of the equations of gas kinetics and hydrodynamics'. *J. Mathematical Modeling* **1**(3), 100–109.

[13] Bocharov, A. V., V. V. Sokolov, and S. I. Svinolupov: 1993, 'On some equivalence problems for differential equations Preprint'. Preprint, WRI/Ufa.

[14] Boisvert, R. E., W. F. Ames, and U. N. Srivastava: 1983, 'Group properties and new solutions of the Navier-Stokes equations'. *J.Eng.Math.* **17**, 203–221.

[15] Brandi, P., A. Salvadori, and Z. Kamont: 2002, 'Existence of generalized solutions of hyperbolic functional differential equations'. *Nonlinear Analysis* **50**(7), 919–940.

[16] Bryant, R. L., S.-S. Chern, R. B. Gardner, H. L. Goldschmidt, and P. A. Griffiths: 1991, *Exterior differential systems*. New York: Springer-Verlag. Math.Sci. Res. Inst. Publ. vol.18.

[17] Bunimovich, A. I. and A. V. Krasnoslobodtsev: 1982, 'Invariant–group solutions of kinetic equations'. *Mechan. jzydkosti i gasa* (4), 135–140.

[18] Bunimovich, A. I. and A. V. Krasnoslobodtsev: 1983, 'On some invariant transformations of kinetic equations'. *Vestnik Moscow State Univ., ser. 1., Matemat. mechan.* (4), 69–72.

[19] Burnat, M.: 1962, 'The Cauchy problem of compressible flows of simple wave type'. *Arch. mech. stos.* **14**(3/4), 313–341.

[20] Bytev, V. O.: 1972, 'Group properties of the Navier-Stokes equations'. *Chislennye metody mehaniki sploshnoi sredy (Novosibirsk)* **3**(3), 13–17.

[21] Cantwell, B. J.: 1978, 'Similarity transformations for the two-dimensional, unsteady, stream-function equation'. *J. Fluid Mech.* **85**, 257–271.

[22] Cantwell, B. J.: 2002, *Introduction to symmetry analysis*. Cambridge: Cambridge University Press.

[23] Cartan, E.: 1924, 'Sur les variétés à connexion projective'. *Bull. Soc. Math. France* **52**, 205–241.

[24] Cartan, E.: 1946, *Les systems differentiels exterieurs et leurs applications scientifiques*. Hermann.

[25] Chern, S.-S.: 1940, 'The geometry of the differential equation $y''' = F(x, y, y', y'')$'. *Rep. Nat. Tsing Hua Univ.* **4**, 97–111.

[26] Chetverikov, V. N. and A. G. Kudryavtsev: 1995a, 'A method for computing symmetries and conservation laws of integro-differential equations'. *Acta Applicandae Mathematicae* **41**, 45–56.

[27] Chetverikov, V. N. and A. G. Kudryavtsev: 1995b, 'Modelling integro-differential equations and a method for computing their symmetries and conservation laws'. *American Mathematical Society Translations* **167**, 1–22.

[28] Chirkunov, Y. A.: 1984, 'Nonlinear viscoelastic one-dimensional Kelvin models'. *Dynamika sploshnoi sredy* (64), 121–131. Institute of Hydrodynamics, Novosibirsk.

[29] Chupakhin, A. P.: 1997, 'On barochronic motions of a gas'. *Dokl. RAS* **352**(5), 624–626.

[30] Clarkson, P. A.: 1995, 'Nonclassical Symmetry Reductions of the Boussinesq Equation'. *Chaos Solitons & Fractals* **5**(12), 2261–2301.

[31] Copson, E. T.: 1953, 'On sound waves of finite amplitude'. *Proc. Royal Soc.* **23**(2), 216.

[32] Courant, R. and D. Hilbert: 1953, *Methods of Mathematical Physics*. New York: Interscience–Wiley.

[33] Davenport, J. H.: 1994, 'Computer algebra - past, present and future'. *Euromath Bulletin* **1**(2), 25–45.

[34] Dorodnitsyn, V. A.: 2001, *Group properties of finite-difference equations*. Moscow: Finalist.

[35] Doubrov, B.: 2001, 'Contact trivialization of ordinary differential equations'. In: O. Kowalski, D. Krupka, and J. Slovak (eds.): *Proceedings of the 8th International Conference on Differential Geometry and Its Applications*. Opava, Czech Republic: Silesian University in Opava, pp. 73–84.

[36] Doubrov, B., B. Komrakov, and T. Morimoto: 1999, 'Equivalence of holonomic differential equations'. *Lobachevskii Journal of Mathematics* **3**, 39–71.

[37] Driver, R. D.: 1977, *Ordinary and delay differential equations*. New York: Springer-Verlag.

[38] Ernst, M. H. and E. M. Hendriks: 1979, 'An exactly solvable the non-linear Boltzmann equation'. *Phys. Lett.* **70A**(3), 183–185.

[39] Erugin, N. N.: 1948, 'On functionally invariant solutions'. *Scientific notes of Leningrad State University, math.science* (15), 101–134.

[40] Erugin, N. N. and M. M. Smirnov: 1981, 'Functionally invariant solutions of partial differential equations'. *Differential equations* **17**(5), 853–865.

[41] Euler, N. and M. Euler: 2004, 'Sundman symmetries of nonlinear second-order and third-order ordinary differential equations'. *Journal of Nonlinear Mathematical Physics* **11**(3), 399–421.

[42] Euler, N., T. Wolf, P. G. L. Leach, and M. Euler: 2003, 'Linearizable third-order ordinary differential equations and generalized Sundman transformations: The Case $X''' = 0$'. *Acta Applicandae Mathematicae* **76**, 89–115.

[43] Ferapontov, E. V. and K. R. Khusnutdinova: 2004, 'On the integrability of (2+1)-dimensional quasilinear systems'. *ArXiv: nlin.SI/0305044*. to appear in Comm. Math. Phys.

[44] Finikov, S. P.: 1948, *Method of Cartan exterior forms*. Moscow–Leningrad: Gostechizdat.

[45] Fusco, D. and N. Manganaro: 1996, 'A method for finding exact solutions to hyperbolic systems of first-order PDEs'. *IMA J. Appl. Math.* **57**, 223–242.

[46] Fushchich, W. I.: 1981, 'On the way of investigation of group properties of integro-differential equations'. *Ukrain. Math. J.* **33**(6), 834–838.

[47] Fushchich, W. I. and R. O. Popovych: 1994a, 'Symmetry reduction and exact solution of the Navier-Stokes equations'. *Nonlinear Mathematical Physics* **1**(1), 75–113.

[48] Fushchich, W. I. and R. O. Popovych: 1994b, 'Symmetry reduction and exact solution of the Navier-Stokes equations'. *Nonlinear Mathematical Physics* **1**(2), 158–188.

[49] Gaeta, G.: 2004, 'Symmetry of stochastic equations'. *ArXiv: nlin.SI/0305044* pp. 1–18.

[50] Gaeta, G. and N. R. Quinter: 1999, 'Lie-point symmetries and differential equations'. *Journal of Physics A: Mathematical and General* **32**, 8485–8505.

[51] Galaktionov, V. A.: 1995, 'Invariant subspaces and new explicit solutions to evolution equations with quadratic nonlinearities'. *Proc. Royal Soc.* **125A**, 225–246.

[52] Ganzha, V. G., S. V. Meleshko, F. A. Murzin, V. P. Shapeev, and N. N. Yanenko: 1981, 'Realization on a computer of an algorithm for studying consistency of systems of partial differential equations'. *Dokl. AS USSR* **261**(5), 1044–1046.

[53] Ganzha, V. G., S. V. Meleshko, and V. P. Shapeev: 1989, 'Intermediate calculations in analytical investigation of differential equations (on a computer)'. *Modelirovanie v mechanike* **3(20)**(4), 49–58.

[54] Ganzha, V. G., S. V. Meleshko, and V. P. Shapeev: 1992, 'Automation of compatibility analysis of quasilinear PDE systems in REDUCE'. In: *Computer Algebra and Its Applications to Mechanics*. pp. 89–95.

[55] Ganzha, V. G., S. V. Meleshko, and V. P. Shapeev: 1993, 'Application of REDUCE system for an analysis of compatibility of partial differential equations'. *Modelirovanie v mechanike* **7(24)**(4), 26–35.

[56] Glassey, R. T.: 1996, *The Cauchy Problem in Kinetic Theory*. Philadelphia: Society for Industrial & Applied Mathematics.

[57] Goursat, E.: 1933, *Course of mathematical analysis*. Moscow-Leningrad: ONTI.

[58] Grauel, A. and W.-H. Steeb: 1985, 'Similarity solutions of the Euler equation and the Navier-Stokes equations in two space dimensions'. *Int.J.Theor.Phys.* **24**, 255–265.

[59] Grebot, G.: 1997, 'The Characterization of Third order Ordinary Differential Equations Admitting a Transitive Fiber-Preserving Point Symmetry Group'. *Journal of Mathematical Analysis and Applications* **206**, 364–388.

[60] Grigoriev, Y. N. and S. V. Meleshko: 1986, 'Investigation of invariant solutions of the Boltzmann kinetic equation and its models'. Preprint of Institute of Theoretical and Applied Mechanics.

[61] Grigoriev, Y. N. and S. V. Meleshko: 1987, 'Group analysis of the integro–differential Boltzman equation'. *Dokl. AS USSR* **297**(2), 323–327.

[62] Grigoriev, Y. N. and S. V. Meleshko: 1995, 'Group analysis of kinetic equations'. *Russian Journal Numerical Analysis and Mathematical Modelling* **9**(5).

[63] Grigoriev, Y. N. and S. V. Meleshko: 1996, 'Lie group and invariant solutions of a system of the kinetic Boltzmann equations'. In: *Modern problems in mathematics and mechanics*, Vol. 1. Moscow: Department of mechanics and mathematics of Moscow State University, pp. 125–128. Proceedings of International conference in honor P.L.Chebyshev's 175 anniversary.

[64] Grigoriev, Y. N. and S. V. Meleshko: 1998, 'Bobylev-Krook-Wu modes for multicomponent gas mixtures'. *Physical Review Letters* **81**(1), 93–95.

[65] Grundland, A. M. and L. Lalague: 1996, 'Invariant and partially invariant solutions of the equations describing a non-stationary and isotropic flow for an ideal and compressible fluid in (3+1) dimensions'. *Journal of Physics A: Mathematical and General* **29**, 1723–1739.

[66] Gusyatnikova, V. N. and V. A. Yumaguzhin: 1999, 'Contact Transformations and Local Reducibility of ODE to the Form $y''' = 0$'. *Acta Applicandae Mathematicae* **56**, 155–179.

[67] Gzigzin, A. E. and V. P. Shapeev: 1975, 'To the problem on continuous joining of particular solutions of systems of partial differential equations'. *Chislennye metody mehaniki sploshnoi sredy (Novosibirsk)* **6**(7), 44–52.

[68] Hale, J.: 1977, *Functional Differential Equations*. New York: Springer-Verlag.

[69] Hearn, A. C.: 1987, *REDUCE Users Manual, ver. 3.3*. Santa Monica: The Rand Corporation CP 78.

[70] Ibragimov, N. H.: 1979, 'On the theory of Lie-Bäcklund transformation groups'. *Matem. Sbornik* **109**(2), 229–253. English translation, Math. USSR Sbornik, 37 (2), 1980, 205–226.

[71] Ibragimov, N. H.: 1983, *Transformation Groups Applied to Mathematical Physics*. Moscow: Nauka. English translation, Reidel, D., Ed , Dordrecht, 1985.

[72] Ibragimov, N. H. (ed.): 1994, 1995, 1996, *CRC Handbook of Lie Group Analysis of Differential Equations*, Vol. 1, 2, 3. Boca Raton: CRC Press.

[73] Ibragimov, N. H.: 1999, *Elementary Lie Group Analysis and Ordinary Differential Equations*. Chichester: Wiley & Sons.

[74] Ibragimov, N. H.: 2002a, 'Invariants of a remarkable family of nonlinear equations'. *Nonlinear Dynamics* **30**(2), 155–166.

[75] Ibragimov, N. H.: 2002b, 'Laplace type invariants for parabolic equations'. *Nonlinear Dynamics* **28**(2), 125–133.

[76] Ibragimov, N. H., V. F. Kovalev, and V. V. Pustovalov: 2002, 'Symmetries of integro-differential equations: a survey of methods illustrated by the Benny equations'. *Nonlinear Dynamics* **28**(2), 135–153.

[77] Ibragimov, N. H. and S. V. Meleshko: 2004, 'Linearization of third-order ordinary differential equations'. *Archives of ALGA* **1**, 71–93.

[78] Ibragimov, N. H. and M. Torrisi: 1992, 'A simple method for group analysis and its application to a model of detonation'. *J. Math. Phys.* **33**(11), 3931–3937.

[79] Ibragimov, N. H. and G. Unal: 1994, 'Equivalence transformations of Navier-Stokes equations'. *Bulletin of the Technical University of Istanbul* **47**(1–2), 203–207.

[80] Kaptsov, O. V.: 1992, 'Invariant sets of evolution equations'. *Nonlinear analysis, theory, methods of applications* **19**(8), 753–761.

[81] Kaptsov, O. V.: 1993, 'Construction of exact solutions of hydrodynamics equations'. Doctoral thesis.

[82] Kaptsov, O. V.: 1995, 'B-determining Equations: Applications to Nonlinear Partial Differential Equations'. *Europ. J. Appl. Math.* **6**, 265–286.

[83] Khabirov, S. V.: 1990, 'Nonisentropic one-dimensional gas motions constructed by means of the contact group of the nonhomogeneous Monge-Ampere equation'. *Matem. Sbornik* **181**(12), 1607–1635. English translation, Math. USSR Sbornik, 71 (2), 1992, 447–462.

[84] Khabirov, S. V.: 1993, 'Partially invariant solutions of hydrodynamics equations'. In: *Exact solutions of differential equations and their asymptotics*. Ufa: Russian Academy of Science, Institute of Mathematics with Computer Center, pp. 42–49.

[85] Kolmanovskii, V. and A. Myshkis: 1992, *Applied theory of functional differential equations*. : Dordrecht: Kluwer Academic Publishers.

[86] Komarovskii, L. V.: 1960a, 'On one exact solution of a space unsteady flow of a gas of double wave type'. *Dokl. AS USSR* **135**(1), 33–35.

[87] Komarovskii, L. V.: 1960b, 'On space unsteady flows of a gas with degenerate hodograph'. *Prikl. matem. and mech.* **24**(3), 491–495.

[88] Kovalev, V. F., S. V. Krivenko, and V. V. Pustovalov: 1992, 'Group Symmetry of the kinetic equations of the collisionless plasma'. *JETF Letters* **55**(4), 256–259.

[89] Kovalev, V. F., S. V. Krivenko, and V. V. Pustovalov: 1995, 'Plasma theory: Vlasov–Maxwell and Related equations'. In: N. H. Ibragimov (ed.): *CRC Handbook of Lie Group Analysis of differential equations*. Boca Raton: CRC Press.

[90] Krook, M. and T. T. Wu: 1976, 'Formation of Maxwellian tails'. *Phys. Rev. Lett.* **36**(19), 1107–1109.

[91] Krook, M. and T. T. Wu: 1977, 'Exact solutions of the Boltzmann equation'. *Phys. Fluids* **20**(10), 1589–1595.

[92] Kucharczyk, P., Z. Peradzynski, and E. Zavistowska: 1973, 'Unsteady multidimensional isentropic flows described by linear Riemann invariants'. *Arch. of Mech* **25**(2), 319–350.

[93] Kuranishi, M.: 1967, *Lectures on involutive systems of partial differential equations*. Säo Paulo: Publ.Soc. Math. Säo Paulo.

[94] Lie, S.: 1883, 'Klassifikation und Integration von gewöhnlichen Differentialgleichungen zwischen x, y, die eine Gruppe von Transformationen gestatten. III'. *Archiv for Matematik og Naturvidenskab* **8**(4), 371–427. Reprinted in Lie's Gessammelte Abhandlundgen, 1924, 5, paper XIY, pp. 362–427.

[95] Lie, S.: 1896, *Geometrie der Berührungstransformationen*. Leipzig: B.G. Teubner. Dargestellt von Sophus Lie und Georg Scheffers.

[96] Lie, S. and G. Scheffers: 1891, *Lectures on differential equations with known infinitesimal transformations*. Leipzig: B.G.Teubner.

[97] Linchuk, L. V.: 2001, 'On group analysis of functional differential equations'. In: V. A. Baikov, R. K. Gazizov, N. H. Ibragimov, and F. M. Mahomed (eds.): *Modern group analysis for the new millennium. Proceedings of the International Conference MOGRAN 2000*. Ufa: USATU Publishers, pp. 111–115.

[98] Lloyd, S. P.: 1981, 'The infinitesimal group of the Navier-Stokes equations'. *Acta Math.* **38**, 85–98.

[99] Ludford, C. S.: 1955, 'Generalized Riemann invariants'. *Pacif. J. Math.* **5**, 441–450.

[100] Ludlow, D. K., P. A. Clarkson, and A. P. Bassom: 1998, 'Nonclassical symmetry reductions of the three-dimensional incompressible Navier–Stokes equations'. *Journal of Physics A: Mathematical and General* **31**, 7965–7980.

[101] Ludlow, D. K., P. A. Clarkson, and A. P. Bassom: 1999, 'Similarity reduction and exact solutions for the two-dimensional incompressible Navier–Stokes equations'. *Studies in Appl. Math.* **103**, 183–240.

[102] Mahomed, F. M. and C. Wafo Soh: 2001, 'Integration of stochastic ordinary differential equations from a symmetry standpoint'. *Journal of Physics A: Mathematical and General* **34**, 777–782.

[103] Manganaro, N. and S. V. M.: 2002, 'Reduction Procedure and Generalized Simple Waves for Systems Written in the Riemann Variables'. *Nonlinear Dynamics* **30**(1), 87–102.

[104] Mansfield, E. L. and P. A. Clarkson: 1997, 'Applications of the differential algebra package diffgrob2 to classical symmetries of differential equations'. *J. Symbolic Computation* **23**(5–6), 517–533.

[105] Mansfield, E. L., G. J. Reid, and P. A. Clarkson: 1998, 'Nonclassical reductions of a 3+1-cubic nonlinear Schrödinger system'. *Computer Physics Communications* **115** (2–3), 460–488.

[106] Martin, M. H.: 1953, 'The propagation of a plane shock into a quiet atmosphere'. *Canadian J. Math.* **3**, 165–187.

[107] Meleshko, S. V.: 1980, 'DP-conditions and the adjoin problem of different DP-solutions'. *Chislennye metody mehaniki sploshnoi sredy (Novosibirsk)* **11**(6), 96–109.

[108] Meleshko, S. V.: 1983, 'Differential constraints and one-parameter Lie-Bäcklund transformations groups'. *Dokl. AS USSR* **271**(1), 42–46.

[109] Meleshko, S. V.: 1989, 'On nonisentropic steady space and plane unsteady double waves'. *Prikl. matem. and mech.* **53**(2), 255–260. English transl. in J.Appl.Maths Mechs, 1989, 53 (2).

[110] Meleshko, S. V.: 1990, 'Double waves in ideal rigid-plastic body with plane deformation'. *Journ.prikl.mech.i techn.phyz.* (2), 131–136.

[111] Meleshko, S. V.: 1991a, 'Classification of solutions with degenerate hodograph of the gas dynamics and plasticity equations'. Doctoral thesis.

[112] Meleshko, S. V.: 1991b, 'On solutions with degenerate hodograph of the quasisteady equations of the plasticity theory with the Mises flow condition'. *Journ.prikl.mech.i techn.phyz.* (1), 82–88.

[113] Meleshko, S. V.: 1994a, 'Homogeneous autonomous systems with three independent variables'. *Prikl. matem. and mech.* **58**(5), 97–102. English transl. in J.Appl.Maths Mechs, 1994, 58 (5).

[114] Meleshko, S. V.: 1994b, 'One class of partially invariant solutions of plane gas flows'. *Differential Equations* **30**(10), 1690–1693.

[115] Meleshko, S. V. and V. V. Pukhnachov: 1999, 'One class of partially invariant solutions of the Navier–Stokes equations'. *Journal of Applied Mechanics and Technical Physics* **40**(2), 24–33.

[116] Meleshko, S. V. and V. P. Shapeev: 1999, 'An application of the differential constraints method for the two-dimensional equations of gas dynamics'. *Prikl. Matem. i Mech.* **63**(6), 909–916. English transl. in J.Appl.Maths Mechs, 1999, 63 (6).

[117] Meshcheryakova, E. Y.: 2002, 'Exact solutions of equations of rotationally symmetric motion of an ideal incompressible liquid'. *Journal of Applied Mechanics and Technical Physics* **43**(3), 397–405.

[118] Meshcheryakova, E. Y.: 2003, 'New steady and self-similar solutions of the Euler equations'. *Journal of Applied Mechanics and Technical Physics* **44**(4), 455–460.

[119] Miller, W.: 1965, *Symmetry and separation of variables*. New–York: Addison-Wesley Pub. Co.

[120] Misawa, T.: 1994, 'New conserved quantities form symmetry for stochastic dynamical systems'. *Journal of Physics A: Mathematical and General* **27**, 177–192.

[121] Myshkis, A. D.: 1972, *Linear differential equations with retarded argument*. Moscow: Nauka.

[122] Neut, S. and M. Petitot: 2002, 'La géométrie de l'équation $y''' = f(x, y, y', y'')$'. *C. R. Acad. Sci. Paris* **335**, 515–518. Ser.I, Équations différentielles/Ordinary Differential Equations.

[123] Nikolskii, A. A.: 1963a, 'The simplest exact solutions of the Boltzmann equation of a rarefied gas motion'. *Dokl. AS USSR* **151**(2), 299–301.

[124] Nikolskii, A. A.: 1963b, 'Three dimensional homogeneous expansion–contraction of a rarefied gas with power interaction functions'. *Dokl. AS USSR* **151**(3), 522–524.

[125] Nikolskii, A. A.: 1965, 'Homogeneous motion of displacement of monatomic rarefied gas'. *Inghenerny journal* **5**(4), 752–755.

[126] Nonenmacher, T. F.: 1984, 'Application of the similarity method to the nonlinear Boltzmann equation'. *J. of Appl. Mathem. and Physics (ZAMP)* **35**(9), 680–691.

[127] Olver, P. J.: 1986, *Applications of Lie groups to differential equations*. New York: Springer-Verlag.

[128] Olver, P. J. and P. Rosenau: 1987, 'Group-invariant solutions of differential equations'. *SIAM J. Appl. Math.* **47**, 263–275.

[129] Ovsiannikov, L. V.: 1965, 'Invariant-group solutions of hydrodynamics equations'. In: *Proceedings of the Second All–Union Congress on Theoretical and Applied Mechanics*. Moscow: Nauka, pp. 302–305.

[130] Ovsiannikov, L. V.: 1978, *Group analysis of differential equations*. Moscow: Nauka. English translation, Ames, W.F., Ed., published by Academic Press, New York, 1982.

[131] Ovsiannikov, L. V.: 1993, 'On optimal system of subalgebras'. *Docl. RAS* **333**(6), 702–704.

[132] Ovsiannikov, L. V.: 1994a, 'Isobaric motions of a gas'. *Differential Equations* **30**(10), 1792–1799.

[133] Ovsiannikov, L. V.: 1994b, 'Program SUBMODELS. Gas dynamics'. *J.Appl.Maths Mechs* **58**(4), 30–55.

[134] Ovsiannikov, L. V.: 1995, 'Regular and irregular partially invariant solutions'. *Dokl. RAS* **343**(2), 156–159.

[135] Ovsiannikov, L. V. and A. P. Chupakhin: 1996, 'Regular partially invariant submodels of gas dynamics equations'. *J.Applied Mathematics and mechanics* **60**(6), 990–999.

[136] Peradzynski, Z.: 1975, 'Hyperbolic flows in ideal plasticity'. *Arch. Mech* **27**(1), 141–156.

[137] Pogodin, Y. Y., V. A. Suchkov, and N. N. Yanenko: 1958, 'On travelling waves of the gas dynamics equations'. *Dokl. AS USSR* **119**(3), 443–445.

[138] Pommaret, J. E.: 1978, *Systems of partial differential equations and Lie Pseudogroups*. New York: Gordon & Breach.

[139] Popovych, R. O.: 1995, 'On Lie reduction of the Navier–Stokes equations'. *Nonlinear Mathematical Physics* **2**(3–4), 301–311.

[140] Pukhnachov, V. V.: 1960, 'Group properties of the Navier-Stokes equations in two-dimensional case'. *J. Appl. Mech. and Techn. Phys.* (1), 83–90.

[141] Pukhnachov, V. V.: 1974, 'Free boundary problems of the Navier–Stokes equations'. Doctoral thesis.

[142] Pukhnachov, V. V.: 2002, 'New class of exact solutions of the Euler equations'. *Docl. RAS* **382**(6), 777–780.

[143] Pukhnachov, V. V.: 2003, 'Exact solutions of the hydrodynamics equations derived from partially invariant solutions'. *Journal of Applied Mechanics and Technical Physics* **44**(3), 317–323.

[144] Rabotnov, Y. N.: 1977, *Elements of hereditary solid mechanics*. Moscow: Nauka.

[145] Rajagopal, K. R.: 1984, 'A class of exact solutions to the Navier–Stokes equations'. *Int.J.Engng. Sci.* **22**(4), 451–458.

[146] Riquier, C. H.: 1910, *Les systemes d'equations aux derivees partielles*. Paris: Gauthier-Villars.

[147] Rozhdestvenskii, B. L. and N. N. Yanenko: 1978, *Systems of quasilinear equations and their applications to gas dynamics, 2nd ed.* Moscow: Nauka. English translation published by Amer. Math. Soc., Providence, RI, 1983.

[148] Schlichting, H.: 1979, *Boundary-Layer Theory*. New York: McGraw-Hill, Inc. Seventh edition.

[149] Sedov, L. I.: 1993, *Similarity and Dimensional Methods in Mechanics, 10th ed.* Boca Raton: CRC Press.

[150] Seiler, W. M.: 1997, 'Involution and symmetry reductions'. *Math. Comput. Modelling* **25**(8/9), 63–73.

[151] Senashov, S. I.: 1990, 'Group classification of viscoelastic bar equation'. *Modelirovanie v mechanike* **4(21)**(1), 69–72.

[152] Shapeev, V. P.: 1974, 'Applications of the method of differential constraints to one–dimensional continuum mechanics equations'. Ph.D. thesis, Computer center, RAS, Novosibirsk.

[153] Sheftel, M. B.: 1982, 'On Lie–Bäclund groups, admitted by the one–dimensional gas dynamics equations'. *Vestnik Leningrad Univ., (ser. mathem., mech., astr.* **7**(2), 37–41.

[154] Shurygin, V. A. and N. N. Yanenko: 1961a, 'On computer realization of algebraic-differential algorithms'. *Cybernetics problems* (6), 33–43.

[155] Shurygin, V. A. and N. N. Yanenko: 1961b, 'On realization of algebraic-differential algorithms on computer'. *Problems of cybernetics* **6**, 33–43.

[156] Sidorov, A. F.: 1959, 'On unsteady potential motions of a polytropic gas with degenerate hodograph'. *Applied Mathematics and Mechanics* **23**(5), 940–943.

[157] Sidorov, A. F.: 1970, 'On exact solutions of the gas dynamics equations of triple wave type'. *Dokl. AS USSR* **194**(4), 782–785.

[158] Sidorov, A. F.: 1987, 'Application of special constructions of series for calculating singularities of generalized solutions of nonlinear equations'. In: *Construction of algorithms and solutions of mathematical physics problems*. Moscow: Institute of Applied Mathematics AS USSR, pp. 79–97.

[159] Sidorov, A. F.: 1989, 'On two classes of solutions of fluid and gas dynamics equations and their relations with the theory of travelling waves'. *J. Appl. Mech. and Techn. Phys.* **30**(2), 34–40.

[160] Sidorov, A. F., V. P. Shapeev, and N. N. Yanenko: 1984, *The method of differential constraints and its applications in gas dynamics*. Novosibirsk: Nauka.

[161] Sidorov, A. F. and N. N. Yanenko: 1958, 'On question about unsteady plane polytropic gas flows with rectangular characteristics'. *Dokl. AS USSR* **123**(5), 832–834.

[162] Smirnov, S. I. and S. L. Sobolev: 1932, 'Sur le probleme plan des vibrations elastiques'. *C. R. Acad. Sci. Paris.* **194**, 1437–1439.

[163] Smirnov, V. I.: 1951, *Course in high mathematics*. Moscow: Gostechizdat.

[164] Sobolev, S. L.: 1934, 'Functionally invariant solutions of the wave equation'. In: *Proceedings of the V.A.Steklov's Institute of Physics and Mathematics*, Vol. 5. Moscow: Institute of Physics and Mathematics, pp. 259–264.

[165] Tanthanuch, J. and S. V. Meleshko: 2004, 'On Definition of an Admitted Lie Group for Functional Differential Equations'. *Communications in nonlinear science and numerical simulations* **9**(1), 117–125.

[166] Taranov, V. B.: 1976, 'On symmetry of one–dimensional high frequency motion of non-collision plasma'. *Journ. techn. physiki* **46**(6), 1271–1277.

[167] Titov, S. S.: 1978, 'On solutions of nonlinear partial differential equations in the form of polynomials with respect to one of variables'. *Chislennye metody mehaniki sploshnoi sredy (Novosibirsk)* **8**(1), 144–149.

[168] Topunov, V. L.: 1983, 'Application of a computer to studying pfaff systems'. *Chislennye metody mehaniki sploshnoi sredy (Novosibirsk)* **14**(3), 150–158.

[169] Topunov, V. L.: 1989, 'Reducing systems of linear differential equations to a passive form'. *Acta Appl. Math.* **16**, 191–206.

[170] Tresse, A. M.: 1896, 'Détermination des invariants ponctuels de léquation différentielle ordinaire du second ordre $y'' = \omega(x, y, y')$'. Preisschriften der fürstlichen Jablonowski'schen Geselschaft XXXII, Leipzig, S.Herzel.

[171] Turchin, V. F.: 1977, *Basic Refal and its realization on computers*. Moscow: CNIIPIASS. English edition: "REFAL-5 programming guide and reference manual", New England Publishing Co., Holyoke, 1989.

[172] Ulianov, O. N.: 1992, 'Two classes of solutions of nonlinear continuum mechanics equations'. Ph.D. thesis, Institute of Mathematics and Mechanics, Ekaterinburg.

[173] Ünal, G.: 2003, 'Symmetries of Ito and Stratonovich dynamical systems and their conserved quantities'. *Nonlinear Dynamics* **32**, 417–426.

[174] Ünal, G. and J. Q. Sun: 2004, 'Symmetries conserved quantities of stochastic dynamical control systems'. *Nonlinear Dynamics* **36**, 107–122.

[175] Wafo Soh, C., F. M. Mahomed, and C. Qu: 2002, 'Contact symmetry algebras of scalar ordinary differential equations'. *Nonlinear Dynamics* **28**, 213–230.

[176] Yanenko, N. N.: 1956, 'Travelling waves of a system of quasilinear equation'. *Dokl. AS USSR* **109**(1), 44–47.

[177] Yanenko, N. N.: 1964, 'Compatibility theory and methods of integrating systems of nonlinear partial differential equations'. In: *Proceedings of the Fourth All–Union Mathematics Congress*. Leningrad: Nauka, p. 613.

[178] Zavialov, Y. S.: 1955, 'On some integrals of one dimensional gas motion'. *Dokl. AS USSR* **103**(5), 781–782.

[179] Zawistowski, Z. J.: 2002, 'Symmetries of equations with functional arguments'. *Reports on mathematical physics* **50**(2), 125–132.

[180] Zubov, E. N.: 1981a, 'On double waves of unsteady space flows of an ideal gas with variable entropy'. *Docl. AS USSR* **263**(5), 1087–1091.

[181] Zubov, E. N.: 1981b, 'On one class of exact solutions of the gas dynamics equations with variable entropy'. In: *Numerical and analytical methods of solving continuum mechanics problems*. Sverdlovsk: Institute of Mathematics and Mechanics, pp. 41–67.

[182] Zubov, E. N.: 1983, 'On some exact solutions of space unsteady flows of an ideal gas with variable entropy'. In: *Exact and approximate methods of study of continuum mechanics problems*. Sverdlovsk: Institute of Mathematics and Mechanics, pp. 53–68.

Index